Mount Mitchell and the Black Mountains

An Environmental History of the Highest Peaks in Eastern America

Mount Mitchell and the Black Mountains

Timothy Silver

The University of North Carolina Press
Chapel Hill and London

Manufactured in the United States of America

Designed by Heidi Perov
Set in Minion and TheSerif by Tseng Information Systems, Inc.

Printed on recycled paper

The paper in this book meets the guidelines for permanence and durability of the Committee on Production Guidelines for Book Longevity of the Council on Library Resources.

Library of Congress Cataloging-in-Publication Data
Silver, Timothy, 1955–
Mount Mitchell and the Black Mountains : an environmental history of the highest peaks in eastern America / by Timothy Silver.
p. cm.
ISBN-13: 978-0-8078-2755-0 (cloth : alk. paper)
ISBN-10: 0-8078-2755-X (cloth : alk. paper)
ISBN-13: 978-0-8078-5423-5 (pbk. : alk. paper)
ISBN-10: 0-8078-5423-9 (pbk. : alk. paper)
1. Mountain ecology—North Carolina—Mount Mitchell—History. 2. Human ecology—North Carolina—Mount Mitchell—History. 3. Mount Mitchell (N.C.)—Environmental conditions. 4. Mountain ecology—North Carolina—Black Mountains—History. 5. Human ecology—North Carolina—Black Mountains—History. 6. Black Mountains (N.C.)—Environmental conditions. I. Title.
QH105.N8 S55 2003
304.2′8′09756873—dc21 2002007158

cloth 07 06 05 04 03 5 4 3 2 1
paper 10 09 08 6 5 4 3

For those who matter most:
Cathia, Julianna, Mom, and Dad

CONTENTS

ILLUSTRATIONS AND MAPS

ILLUSTRATIONS

MAPS

PREFACE

"Write what you know." It is an author's truism and one that I took to heart in writing this book. I cannot remember a time when I did not know something about North Carolina's Black Mountains. As a toddler I spent a restless night bundled in blankets on the backseat of a 1953 Ford sedan, camping (as we called it then) with my family at Carolina Hemlocks, a U.S. Forest Service campground along the eastern flank of the range. Family lore has it that I awoke before daylight and demanded that my sleepy parents take me to see the South Toe River that flowed nearby. On that or some other such outing, my parents probably told me that Mount Mitchell, one of the peaks looming over the campground, was the highest mountain east of the Mississippi River. If they said that, I have no recollection of it. My earliest Black Mountain memories are of summer afternoons spent wading in the South Toe, the distinctive crackle of campfires at twilight, and the not-quite-musty smell of our gray-green canvas tent.

For my family a summer trip to Carolina Hemlocks also constituted a homecoming of sorts. Our German ancestors (who first went by the name Silber) migrated into western North Carolina from Pennsylvania and settled near Kona, a tiny community ten miles or so downriver from the campground. The clan gained statewide notoriety—some might say infamy—in the early 1830s when one of the in-laws, Frankie Stewart Silver, murdered her allegedly drunken, philandering husband, Charlie. After splitting his skull with an axe, she dismembered his body, burned the remains in a fireplace, and hid the ashes beneath the floor of their Toe River cabin. Convicted of the crime in 1832, she was hanged a year later, an unusual fate for a white woman in North Carolina.

Although I grew up hearing my grandparents tell that story and visited the Black Mountains often as a child, it was as a college student in my twenties that I really began to explore the region. By then I had given up car camping and river watching in favor of backpacking and trout fishing, two activities well suited to the area's steep trails and swift-flowing waters. In later years, when I ventured off to graduate school to study environmental history, my visits to the mountains became more sporadic, but I still returned at every opportunity, lured back again and again by the prospect of watching a sunrise from a secluded campsite or taking an eastern brook trout from a high mountain stream.

Considering my family background and abiding interest in the outdoors, one might conclude that a book like this was inevitable. Perhaps so. For a long time, however, I hesitated to take on what now seems an almost made-to-order topic. As a professor at a university where Appalachian studies is among the most visible graduate programs, I was well aware of the impressive body of work produced by Appalachian historians. I also knew that the Black Mountains had attracted attention from several writers, some of whom had already investigated the region's past.[1] Could a new book add anything significant to our understanding of the region? More to the point, would anyone not familiar with this particular landscape care about what another regional history might reveal?

After nosing around in the sources (at the time as much from personal as professional interest) I answered "yes" to both questions, not because what I discovered was completely new, but because my initial research suggested that as an environmental historian I might use the Black Mountains to offer a fresh perspective on the Appalachian past. Environmental history is a discipline dedicated to exploring relationships between people and nature, to discovering how humans have affected the natural world and, in turn, been affected by it. Getting at those relationships requires what one scholar calls an "earth's-eye view of the past," a view in which trees, crops, weeds, wildlife, and microorganisms are every bit as important as governments, wars, economies, and other human institutions.[2]

For almost three decades environmental historians have been writing those kinds of books about various regions of North America, including New England, the Great Plains, California, the Everglades, and the Pacific Northwest. Yet only within the last five years—in a study of deforestation in West Virginia, in one general history, and in two volumes dealing with the Great Smoky Mountains—have scholars taken the first tentative steps toward adopting an environmental perspective on the southern Appalachians.[3] Even in those fine works the things that people do—building railroads, cutting timber, and setting aside land for national parks—get most of the attention. When it comes to writing the history of North America's oldest mountains, nature has been little more than a supporting actor in a distinctly human drama. The more I learned about the Black Mountains, the more I became convinced that we need a different and, I think, more true-to-life chronicle of the southern Appalachians, one in which nature gets equal time with people.

This book is an attempt to write such a history of a single Appalachian range. Technically the story begins nearly a billion years ago, with the formation of rocks that now lie buried deep beneath the Black Mountains. But my primary focus is the relatively short period during which people have lived on the land, starting some 10,000 years ago and concluding in the present. I devote much of my attention to the last 100 years, during which these peaks, like the Appalachians in general, underwent rapid development and change.

My thesis is simple. I argue that human perceptions of nature—how people thought about the natural world and envisioned themselves in it—dictated most of their activities in the region. However, even as humans confidently went about their business, nature moved to its own peculiar rhythms, sometimes changing the land in ways that people never imagined and often could not fathom, thereby helping to create the Black Mountains that we know today. Because neither people nor mountain ranges exist in isolation, I have tried throughout to show how events on this single landscape also reflected broader trends in North Carolina, Appalachia, the South, and the nation as a whole.

For me, though, this was always more than a scholarly work, not only because of my genealogy, but also because I, like many people in North Carolina, find the Black Mountains interesting enough in their own right. Their summits, including the preeminent Mount Mitchell, comprise some of the most unusual natural environments in the American South, habitats that at first glance seem more like those of southern Canada and northern New England. Like most North Carolinians I take unabashed pride in that particularity. On more than one occasion I have explained to misguided visitors that we—not our parvenu neighbors in Tennessee or Virginia and not the genteel citizens of New Hampshire, Vermont, or Maine—hold clear title to the highest ground in the East.

But that is not the only thing about this landscape that piques our curiosity. In 1857 Elisha Mitchell, the University of North Carolina professor for whom the tallest mountain is named, fell to his death while attempting to measure the summit. That tragedy, which claimed the life of one of the state's eminent men, imbues the land with mystery, danger, and a sense of intrigue that only heightens its distinctiveness. North Carolina acknowledged as much in 1915 when it chose Mount Mitchell as the site for its first state park. In recent years the Black Mountains have become famous for another reason. Over the last two decades the region's dying spruce-fir forests have spawned an intense debate about the effects of air pollution in the southern Appalachians. Indeed, some environmentalists believe that these mountains have been the victim of another, far more serious tragedy: our complacency about the toxins that spew from our power plants and automobiles. According to some clean-air advocates, we are now destroying one of the natural and cultural landmarks we hold most dear.

Though I, perhaps as much as any North Carolinian, think of the Black Mountains as a special place and worry about their future, I initially tried to write about them as if I were a stranger to the region. When I first put my fingers to the keyboard, I adopted the voice of an unobtrusive expert, a scholarly narrator carefully laying out my research without revealing much of myself in the process. But after several ponderous drafts of an early chapter ended up in the recycling bin, I determined that to continue in that vein would be not only counterproductive but also dishonest. Whether I said so openly or not, my affinity for the region and my experiences there would inevitably influence my research and writing. It would be better, I decided, to let those connections show, to draw on them openly, and in essence, make my experiences part of the region's history.

Having come of age intellectually in the 1980s, I knew that such notions had sparked endless debate among historians and their postmodern critics over the proper relationship between writers and their subjects. But truth be told, I did not delve deeply into the abstract world of literary theory. I was simply looking for a way to breathe life into the story that I was trying to tell.

Taking a cue from environmental historian Donald Worster, who once urged his colleagues to buy "a good set of walking shoes" and get "some mud on them," I put aside my note cards and half-finished outlines, got up from the computer, and went back to the Black Mountains.[4] I visited often and at all seasons, hiking, camping, fishing, and rambling through the state park and Forest Service lands—in short, doing all the things that attracted me to the area in the first place—and recorded my observations in a loosely organized journal. I kept similar notes when research took me to Asheville, Chapel Hill, Raleigh, and other pertinent sites.

In time what began as a rather haphazard travelogue became an integral part of the finished book, a book that is—at least for a professional historian—somewhat unconventional in its organization and narrative style. Every major chapter revolves around four expanded entries from that original journal. I have arranged the entries according to the seasons so that each chapter takes readers through a calendar year in the Black Mountains, allowing them to experience the natural world as I saw it and to understand that now, as in times past, nature remains an active, turbulent, and occasionally violent agent of change. As I traveled in the region, I also tried to sense what earlier visitors might have experienced. More important, I sought to use the modern landscape as a text, to read it in much the same way as my printed sources, constantly searching for patterns, distinguishing features, or anything else that offered clues about the past. I now regard those observations as crucial empirical evidence, as vital to this history as any document uncovered in a state archive or university library.

The narrative that follows, then, can be read in two ways. Scholars may wish to think of it as local environmental history, a case study of people and nature in the southern Appalachians. As such, it adheres closely to one of the guiding principles of regional and local studies, namely, that the story of one small place can contribute to our understanding of the wider world. More general readers interested in the Black Mountains for their own sake—including those who have hiked the same trails, slept in the same woods, and fished the same streams as I—may wish to think of the book simply as the story of a unique and wonderful place, one that is, regrettably, very much at risk. As a writer with one foot planted firmly in both camps, I ask each for a measure of forbearance. Academic experts, accustomed to more orthodox histories, will need to accept (or at least tolerate) my presence in the narrative. Likewise, nonprofessionals will have to live with a scholar's idiosyncrasies, including my conventional commitment to documentation and lengthy endnotes. With those and other minor indulgences, I trust, all readers will join me in recognizing that the Black Mountains have much to teach us, not only about nature and history, but also about ourselves.

ACKNOWLEDGMENTS

Not long after I began work on this book, I planted a Norway spruce, an exotic evergreen that figures in the Black Mountain story, in my front yard. At the time the tree was little more than a seedling, standing just over twenty inches tall. Now it tops out at more than twenty feet. It is nature's way of reminding me that I have been at this task for well over a half-decade. During that time I have accumulated an embarrassingly long list of scholarly debts, many of which can never truly be repaid. The best that I can do is to express my appreciation to those institutions and individuals who have made it possible for the book, like the tree, to reach maturity.

Thanks to the history department at Appalachian State University, I have gainful employment in a place that I love with a teaching schedule flexible enough to allow for research and writing. My work on the Black Mountains began with a summer grant from the university's Office of Graduate Studies and Research and a department-sanctioned off-campus scholarly assignment, both of which helped facilitate early trips to various archives. The history department also awarded me its I. G. Greer Distinguished Professorship in History for 1997–99. That signal honor, made all the more meaningful because it came from colleagues, brought some very practical benefits, including a generous research stipend and a semester's leave. In addition I held a National Endowment for the Humanities Fellowship for College Teachers that afforded a luxurious yearlong sabbatical during which I wrote most of a first draft.

Even if blessed with time and resources, a historian cannot work without librarians. At the Southern Historical Collection and at the North Carolina Collection, a host of dedicated professionals made my research pleasant and productive. The staff of the North Carolina Division of Archives and History in Raleigh guided me safely through the state's labyrinthine bureaucracy and the blizzard of paper it generates. Michael Hill, of the archive's research branch, deserves special thanks for finding some crucial documents relating to segregation in the state parks. Cheryl Oakes at the Forest History Society answered many e-mail queries about Forest Service policy. Ann S. Wright and Zoe Rhine of Asheville's Pack Memorial Library were especially helpful during my search for illustrations. And I would have been lost—not to mention deeply in debt for overdue fines—without ongoing assistance from Jack Love, Dianna Johnson, Dean Williams, and Susan Jennings at Appalachian State's Belk Library.

Others helped keep me from stumbling when I ventured onto unfamiliar scholarly terrain. Formal interviews with Virginia Boone, Carolyn Marlowe, Robert Bruck, A. E. Ammons, Joe Scarborough, and Thomas Ellis added immeasurably to my work. David Moore showed me around the Warren Wilson archaeological site, and Charles Hudson helped me puzzle out the route Hernando de Soto probably took through the Appalachians. I discussed Elisha Mitchell's travels and the finer points of trigonometry and long-distance surveying with Perrin Wright. Michael Schafale shared his ideas about land use in spruce-fir forests; John Sharpe and Lea Beazley talked with me about Mount Mitchell State Park. Matthew Rowe, Gary Walker, and Howard Neufeld sharpened my understanding of ecology and chaos theory. Coleman Doggett, of the North Carolina Division of Forest Resources, provided a wealth of useful documents and photographs.

Many ideas that eventually found their way into this book grew out of lively debates in my classes and seminars at Appalachian State. My students—graduate and undergraduate alike—will no doubt see themselves in the following pages. In the summers of 1996, 1998, and 2000 Carolyn Merchant, Shepard Krech III, and I led "Nature Transformed: Imagination and the North American Landscape," a National Endowment for the Humanities Institute at the National Humanities Center. Each year twenty dedicated high-school teachers (joined in 1998 by a National Park Service ranger) braved sweltering heat—and worse, relentless air conditioning—to ponder the elusive meanings of ecology, conservation, nature, and wilderness. Those discussions contributed much to my thinking. But more than that, I appreciate the participants' willingness to trek with me through the woods and fields of the North Carolina Piedmont. It was during those excursions, so efficiently organized by seminar administrators Richard Schramm and Crystal Waters, that I first began to think about landscapes as texts.

A number of colleagues around the country read and commented on parts or all of my manuscript. For their interest and insight I thank Margaret Lynn Brown, Daniel S. Pierce, John Inscoe, Paul S. Sutter, Shepard Krech III, Harry L. Watson, Ruth Currie, Lynn Nelson, Antoinette Van Zelm, Jack Temple Kirby, Tom Hatley, Richard Schramm, Robert Griffith, and Edward Nickens. Ted Fitts, chair of the history department at Moses Brown School and a 1998 participant in "Nature Transformed," merits special consideration, not only for his painstaking attention to my chapters but also for his warm friendship. I received similar aid and encouragement from many individuals in my own department, notably David White, Jim Goff, Steve Simon, John Alexander Williams, Karl Campbell, Sheila Phipps, Janine Lanza, Michael Krenn, Brenda Greene, and Donna Davis. Mike Wade, who read the entire work, also served as department chair while the book was in progress. He is perhaps the most supportive and able administrator a scholar could ever hope to know. He remains a valued colleague and friend.

David Perry, my editor at the University of North Carolina Press, deserves far more acknowledgment than space permits. Suffice it to say that he believed in the book and my methods so much that he once accompanied me on an overnight hike to a 6,000-foot Black Mountain peak—in the middle of January. For those and the other miles we have traveled together, I am grateful. Mark Simpson-Vos, Paula Wald, Stephanie Wenzel, Heidi Perov, and several others at the press extended me many kindnesses and saved me from numerous mistakes.

Because my research often took me to Chapel Hill, I had the pleasure of renewing friendships with Raymond and Judith Pulley, mentors from my undergraduate days, whose hospitality, especially when it comes to entertaining wayward historians, knows no bounds. James Axtell and James P. Whittenburg, mentors during my graduate studies at the College of William and Mary, also provided long-distance advice and counsel. Years ago (more than any of us would care to admit) they, along with Edward P. Crapol and Michael McGiffert, instilled in me a passion for American history and a keen appreciation for the scholar's craft.

On some of my excursions into the Black Mountains I enjoyed the company of some truly expert fishermen and backpackers, including Richard Tumbleston, Al Hines, Drew Sumrell, Kevin Hartley, Wes Waugh, and Charles Miller. A number of friends from St. Mary of the Hills Episcopal Parish, including Father Rick and Beth Lawler, Bill and Donna Devereux,

Alex and JoAnn Hallmark, and Lynn Coulthard, provided fellowship and welcome distractions from my work, all the while reminding me that I was writing about real issues that mattered to real people.

Finally, I thank my family. My parents, Howard and Mona Silver, taught me at an early age to appreciate nature and to love the wild. For those and many other lessons, I have long been grateful. My in-laws, David and Maria Tribby, Ned and Elena Taylor, and Mark and Catherine Tribby, have been unwavering in their support and affably tolerant of a liberal academic's eccentricities. But I owe the greatest debt to my wife, Cathia, a fine university teacher in her own right, and to our eternally effervescent daughter, Julianna. With infinite patience, happy hugs, and more than a modicum of grace, they daily grant me those most precious of familial gifts: unconditional love, acceptance, and the freedom to be who I am. For them, I live.

Mount Mitchell and the Black Mountains

ONE

Origins

JANUARY

UPPER SOUTH TOE RIVER

At 2:00 A.M. I hear the wind. It begins as a murmur on the lower slopes, gains strength as it moves closer, and finally rushes past in a whistling roar. Yellow birches and white oaks bend. Half-frozen hickories pop and squeak. Hemlock boughs crack and fall, spraying twigs and bark across crusty, week-old snow. Close to the ground, dark green rhododendron leaves, shriveled and stiff from cold, rattle lightly as the gust moves on. Higher up the mountain, almost out of earshot, the blast twists and bends through red spruce and Fraser fir. Then, for the briefest instant, calm. In the silence I can pick out the soft tenor of the river moving beneath the ice until, somewhere down the ridge, the murmur begins again.

On this night my tent sits in a sheltered cove high on the southeastern face of the Black Mountains, 4,000-odd feet above sea level. Eighteen inches of snow cover the ground. Three and a half miles north and 2,500 feet up, on the highest mountain in eastern America, snow depths exceed two feet. At my camp the temperature hovers around twenty degrees Fahrenheit; at the summit it will drop into the single digits, typical readings for this time of year.

Kept awake by the wind, I tune a transistor radio to a weather forecast from the North Carolina Piedmont. There, where the hills top out at a mere thousand feet, a meteorologist reports current temperatures in the forties. Fed by warm southern breezes, daytime highs will rise to near sixty. Farther east, where North Carolina's coastal plain dips toward the Atlantic Ocean, cities such as Wilmington and New Bern may see seventy degrees by midafternoon. "Springlike," the announcer calls it, a "January thaw," the perfect time to enjoy the "mild Carolina winter." I switch on a flashlight and watch my breath condense into fog.

Tucked away on the headwaters of this mountain river, I am less than 100 air miles from the central Piedmont and barely twice that distance from the eastern coastal plain. But later today, as I break camp and head toward the mountaintop, the wind will blow hard from the northwest. Thirty degrees will feel like ten. The sun will shine; but the snow will stay where it fell eight days ago, and the ice along the river's edge will only thicken when daylight fades and temperatures drop. This is deep winter. At its most severe it mimics the climes of New England and southern Canada. Yet in the Black Mountains, as in other parts of the high Appalachians, the cold is a curious local anomaly, existing only a geographic stone's throw from some of the South's most temperate lowlands.

Winter is also a climatic artifact, a souvenir from the deep recesses of geologic time. These high peaks emerged long ago amid surging seas and colliding continents. The icy South Toe sprang from ancient weather systems that hammered the slopes with eroding rains. And knee-deep snow is a seasonal remnant of earth's most recent cold snap, the Great Ice Age, a time when the Appalachians were more like the Arctic and spruce and fir spilled out of the mountains onto the Piedmont and coastal plain. Now isolated on the highest ridges, in stark defiance of the warmth below, Black Mountain winter—more than any other season—inclines us to take the long view of the region's history, to think about prehistoric landscapes with odd-sounding names, to consider the transforming power of cold and ice and the remarkable ability of a mountain range to reshape itself, even in the absence of human beings.

A billion years ago earth was a different place. It had no Black Mountains, no Appalachian Range, and no North America. The planet's vast oceans circulated around a huge single slab of sand and rock. About 750 million years back the hard plates in the crust of this megacontinent slowly shifted and separated, floating into the seas like giant icebergs, splitting the land mass into several small fragments and larger continents. Among them were Laurentia, or proto-North America, and Gondwana, or proto-Africa. For more than 200 million years the vast seas between these lands served as catch basins for sand and clay washing from the surrounding terrain. On the floor of those ancient oceans, where such continental debris mixed with volcanic material, the rocks that now form the South's highest landscape began to take shape.[1]

Turning those rocks into mountains was slow business—so slow that human minds can barely fathom it. The best that geologists can do is to break down the process into a series of shorter, more comprehensible episodes that pinpoint key changes in the planet's landforms. The ceaseless saga probably began 500 million years ago, when some of the continental fragments drifted toward Laurentia and collided with that continent's eastern shore. As the seas between the land masses slowly closed, rocks on the ocean floor surged up and over other formations to the west, creating the base of the southern Appalachians. Buried for millions of years beneath Laurentia's surface, those deposits solidified into extremely durable metamorphic rocks layered with quartz, feldspar, and mica. Known as gneiss and schist, the massive formations had a striped appearance, the result of dark bands of mica mixing with lighter strips of quartz and feldspar.[2]

As the rocks melded, the inexorable movement of continental plates again pushed land masses toward one another until, roughly 300 million years ago, Laurentia and Gondwana merged into a second supercontinent called Pangaea. When lands collided, the continental crust folded and wrinkled—like a giant rug pushed up from one end—driving more rocks across the existing uplifts until the mountains of Pangaea resembled today's Alps. For another 100 million years wind and rain battered the primeval range, pounding it into a great plateau. Then, some 180 to 200 million years ago, the agonizingly slow, incessant movement of continental plates began to split Pangaea into the continents recognizable today. The weathered mountains, known as the Appalachian Plateau, eventually became part of North America, breaking apart from similar uplands in Europe and Africa. In the region that is now North Carolina, the Great Smoky Mountains formed the western boundary of the plateau, while the narrow band of mountains known as the Blue Ridge stretched in a long diagonal line southwest to northeast across the eastern edge of the infant continent.[3]

Long after Pangaea divided, periodic episodes of folding and uplift continued to push ancient rock formations to the surface. About 65 million years ago one such geologic event exposed a jagged mass of gneiss and schist that had been buried since the earliest phases of continental movement. Between 39°5′ and 39°7′ north latitude (just northeast of the modern city of Asheville), some of the new mountains formed a fifteen-mile-long semicircle that vaguely resembled a fishhook. The odd formation did not conform to the general southwest-northeast sweep of the Appalachians. Instead, the peaks emerged as a cross range, extending at a sharp angle north and slightly west from the Blue Ridge.[4]

For tens of millions of years, while nearby mountains on the Appala-

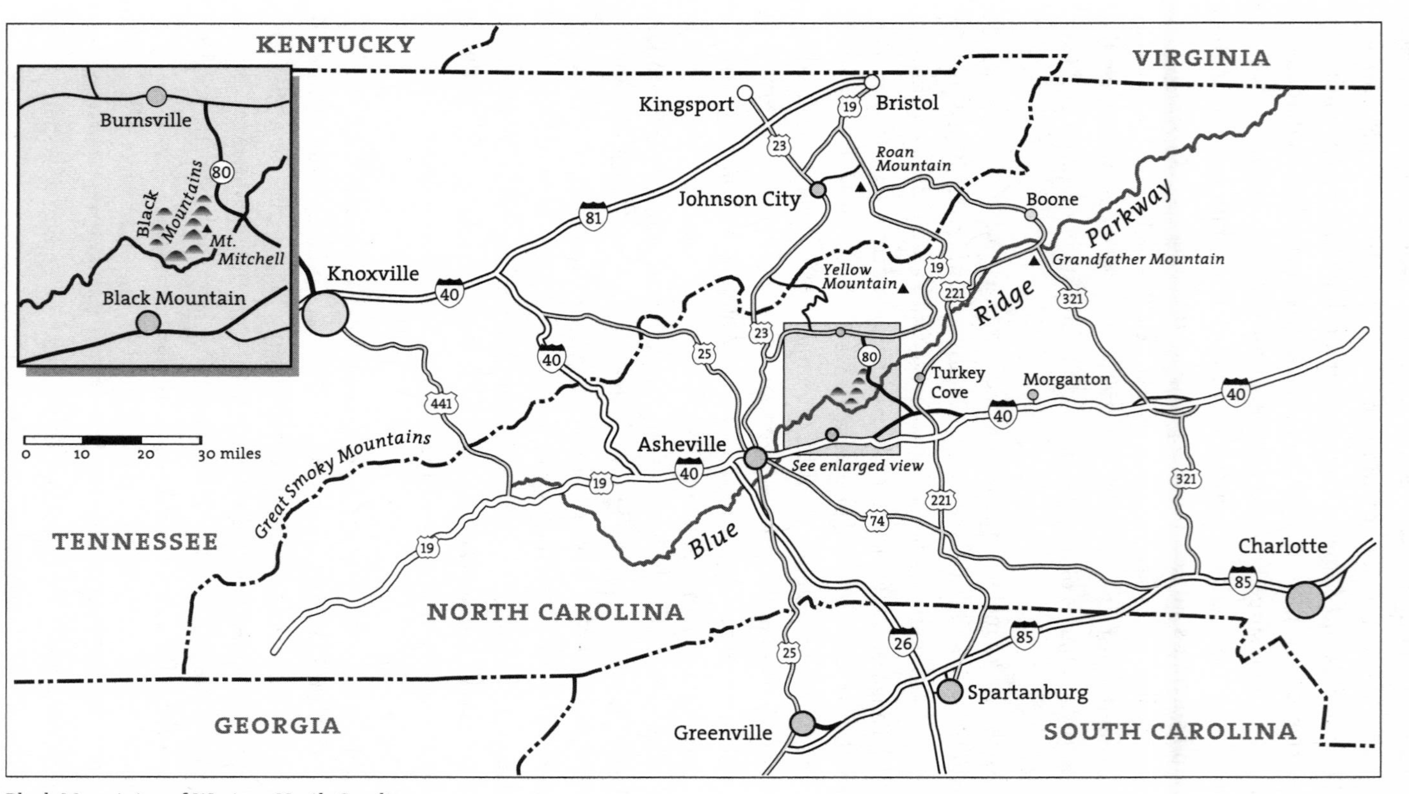

Black Mountains of Western North Carolina

A view of the eastern flank of the Black Mountains showing their domelike character and the dark crown of spruce-fir that gives the range its name. U.S. Forest Service.

chian Plateau weathered into rounded hills, the compressed, superhardened rocks in this cross range proved highly resistant to wind and erosion. By the time humans got around to charting the region, it still contained eighteen peaks over 6,300 feet in elevation. Six of those were (and are) among the ten tallest mountains in the Appalachian chain. Just slightly south of the cross range's midpoint, one especially durable chunk of mica gneiss stood as the highest spot of ground east of the Mississippi River. Because the peaks were covered with thick, dark forests, early white settlers called them the Black Mountains, or simply, the Blacks.[5]

The Blacks survived prehistoric water and wind better than other mountains, but they did not escape unscathed. When the range was young, rains raced from the summits in thundering torrents, carrying sediment that slashed deep grooves in softer material along the steep, rocky slopes. Over time, however, the plunging falls and narrow gorges typical of young landscapes gradually gave way to the "gently elevated coves" and "smoothly rounded valleys" of the modern mountains. Compared with the Rockies or other high peaks in western North America, the East's most formidable mountains do not appear especially rugged. Indeed, other peaks in the adjacent Blue Ridge (those surrounding Grandfather Mountain, for example) stand out in much sharper relief against the nearby terrain. Viewed from the air or from a distance, Black Mountain summits cluster together, conic and domelike, giving the range a crumpled, corrugated appearance. Myriad springs and creeks flow from virtually every furrow and create the river systems that drain the region. For the most part they move with relative ease now, only occasionally diving across some remnant precipice into small plunge pools. But high in the Blacks, where creeks and rivers surge after every spring rain and where sizable boulders tumble down the streambeds, it is still possible to witness the mountain-sculpting power of moving water.[6]

In the Blacks, as elsewhere in the southern Appalachians, the general course of such runoff is determined by the Blue Ridge. As a boundary between the flatter Piedmont and more mountains to the west, the Blue Ridge serves as the Eastern Continental Divide. Rains that fall on its eastern slopes find their way to the Atlantic Ocean, while those that land on its west side wind up in the Gulf of Mexico. Because the Blacks extend north from the western edge of the divide, they give rise to several important rivers that eventually feed the gulf. But on the cross range's extreme southeastern flank, where the fishhook curves into the Blue Ridge, lies a geographic no-man's-land. It is created by two mountains, today known as Graybeard and the Pinnacle of the Blue Ridge (Pinnacle for short). Those peaks form a bridge between the Blacks and the eastern side of the divide. Several creeks that begin here and that, by custom if not by strict geographic placement, have always been linked to the Black Mountains actually flow toward the Atlantic. From such elevated and varied terrain, water is slow to make its way to the lowlands.[7]

A raindrop falling on the long north-south part of the range (the shank of the fishhook) usually trickles into either the South Toe or the Cane River. The South Toe originates on the range's eastern slopes, while the Cane forms high on its western flank. As rivers go, both are narrow and shallow, typical boulder-strewn, high-gradient Appalachian streams. The Cane and the South Toe flow along the base of the mountains, creating long, level valleys that extend the entire length of the range. The South Toe eventually intersects the larger North Toe River, which originates well north of the Blacks on the western slope of the Blue Ridge. Though local residents and most scholars have traditionally described the region where the two streams meet as the Toe River Valley, that is a misnomer. The stream retains the name North Toe even after it joins the south branch. The Cane, too, eventually meets the North Toe northwest of the Blacks, forming a mighty stream known as the Nolichucky River. The Nolichucky

The boulder-strewn South Toe River with the East's highest peak in the background. N.C. Division of Archives and History.

then takes a twisted path out of the Appalachians to the French Broad, the Tennessee, the Ohio, and the Mississippi.[8]

But if a raindrop falls elsewhere in the Black Mountains, it follows a different path. If it comes down in the extreme west (at the hook's point), it probably journeys to the French Broad via Dillingham Creek or the Ivy River, two streams that drain this portion of the range. South of there (where the hook starts its curve) falling water first makes its way into the North Fork of the Swannanoa River, then to the Swannanoa proper, and on to the French Broad and points west. At the extreme southern end of the mountains, where the range curves into the Blue Ridge, a raindrop might slip off the fishhook's bend into Mill or Curtis Creek. From there it travels into the Catawba River, on to the Santee River, and through the

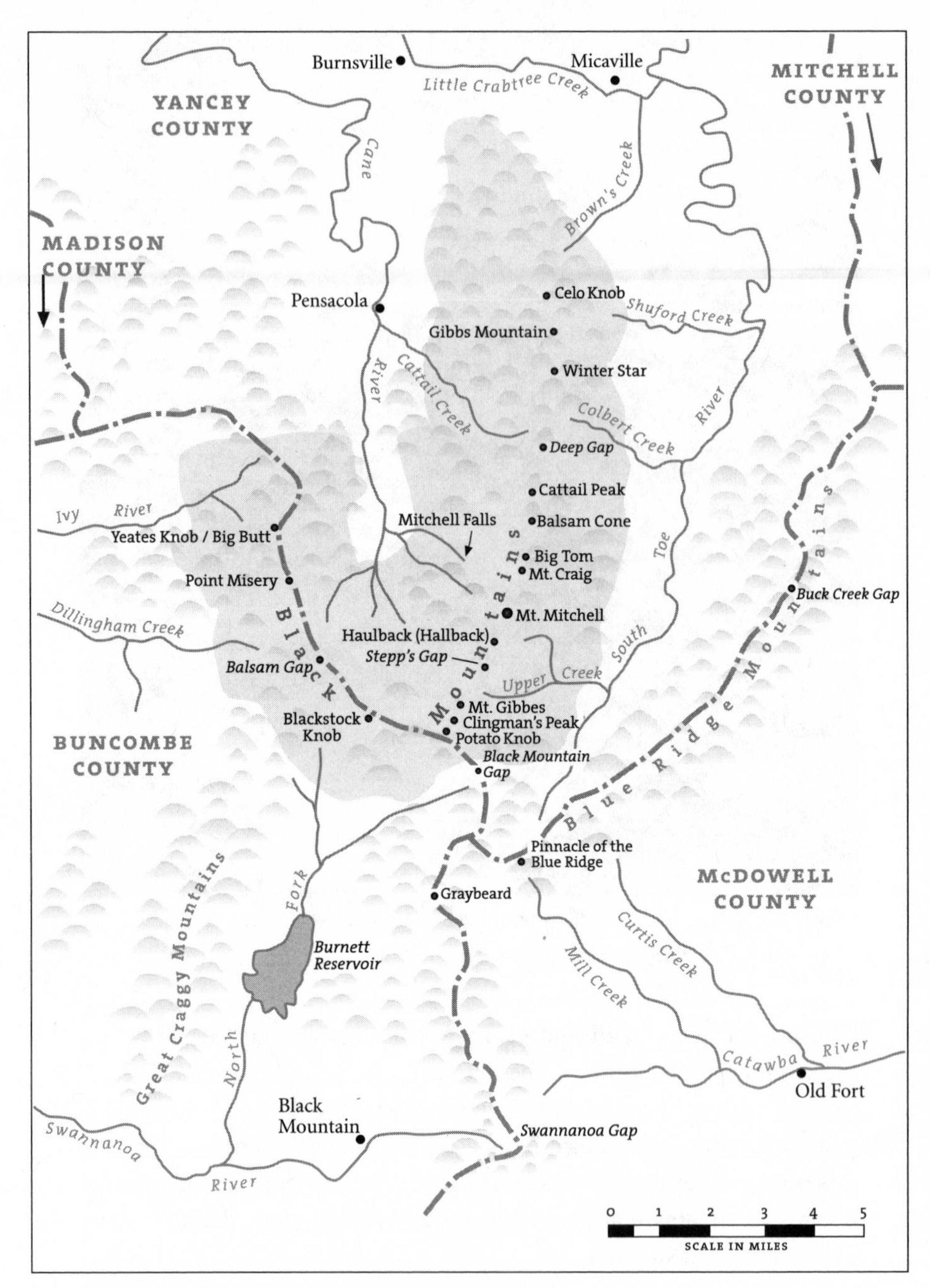

Prominent Geographic Features of the Black Mountains

South Carolina lowcountry to the Atlantic. Due to their topography and unique geographic placement, the East's highest peaks shunt rainwater to all points of the compass. Indeed, a single summer storm hovering over the Blue Ridge at the southern end of the Black Mountains could easily deposit half its precipitation in the Atlantic Ocean and the remainder in the Gulf of Mexico.[9]

Rain falling on the Blacks in ancient times helped grind hard rocks into soil that eventually sprouted the forests for which the range is named. Condensed into an episodic, time-lapse narrative, the history of the region's trees begins with the breakup of Pangaea, when conifers dominated the infantile North American woods. A hundred million years later the evolution of flowering plants and a diversified insect kingdom allowed pollination and the emergence of deciduous trees. Between 65 million and 25 million years ago—as birds and mammals replaced dinosaurs and the Blacks began the long process of weathering and erosion—a huge forest of broad-leafed and coniferous trees seems to have stretched across much of the northern hemisphere. California redwoods, sequoias, southern magnolias, and sweet gums may be vestiges of these early woods, modern reminders of the warm, wet climates in which they originated.[10]

If those trees ever existed in the high Appalachians, however, they are long gone. Like forests across North America, those that have covered the Blacks in recent times are the fresh offspring of an aging planet, spawned not in tropical heat but under the chilling influence of glacial ice. The Great Ice Age or Pleistocene Epoch began about 2 million years ago. Some twenty-five times during this period, glaciers—some small, some gargantuan—slid out of the Arctic, intermittently advancing and retreating across much of North America. Each time the glaciers receded, average temperatures rose significantly, creating what climatologists call an interglacial and providing a geologically short (10,000 years or so) break from the cold until the ice again crept south. The last of the big glaciers, known in eastern North America as the Laurentide ice sheet, receded some 18,000 to 14,000 years ago. That glacier never made it into the southern Appalachians; its lower boundary extended only to northern Pennsylvania. But high mountain forests, including those in the Blacks, did not escape the effects of the cold.[11]

In front of the ice a narrow band of tundra stretched across northern Virginia and down the spine of the central Appalachians toward North Carolina. It also seems likely that an alpine tundra, created by the colder temperatures common to higher elevations, existed on mountains above

3,500 feet. If so, the upper reaches of the Blacks must have been covered with spongy soil that never completely thawed, even during the warmest days of summer. It sprouted few trees worthy of the name, only mosses, lichens, shrubs, and dwarf willows like those now common in the Arctic. Real woods began farther down the slopes where those same plants mixed with spruce, fir, jack pine, and tamarack to form an open, boreal forest. With some variation this coniferous vegetation extended east to the Carolina coastal plan and west to the Mississippi River and beyond.[12]

The animals that inhabited the mountains in those late Pleistocene days were something to behold. Each spring the alpine tundra came alive with a variety of lemmings, voles, shrews, and jumping mice. The rodents furnished food for ermine, marten, and numerous other small predators that hunted the boggy habitat. Where the woods thickened, ptarmigans and spruce grouse shuffled across the forest floor, while gray jays screeched from the lowest branches of the evergreens. Although fossil evidence remains fragmentary, the birds probably shared their boreal surroundings with an awe-inspiring array of large mammals. Perhaps caribou browsed the tender vegetation along the tundra-forest boundary. Mammoths and mastodons may have roamed the spruce-fir region with musk oxen, bison, and moose. Three types of bears (possibly including grizzlies) hunted and scavenged alongside the now-extinct dire wolf. And there were oddities: ground sloths, giant beavers (350-pound rodents that, at first glance, looked more like bears), and piglike long-nosed peccaries. It was a strange world, one that the Appalachians have largely lost. Yet even as it disappeared, it gave rise to many of the plants and animals that still survive in the Black Mountains.[13]

As the ice sheets abated, the Blacks, like the rest of North America, entered the continent's most recent (and current) interglacial, the Holocene Epoch, a division of time that includes the brief (geologically speaking) moment that humans have inhabited the Appalachians. The Holocene brought warmer weather to North America, and as the climate shifted, habitats that had sustained the Pleistocene menagerie began a hasty retreat to the north and west. Natural tundra vanished from the highest ridges 13,000 to 12,000 years ago. In protected river valleys, like those along the Cane and South Toe, where temperatures were milder, the open spruce-fir woodland gave way to mixed forests of pine and hardwoods. Within another millennium, many of the larger mammals had either become extinct or relocated. Three thousand years later most of the boreal rodents had vanished from the lower slopes. The progression continued throughout the first eight millennia of the Holocene. In fits and starts cold-adapted plants and animals moved out; those that had evolved in warmer regions moved in. As they did, the modern forests of the Black Mountains slowly took shape.[14]

MAY

GREENKNOB FIRE TOWER

Spring does not come easily to the Blacks. Today afternoon temperatures will struggle to reach the mid-fifties. Even so, I sweat, wheeze, and unzip my jacket as I work my way through the final switchback toward the top of a lofty promontory called Greenknob. At the top of the ridge, 5,000-plus feet above sea level, the U.S. Forest Service maintains a fire tower that is a favorite stop for hikers. The view from the tower is spectacular; one wonders how the fire watchers concentrate on their work.

I remind myself that I am here on business. I have come to see the forests from above, to seek their patterns, and to imagine the postglacial march of trees that populated the steep slopes. Greenknob is situated high on the Blue Ridge, just east of its intersection with the Blacks. As I make my way to the western side of the tower, I can gaze across the South Toe Valley onto the eastern face of the cross range. In the clear air of early spring, as a climbing sun illuminates the mountains, the forests seem divided into strips of dull color. At the lowest elevations where the trees have leafed out, the woods are pale green. Slightly higher, budding maples are tipped with claret. They mix with hickories, yellow with new shoots but still a long way from full leaf. White-blooming downy serviceberries—locally known as "sarvis"—appear here and there, small clouds floating in a pastel sky. From there the still-wintry deciduous forests turn a torpid gray before giving way to the dark evergreens that blanket the summits.

These seemingly distinct bands of vegetation stem from the relationship between elevation and climate. Generally speaking, each thousand feet of elevation brings a drop in temperature of three degrees Fahrenheit. Consequently, when spring comes to the mountains, the sheer height of the Blacks momentarily blocks its progress. Warm weather arrives first and lingers longest in the river valleys. From there it advances slowly up the slopes, turning gray woods green, until it reaches the conifers at the top. It is not unlike what happened at the end of the Pleistocene. As the overall warming trend moved across the Blue Ridge, colder temperatures in the high mountains interrupted forest migrations. The climate in the upper elevations remained cold enough to sustain evergreens and hardwoods that normally would have settled farther north, while trees more suited to southern latitudes moved onto the warm lower slopes. From the fire tower it all seems eminently logical: trees advancing step by step to take their rightful places on the mountains. Today's forests are much dif-

ferent from those that once covered these slopes. But for the moment, the late Pleistocene seems close. I can see glaciers sliding backward, boreal forests retreating, soils warming, and deciduous trees sprouting.

The great tree migration began with spruce and fir. Eighteen thousand years ago—when Pleistocene temperatures were at their lowest—several varieties of cold-adapted spruce grew as far south as central and lowland South Carolina. Over the next 12,000 years, as temperatures began to warm, the trees became increasingly concentrated in the cooler uplands until, at a point 6,000 years past (when the Holocene climate was at its warmest), most southern spruce populations became isolated in the Appalachians. Four millennia later one species, now known as red spruce, had formed forests that covered the highest peaks and extended downslope to the mid-elevations of the Blacks. Fir moved even faster, probably taking refuge in the Blacks 10,000 years ago and moving north from there. The trees may have disappeared from the Blacks during the warmest millennia of the Holocene, but they returned 2,000 to 5,000 years ago as the interglacial climate again began to cool. During that period a new and distinct species, today called Fraser fir, rose to prominence alongside red spruce in the southern mountains.[15]

The spruce-fir vegetation that covers the summits is, like the winter season that sustains it, a relic of the ice age, having survived in one form or another since the retreat of the last big glacier. In modern forests red spruce appears sporadically in small stands above 4,500 feet. A thousand feet farther up it becomes more evident and mixes with Fraser fir. Above 6,000 feet fir predominates. The evergreens cling precariously to the highest slabs of gneiss and schist, isolated from the surrounding hardwoods, awaiting, perhaps, the inevitable return of the ice.[16]

Downslope, spruce and fir begin to give way to what some naturalists call a northern hardwood forest. It is most prominent between elevations of 4,500 and 5,500 feet, though its remnants sometimes extend to 3,500 feet and below. The gray trees, so slow to green up in spring, are primarily American beech, yellow birch, and buckeye, all of which mix with red maple at lower elevations. Native rhododendron thrives in the forest understory. In the wake of the glacier, forests like this generally moved to northern Pennsylvania, New York, and New England. Once into the Blacks, however, the trees remained trapped on the range, caught in the vagaries of elevation and temperature.[17]

Below 3,000 feet, where steep slopes fade into river valleys, mid-eleva-

The indistinct ecological boundary between Black Mountain spruce-fir and northern hardwood forests. U.S. Forest Service.

tion hardwoods give way to a mixed deciduous or Appalachian hardwood forest. Here, where spring arrives first, the modern forests consist of yellow poplars, pignut and mockernut hickories, and a few remnant maples from the upper elevations. Oaks are plentiful, with white and northern red oaks most visible. Seventy-five years ago the forest was also home to the American chestnut, whose narrow, saw-toothed leaves shifted from pale green to amber as summer slipped into fall. Standing a hundred feet tall and twelve feet in diameter, a single large chestnut might cast a shadow that covered forty yards of forest floor. Such noble trees were once common on elevated ridges in the lower Blacks, perhaps accounting for 75 to 90 percent of the timber in those forests.[18]

All these trees began their postglacial march from refuge forests (so named because they were far enough from the glacial cold to support hardwoods) to the east and south. It was a haphazard journey. Because squirrels and other animals carried acorns great distances, oaks moved quickly. Maples, which dispersed their seeds in the wind, were also early arrivals. American chestnuts probably endured the last glaciation far out on the continental shelf east of the Carolinas, on ground exposed when the glaciers locked up North America's water supply and ocean levels dropped. (That land disappeared into the sea as the glaciers receded.) From there, chestnuts moved slowly inland, migrating at half the speed of maples and only a third as fast as oaks. Chestnuts and the other new arrivals had to compete with and replace existing trees, a process that continued piecemeal for thousands of years. Unlike the boreal and northern hardwood forests farther up, those at the base of the mountains are typical southern woods. Similar vegetation thrives in upland regions across the Carolinas, Georgia, and Alabama, where it also settled as the climate warmed.[19]

Spruce-fir, northern hardwoods, Appalachian hardwoods—it all seems so neat and organized. But perfunctory forest labels, like the woods themselves, have long histories, defined as much by human thought as by nature's processes. This method of categorizing forests—including those in the Blacks—is intricately connected to the evolution of ecology as an academic discipline and the scientific controversies that swirled around the field. That story begins with Frederic Clements, a pioneer ecologist whose ideas have had a profound influence on how most Americans think about nature.[20]

Clements was born a long way from the Black Mountains, on the Nebraska grasslands, in 1874. An avid collector of plants as a youth, he studied botany in college and took a Ph.D. in that field in 1898. Throughout his career Clements concerned himself with one of the biggest questions of vegetation science: how to explain the various patterns of plants that covered Earth's surface. In short, he believed that vegetation organized itself into distinct "associations or formations" that were "uniform over large areas." Such formations (later scientists called them "communities") achieved that uniformity through a process called succession. In any plant community, Clements thought, one type of vegetation replaced another, or *succeeded* it, until that particular piece of ground produced growth ideally suited to the climate.

Much of Clements's work concerned the grasslands of his native plains. But succession, he argued, was a "universal process," occurring over and over "whenever proper conditions" arose. In upland forests, like those of the lower Blacks, succession might begin with pioneering weeds and grasses, then proceed to some sort of early tree cover, such as pines, which give way to oaks and chestnuts and then to red and sugar maples. Forests grew up much like human beings, passing from the instability of youth to the steadiness of old age. At maturity an association reached a state Clements called "climax," a point at which it was "able to reproduce itself, repeating with essential fidelity the stages of its development." Each of those stages was easily identifiable if one looked for certain dominant plants or trees typical of that particular community.[21] The names commonly given to Black Mountain forests—spruce-fir, northern hardwoods, or Appalachian hardwoods—still reflect Clements's method of describing communities by identifying major species.

In the 1930s Clements's view was challenged by Arthur Tansley, a botanist working at Oxford. Seeking an alternative to Clements's notion that plant associations were humanlike "living organisms," Tansley argued that it was better to think of such communities as larger entities that included not only plants and animals but also the surrounding air, water, and physical environment—in sum, everything it took to sustain life. The key to that sustenance, Tansley believed, was energy. It moved through living things by way of various food chains or food webs, concepts originally developed by British zoologist Charles Elton in 1927. Energy came first from the sun. It was captured by plants, which were consumed by animals, which in turn became food for other animals. As long as energy flowed, the dynamics continued. Seeking a name for this setup, Tansley turned to his colleagues in physics, who by the 1930s were speaking of such relationships in terms of systems. Borrowing that appellation and blending it with ideas in his own discipline, Tansley coined a term that would influence both scientific and popular conceptions of the natural world—including that of the Black Mountains—for decades to come. He called his concept the ecosystem.[22]

Tansley's ideas became influential, in part, because they eventually proved easy to blend with Clements's seemingly contradictory theories. The intellectual synthesis took time. A number of prominent ecologists had a hand in perfecting it, testing and correcting some of Tansley's notions about energy flow, and explaining the workings of various food webs. But no ecologist was more important to the fusion of ideas than Eugene P. Odum, who in 1953 authored *Fundamentals of Ecology*, one of the best-known texts in the discipline. He, like Tansley, defined an ecosystem as a community of living things functioning within its physical environment. But in those communities, Odum thought, plants and animals sought stability, balance, and equilibrium. Each depended on the other for life-sustaining energy. In such an ideal world, newly sprouting trees fed deer, deer fed mountain lions and other predators, and waste left by the carnivores, in turn, replenished the trees. The ecosystem's cycle was logical, endless, and perfect. Odum also implied that ecosystems were at their best when they were mature and included a wide diversity of species capable of keeping one another alive. In that state they were, like Clements's climax plant associations, self sustaining and self perpetuating.[23]

Notions of succession, climax forests, balanced ecosystems, and harmonious nature remain important fixtures in the environmental thinking of most Americans. Such concepts echo through popular descriptions of Black Mountain forests. On the slopes of the Blacks, it seems, nature's grand scheme has been to produce, as perfectly as possible, communities suited to all variations of elevation and temperature. Nature appears so systematic here that the Blacks have often been portrayed as an ecomuseum, a place where odd and otherwise distant habitats can be viewed in local microcosm. Guidebooks suggest that the high Appalachians exhibit "almost every forest type that occurs in the eastern half of the continent." Climbing the Blacks is often compared to walking from Georgia to Newfoundland. To find all these forests at sea level, the books suggest, one would have to travel north by car for three days. As one moves from the lowest to the highest reaches of the Black Mountains, however, "all these habitats can be visited on a single three-hour hike." At the top the spruce-fir region, the most exotic ecosystem, is "Canada in Carolina."[24] In this scheme the world of the Blacks is highly stratified, ordered, logical, and easily understood by humans.

But even as Clements, Tansley, and Odum constructed models of climax forests and balanced ecosystems, other scientists dissented. Henry Allen Gleason, a contemporary of Clements, did his boyhood plant collecting in Illinois farm country, took his Ph.D. at Columbia University, and later served as head curator of the New York Botanical Garden. Between 1917 and 1939 he developed the "Individualistic Concept of the Plant Association." On landscapes where Clements found succession, climax, and order, Gleason saw a more random, transient, and competitive hodgepodge of plants. He explained it this way: "Every species of plant is a law unto itself, the distribution of which in space depends upon its individual peculiarities of migration and environmental requirements." A given tree might "migrate everywhere, and grow wherever [it found] favorable conditions." Expanding on Gleason's theories, other ecologists argued that the makeup of plant communities was determined not by hard and fast rules of climate and topography but by the chancy process of seed distribution, an operation that depended on, among other factors, the whims of nut-gathering animals and circuitous wind currents. Stable, climax forests, easily identifiable by their dominant trees, were mostly creations of the human mind. "The behavior of the plant," Gleason declared, "offers in itself no reason at all for the segregation of definite communities."[25]

Although never completely discounted, even during the heyday of

Clements and Odum, Gleason's ideas have enjoyed a recent revival within the scientific community. Today many forest ecologists no longer believe that nature moves inexorably toward climax and balance. Change is simply too common; the chance for disruption of any successional process, too great. A key concept in this more modern view is the idea of natural disturbance, defined as any event, not attributable to human action, that affects forest patterns and growth. The advancing glaciers that turned mountaintops into tundra were agents of natural disturbance, as were the rushing waters that carved riverbeds and built Black Mountain soils. But disturbance does not always take place on such a grand scale. Local winds scattering maple seeds and animals burying acorns can also bring change. As a result, forests that appear perfectly stable are in constant upheaval. Plants move in and out; animal populations wax and wane; general labels for climax communities no longer apply. Some ecologists now find it more accurate to think in terms of smaller, irregular "patches" of vegetation shaped as much by extreme local conditions as by general climate and topography. This theory of "patch dynamics" has become especially important in describing Appalachian forests. Plants growing in a sheltered cove may be strikingly different from those sprouting on a rocky outcrop a few hundred feet away. Like designs in a kaleidoscope, mountain plants and trees form a "shifting mosaic." Vegetation may combine briefly into some identifiable configuration. But the pattern can change at the impulse of a passing storm, a swarm of insects, an ambitious squirrel, or any of a thousand other influences.[26]

The belief that plant communities might not be as constant and uniform as Clements and Odum once suggested has also drawn impetus from the study of chaotic theory, or as it is commonly called, chaos. Chaos is a fairly new branch of mathematics and science that today influences many academic disciplines. Ideas about chaos stem from the belief that, in any system, slight changes in input can drastically affect final outcomes. Many occurrences in nature that appear random really are not. They are caused by events that seem innocuous but that, when multiplied within a system, have dramatic results. Scientists call this phenomenon "sensitive dependence on initial conditions," or how "small scales intertwine with large."

A popular example used to illustrate chaos theory is a falling snowflake. Snowflakes result from water freezing into ice, a common process that begins at thirty-two degrees Fahrenheit. In a stable artificial environment such as a tray in a freezer, water crystallizes from the outside of the tray in, forming a smooth ice cube. Outside the laboratory, however, in the real world of nature, water droplets freeze from the inside out. Ice crystals send out tips, the tips expand their boundaries, the boundaries become unstable, and "new tips shoot out from the sides." The result is a six-sided snowflake. But as the flake floats through the air, the particular ways in which its tips form depend on a profusion of variables, including slight changes in temperature, humidity, wind currents, and atmospheric impurities. Because each drifting snowflake takes a slightly different path, it experiences those variables in minutely different ways and ultimately forms a pattern different from every other snowflake on earth. This process is the "essence of chaos." It involves a large-scale operation—water crystallizing into ice—that in a controlled domain of pure science is predictable, mechanistic, and linear. But in the everyday world of nature the process is small scale, individual, nonlinear, and uncertain. Many paths are open to each freezing water droplet. Which one it takes depends on all the changing conditions it experiences, and "the combinations may as well be infinite."[27]

Trees are not snowflakes. When it comes to forest ecology, however, chaos theory suggests that the woods of the Black Mountains that appear so stable are not just subject to occasional disruption but are, in fact, more complex and changeable than we can imagine. Patterns of succession and climax—the universal, linear mechanisms of Clementsian ecology—are not immutable rules but, rather, just a few of the myriad possibilities open to a given patch of woods. Like a falling snowflake, a forest in the Black Mountains is the result of all the large and small forces to which it has been exposed, from the continental movement of glaciers to the stirring of local winds. Each group of trees (indeed each tree, as Henry Gleason might say) is the product of its own unique history. As one careful student of ecology has written, chaos is "not a complete surrender to disorder." Instead, it implies that if order exists, it is "going to be much more difficult to locate and describe" and will "always have an unruly element of indeterminacy in it."[28]

To witness the effects of chaos and disturbance in the Black Mountains, one need look no further than the region's turbulent weather. Indeed, weather itself is a prime example of chaotic nature at work, as it brings together large regional influences of climate with distinctly local fluctuations of temperature, barometric pressure, and precipitation.[29] On and off since the 1870s the U.S. government has collected weather data from the highest peaks of the Blacks. Thousands of local observers have kept their own extensive records. Such observations indicate that temperatures in the Blacks generally run 25 to 30 degrees cooler than those at sea level. In Asheville, fifteen air miles away at 2,210 feet above sea level, readings are 12 to 15 degrees warmer than they are on the summits. In January, the coldest month in the Blacks, the average high is about 34 degrees Fahrenheit; the average low, 18. In July, the warmest month, highs hover around 67; lows average 54. The frost-free period usually falls between late May and early October. Annual precipitation, the records suggest, is plentiful and fairly evenly distributed throughout all seasons, totaling roughly 80 inches a year.[30]

But like forest labels, average annual readings are idyllic abstractions spawned by the human need to create order where little exists. As anyone who spends much time in the Blacks knows, mountain weather is as erratic as any on the planet. Borne on the jet stream, cold dry air from polar ice caps frequently sweeps south into the cross range. Warm waters in the distant Pacific Ocean breed storms that move across the continent and blow into the Blacks from the northwest. Moisture-laden clouds invade from the Gulf of Mexico. Nor'easters whip up the Carolina coastal plain, pelting the Blue Ridge with a counterclockwise backlash of wind and precipitation. The odd Atlantic hurricane churns across the Piedmont, venting its last fury as it breaks apart over the high ridges. In late summer large high-pressure systems known as Bermuda highs anchor off the South Carolina shore, bringing clear days for weeks at a stretch. In such a volatile atmosphere bizarre, even cataclysmic, weather events are common.

On New Year's Eve 1927 a group of hikers left Marion, North Carolina, to spend the night in a cabin near a summit in the Blacks. During the first hours of 1928, the temperature outside their shelter plummeted to twenty-five degrees below zero. The hiking party survived, apparently without incident, and proudly reported a new record low temperature for North Carolina. But a half-century later, in January 1985, a sudden dip in the jet stream sent a frigid mass of Arctic air knifing into the American South. During the worst of the cold the official weather service thermometer on the East's highest peak read thirty-five below zero, a temperature that shattered all previous records for the mountains and the state.[31]

Around 1:00 P.M. on a seemingly normal Wednesday in August 1931, a fast-moving cold front roared into the Appalachians from the northwest, sending temperatures in Asheville plunging to fifty-seven degrees. As cold air mixed with warm, high winds, thunderstorms, and intense hail raked western North Carolina, shredding crops and damaging farm buildings. By dusk, temperatures in the Blacks were in the thirties, and snow fell during the night. Tourists visiting the high peaks built huge fires and wrapped themselves in blankets to keep warm. Yet in July 1949, under the influence of lingering high pressure, official readings on the East's tallest mountain soared to a balmy eighty-one degrees, one of the highest marks ever recorded there.[32]

Precipitation, too, is highly irregular. In November 1977 a sluggish low-pressure system from the Gulf Coast stalled over the southern Appalachians and brought nine inches of rain and devastating flash floods to the mountains. During the last night of heavy storms, a man living on Shuford Creek (a South Toe tributary near the northern end of the Blacks) awoke to a thunderous racket outside his house. "It sounded like trains and jet planes and loud pops like rifles being fired," he later recalled. "I never heard anything like it before—not in my life." The next morning he found a giant tangle of spruce and beech trees lying inches from the back wall of his home. Shattered, twisted, and almost entirely stripped of bark, the trees had been swept down Shuford Creek basin by a 200-yard-wide landslide that originated on a 6,000-foot peak far above the residence. Scars from the slide are still visible on the eastern face of the Blacks.

A decade later, in 1986, a severe seasonal drought—in part the product of a stalled Bermuda high—severely reduced levels of Black Mountain streams, including Shuford Creek, threatening fish populations and depleting the forest water table. The next winter, however, when precipitation returned, several brutal ice storms ripped through the Blacks, sporadically breaking and killing trees throughout the range. The now infamous blizzard of March 1993, spawned in part by a nor'easter and dubbed the Storm of the Century, dumped five feet of snow on the highest mountains. It took a month to melt.[33]

Each of these chaotic weather events sprang from atmospheric disturbances that began hundreds, even thousands of miles from the North Carolina mountains. Yet each had a profound effect on local landscapes. If it is possible to pinpoint such dramatic episodes in the recent past, how many more such occurrences—floods, droughts, winds, landslides, and ice storms—might have had a hand in determining the composition of Black Mountain forests? A complete answer is probably beyond human comprehension. But to make a start at it, one must abandon the logical, abstract, stratified realm of the fire tower and visit the everyday, patchy, chaotic world of the woods.

JULY

BALD KNOB RIDGE

In early summer Bald Knob Ridge is a glorious place. As I step from the sunlight into the trees at the trailhead, the drop in temperature is palpable. Under the deciduous canopy the morning air is calm, wet, cool, and full of birdsong. Native rhododendron blooms in the understory. Wildflowers dot the forest floor. I walk unencumbered in shorts and T-shirt. In the humidity my clothes are soon soaked. My glasses quickly fog.

Farther up the trail, toward midmorning I notice clouds building in the west. I cannot see it happening, but warm air is rising off the valley floor west of the Blacks. As it slips up and over the high mountains, it cools and becomes unstable, spawning thunderheads. At first I ignore the distant rumbles, believing they will pass. But within an hour the sun disappears; the woods turn dark, and birdsong ceases. After another fifteen minutes the storm breaks full force, and I scramble for rain gear. Lightning crackles. Small pellets of hail fall with rain in the downpour. Although intense, the storm is short lived. After twenty minutes the sun returns. A thin white mist rises through the woods. I keep walking, climbing.

Bald Knob Ridge is a smooth, rounded hump of a mountain near the southern end of the Blacks, at the bend in the fishhook. Its name notwithstanding, the ridge is heavily forested, and the trail that ascends it is one of the most picturesque in the Blacks. Because of its relative isolation, the region still has old-growth timber and areas where human influence has been minimal. The trail begins at 3,900 feet and climbs to 5,200. It starts in an Appalachian hardwood forest and ends amid red spruce. By all accounts it provides one of those celebrated short hikes on which one can view forests like those that stretch from Georgia to Canada.

Yet as the experience of this day suggests, Bald Knob Ridge is a place of chaos, of quick-breaking storms and summer showers. Even on this relatively even terrain, where the ascent is gradual on mountains worn smooth by eons of erosion, evidence of disturbance and patchwork change abounds. For the most part the forest types frequently used to explain vegetation in the region are barely visible here. In defiance of any systematic arrangement, trees pop up at random: a hickory here, a maple there, an oak in between. Conifers mingle with hardwoods; shrubs and plants crowd one another on the forest floor. Vegetation erupts thickly in some places, thinly in others. Nature's method has given way to nature's madness. Even in those tiny pockets that have remained relatively free of human activity, the woods seem imperfect and fractured. It is unsettling. It shakes my already fading faith in ordered nature. But it is the way of forests and, consequently, the way of the Black Mountains.

Water is a key mechanism of change in these woods. Walk along Cane River near the communities of Murchison and Pensacola at the western base of the Blacks. Explore the lower South Toe Valley past the villages of Celo and Bowditch on the eastern side, or venture into the town of Swannanoa, where the river that drains the southern end of the range flattens and winds through a sheltered valley. In terms of elevation these regions are far removed from high places such as Bald Knob Ridge. But when heavy precipitation falls on the summits, floodwaters roll down the mountains. Rivers surge out of their banks and deposit alluvial soil along the edges of the streambed. Deciduous trees abound, but they are lowland species that can adapt to a wet environment. Sycamores and box elders grow thick and fast here. At 2,500 feet the river birch, the only member of the birch family that grows at low elevations in the Southeast, is sometimes evident in sporadic stands. Although this floodplain forest exists only in the lowest elevations of the Blacks, it is an environment every bit as extreme as any high-altitude forest. Periodic flooding regularly uproots trees and alters the courses of streams, forcing the vegetation that survives here constantly to adapt to a changing landscape.

A lack of water can be as important as an abundance of it in determining forest composition. Stand on a southern hillside, either on Bald Knob Ridge or elsewhere at lower or middle elevation. By rights this should be a deciduous forest. But here the soil loses moisture quickly due to constant exposure to the sun. Instead of mixed hardwood vegetation, the lowest exposed patches frequently sprout white, shortleaf, pitch, Virginia, or Table Mountain pines, which can survive when conditions turn hot and dry. Rocky slopes and ledges are also home to the Carolina hemlock, a rare local evergreen discovered in 1850.

Walk farther up a particular ridge, and you will discover that such dry regions can be virtually treeless, populated instead with mountain laurel and blueberry, two notoriously drought-resistant shrubs. Venture into a sheltered cove at the same elevation where temperatures are less variable and where the soil retains moisture. Here, in contrast, huge trees grow in profusion: the yellow buckeyes and maples so visible in early spring and the eastern hemlock that prefers valleys and ravines. Slight variations, minute changes in topography, scratches and dents in ancient gneiss and schist—of such minor vicissitudes are Black Mountain forests made and maintained.[34]

The thunderstorms that roll across the high ridges have the potential to unleash one of nature's most powerful tools for shaping forests: a lightning-set fire. But while thunderstorms are common in the Blacks, extensive natural burns are rare. In deciduous forests the thick canopy of trees keeps the forest floor shaded in summer, maintaining cool, damp conditions near the ground. Only in early spring and late fall and only if the weather is dry are such woods susceptible to lightning fires. Dry thunderstorms, which bring lightning but no significant rain, are not unheard of in the Appalachians. But they are atypical. Even during the driest springs, lightning is almost always accompanied by precipitation that dampens dead limbs, leaves, and other combustible litter. As a result fires started by lightning tend to smolder and creep along the forest floor. Most extinguish themselves quickly as the rains intensify. Only on the driest, most exposed southern slopes is it likely that such burns have played a role in maintaining Black Mountain vegetation. Some of the pines that grow at lower elevations—shortleaf, pitch, and Table Mountain—are fire-adapted trees that can give way to hardwoods in the absence of periodic burns. But fires kindled by people have been common in the Blacks during recent decades; whether lightning fires could have maintained the pines without human help remains an open question.[35]

Nowhere is this resistance to fire more apparent than at the highest elevations. At 5,000 feet on Bald Knob Ridge, where spruce becomes evident, trees grow tall and the woods turn dark. Some of the red spruces here may be 250 to 300 years old. On the clearest summer days, sunlight barely oozes through the thick interlocking branches of the evergreen canopy. Rainwater is slow to evaporate and snows are slow to melt. The forest floor is constantly damp and covered with moss. The soil is peatlike, with a consistency of wet sawdust. Dead limbs, litter, and other potentially combustible materials that fall to the ground are soon soggy with rot.[36]

Even when there is no rain, the heating and cooling of air over the Blacks brings clouds and fog to the high mountain forests. On seven of every ten days, significant fog moves across the upper ridges. In warm seasons clouds swirl out of the western valleys and hang on the peaks, sometimes extending 1,500 feet or farther down the slopes. Usually the fog passes quickly. On other occasions it may linger for days, even a week. When it does, the evergreens become saturated with cloud moisture. Water drips from their branches and showers the forest floor. In winter, clouds blowing across the mountains cool and freeze, forming rime ice that sticks to trees like cake frosting. Fog and rime ice are important sources of moisture in the highest elevations and help keep the evergreen forests free of fire. If subjected only to lightning-set fires, the highest stands of red spruce and Fraser fir might go for a thousand years or longer without a significant burn. Foresters tell of blazes kindled at lower elevations that swept up various mountain slopes only to disappear when they entered the dank realm of fir and spruce. Individual trees are flammable, but "the forest as a whole is nearly immune to fire."[37]

Fire is not the only potential disturbance here. Climb a thousand feet above Bald Knob Ridge and it is impossible to ignore the wind. In summer it comes from the south and east, bringing the season's mildest temperatures. But in the cold months it blasts in from the north and northwest. February, March, and April bring the strongest gusts. Efforts to measure maximum wind speed in the Blacks have met with mixed results. Before the mid-1970s, violent gusts routinely destroyed government anemometers before exact measurements of the highest winds could be made. However, during winter storms, winds frequently reach hurricane strength. Studies conducted across the southern Appalachians show that the highest mountains average twenty-four days a year when winds exceed 75 miles per hour. Gusts of more than 100 miles per hour are fairly common in the Blacks. On the coldest winter days, windchill estimates typically range from 30 to 60 degrees below zero. Stories vary regarding the strongest winds ever recorded on the East's highest peak. But one reading of 189 miles per hour, registered just before an anemometer shattered, at least suggests maximum velocity.[38]

The effects of such winds are visible everywhere in Black Mountain forests. At lower elevations hemlocks and pines are especially prone to blowdown and breakage. Where large limbs or whole trees are knocked down by wind, the forest canopy opens, giving species adapted to open, dry terrain a chance to move in. Higher on the slopes, in the upper reaches of the deciduous forests, trees are often twisted and stunted by wind and cold. On the most elevated ridges these patchy woods, typified by small, tightly clustered stands of trees sometimes no taller than twenty feet, look more like orchards.

Winds are most disruptive among spruce and fir. Soils here are thin; in places less than a foot of dirt and organic matter covers the bedrock. Fraser fir and red spruce are, of necessity, shallow-rooted trees that can fall any time winds exceed fifty miles per hour. Moreover, some of the windiest days occur in late winter and early spring just as the forest floor thaws and becomes muddy from winter snowmelt. Once bent by a strong gust, a mature spruce or fir can easily be pried loose from its moorings in the spongy soil. Farther north, in the mountain forests of New York and New

Mountain ash growing in a disturbed area of spruce-fir forest. Photograph by the author.

England, winds create long, narrow bands of dead trees called fir waves. But in the southern Appalachians, and especially in the Blacks, wind tends to blow down trees on exposed slopes where the evergreen canopy is already thin. As a result wind-damaged Fraser firs usually die in irregular patches, creating small random breaks in the high-altitude forest.[39]

In the Blacks such blowdowns are most visible on western slopes exposed to the prevailing winter winds, particularly in the Cane River, Ivy River, and Dillingham Creek watersheds. Once a patch of evergreens is cleared by wind and weather, other plants compete for the vacated territory. Blackberries or similar shrubs sometimes invade the plots. Yellow birch also takes advantage of such clearings. Its seeds are plentiful and light, easily scattered on the wind and quick to germinate in the thin moist soils. The pin cherry is a small tree that often resembles a bush. In summer it produces a tiny, quarter-inch cherry with a large stone. Birds feeding on the fruit often carry it many miles. Because the seeds have an almost impermeable coating, they can lie dormant on the forest floor for a half-century. When wind levels a spruce or a fir, pin cherries spring to life, flourishing on the open, sunny ground. Mountain ash, too, is a showy hardwood that in autumn sprouts bright red berries. It produces fewer seeds than either the yellow birch or the pin cherry. But its fruit is an important food for various birds, which deposit the seed-bearing berries in open areas at high elevations. Well adapted to the moisture-rich, fog-laden world of the highest elevations, mountain ash frequently sprouts in irregular stands alongside the evergreens.[40]

So it went for many millennia in the Blacks. Stirred by forces distant and local, glaciers moved, trees migrated, weather systems blew in and out, and plants and trees fought one another for vacated space. Mountain forests remained in constant flux. And so it continues, unnoticed by most humans. Continental plates still shift and grind away in relentless methodical movement. Long-term studies of climate suggest that North America's most recent interglacial, the period that produced the modern Black Mountain forests, is in fact an aberration. Viewed within the enormity of geologic time, the approximately 10,000 years since the last big glacier constitute only a momentary warm spell on a continent that is normally much colder. Indeed many climatologists believe that for the last 5,000 years or so, overall temperatures in North America have been dropping, perhaps propelling the continent toward another ice age (though lately scientists have uncovered considerable evidence that global warming spawned by a depleted upper ozone layer and the accumulation of certain gases in the upper atmosphere may be altering this trend). The effects of such cooling have not been lost on the East's highest peaks. During much of that time red spruce expanded its range in the Blacks, taking back some of the territory lost to hardwoods at the end of the Pleistocene.[41]

In forests as dynamic as these, labels such as spruce-fir, northern hardwoods, and Appalachian hardwoods now seem woefully inadequate. They are perhaps most useful as mental constructs, a severely limited human way of describing the indescribable. Likewise the once-hallowed concept of the ecosystem has also become something of an abstraction. Scientists studying Black Mountain forests still make use of the term, speaking and writing, for example, of the "spruce-fir ecosystem." But today many ecologists would agree that it is an imperfect concept and simply a useful way of thinking and talking about an infinitely complex world.[42]

Admitting that nature in the Black Mountains is less predictable and stratified than we once thought also requires that we reexamine those long-held popular perceptions of the region's landscape. Only in the most vague and general sense have the Blacks produced miniature replicas of other eastern forests. No matter what the guidebooks suggest, the spruce-fir region is not really a piece of Canada preserved in North Carolina. Fraser fir is, after all, a species that evolved in lower latitudes and remains unique to the southern Appalachians. Though red spruce grows farther north, it is the only spruce that has successfully adjusted to life in southern climes, an adaptation accomplished during its ice-age migrations. Warmer southern temperatures have also made yellow birch more common than in similar forests farther north. From top to bottom, other forests are not reproduced here but, rather, are altered and re-created to suit local conditions.[43]

Fragmented forests punctuated with sunny clearings, rocky outcrops, and small patches of uprooted trees may perplex humans and confound traditional schemes of classification. But other creatures find the kaleidoscopic mixture of vegetation more to their liking. The exotic beasts of the Pleistocene have disappeared, but the Blacks still provide habitats for a stunning array of animals, some of which are found nowhere else in the South and a few nowhere else in the world. They are visible at all seasons, though conscious efforts to find them often meet with frustration. The woods are too complex, the habitats are too varied, and the mountains are too difficult to traverse quietly. An old hunter's adage is especially appropriate in these forests: Look for animals and you never find them; wait long enough and they find you.

OCTOBER

BEAR WALLOW RIDGE

High on this south-facing slope the omnipresent wind has, for once, disappeared with the daylight. Two hundred feet below my backcountry campsite, Upper Creek, a tributary of the South Toe, running low in this dry season, splashes down its streambed with the muffled cadence of distant static. Otherwise an eerie silence pervades the woods. So deep is the quiet that even insensitive human ears can pick out the tick-tick-tick of yellow birch leaves dropping from the forest canopy. Somewhere the buzzing of a lone katydid starts and stops short, a reminder that summer is past and colder temperatures are in the offing. I lie awake, listening.

As its name suggests, Bear Wallow Ridge is an area favored by wildlife. Earlier, along Upper Creek, I happened upon an enormous pile of bear dung, full of berry seeds. A short time later, in the shadows of late afternoon, two deer vanished into a streamside stand of yellow birches. Now, as darkness envelops the mountain, I hear the rustling and shuffling of tiny feet in the forest litter outside my tent. Shrews, moles, voles, woods mice, and other small rodents are making their rounds. A bit later I sense more pronounced movements in the fallen leaves. I refuse to look but assume this is a striped skunk, a notoriously myopic animal, searching for food. Then, from a nearby ridge, comes the hooting of a great horned owl. I recall that the owl is one of the few predators that ever threatens a skunk. Both hunter and hunted are abroad this evening.

I remember a similar fall night several years ago spent in a public campground on the eastern edge of the Blacks. On that occasion I and the other campers awoke in the wee hours to the musky odor, heavy footfalls, and unmistakable panting of a prowling black bear. Having spent several anxious nights with bears in the Great Smoky Mountains, I tried to tell myself that such encounters are usually—though not always—harmless. Conventional wisdom holds that black bears rarely bother humans if the latter have no food, and I had disposed of anything edible long before dark. Still, the thought of the curious creature sniffing around my tent, sometimes only a few feet away, was unsettling. That night no carefully considered logic could squelch my fear. Even after the bear moved on, I never slept.

That recollection and memories of the scat on the trail provide little comfort on this night either. But this time, even on Bear Wallow Ridge, no animals disturb me. Sleep comes sporadically. Awake at dawn, I hear the telltale "yank-yank-yank" of a white-breasted nuthatch flitting about on a nearby deadfall. Somewhere in the distance a woodpecker drums incessantly. At the creek, where I filter water for morning coffee, a ruffed grouse, beating its wings in thunderous rhythm, explodes from a rhododendron thicket. I start at the sound, and my water bottle tumbles downstream. Even in daylight, bears, skunks, and other denizens of the night remain a disquieting memory. I wonder about the creatures that amazed and unnerved earlier travelers. Perhaps their perceptions of Black Mountain forests and wildlife were, like mine, shaped by the powerful interplay between human imagination and the natural world.

In the low-elevation forests of the South Toe, Cane, and other valleys, life revolves around the rivers. As they tumble downslope, tributaries like Upper Creek surge and churn around ancient rocks, creating small waterfalls and riffles that keep their waters heavy with dissolved oxygen. Because the streams originate in the highest ridges, the water starts out cold and remains so even as it forms larger rivers on the valley floors. In summer thick vegetation shades the stream banks, keeping water temperatures cool.

Such fast-flowing rivers are prime habitat for an infinite variety of insects and crustaceans that provide sustenance for cold-water fish, including trout. Today Black Mountain fish populations are more eclectic than in times past, the result of extensive stocking programs and accidental introductions of nonnative species. But before human influence was conspicuous in the range, the fish most prominent in its streams were eastern brook trout, the only trout indigenous to the region.[44]

At one time in the distant past brook trout were native only to the northeastern quadrant of North America. Their range extended south to New Jersey, north to Hudson Bay, and west to the Great Lakes. Like other animals, trout migrated south during the Pleistocene and established themselves in the waters of the southern Appalachians. They stayed only because the mountain streams remained cold enough to support them. Strikingly beautiful fish, with dark green vermiculated backs, red spots, and orange fins, brook trout are, like red spruce and Fraser fir, holdovers from the mountains' ice-age past.[45]

Since then brook trout have figured prominently in the river environments of the Black Mountains. An important food for raccoons, river otters, minks, and other animals common to the region, brook trout helped establish a key ecological link between river and woods so that life in the water helped sustain life on land. And a rich life it was. The small mammals shared their valley habitat with modern beavers, very distant and much smaller relatives of the giant Pleistocene rodents. Away from the rivers in sporadic clearings, small woodland predators held sway. The least weasel, the smallest carnivore in North America, fed on native mice and moles, while its larger cousin, the long-tailed weasel, preyed on shrews, eastern chipmunks, and eastern cottontail rabbits.[46]

Major river valleys in the Blacks have also provided important habitat for larger mammals, including white-tailed deer. Primarily browsers, deer frequently feed on newly sprouting trees and plants. At the onset of warm weather such forage can often be found at lower elevations, where the Appalachian spring begins. Later in the year deer tend to move farther up the slopes to open areas in the hardwood forests. The animals continue to feed there well into autumn, especially if acorns are plentiful, before they move back toward the rivers during winter. Until the late 1700s deer might (ecologists are not yet ready to say for sure) have shared Black Mountain forests with wapiti or American elk. Like deer, elk prefer high altitudes in summer and lower-elevation woods in winter. They frequently congregate in cleared patches. Given the effects of wind and other disturbances at high elevations, the Black Mountains might have offered enough open range on the highest peaks to sustain a small population of these large-hoofed mammals. Some ecologists argue that bison, more commonly called buffalo, were also once native to North Carolina. But archaeological evidence of their presence is meager. Besides, buffalo require even more cleared space than elk. If the animals did live in the state, they were much more prominent in the Piedmont and points east than in the high mountains.[47]

Exactly how many white-tailed deer and elk once lived in the Blacks is difficult to gauge. Today's deer populations are probably much larger than those in the past, primarily because a number of large predators, some of which are now gone from the region, once kept the herds in check. Not the least of these was the mountain lion, or catamount. A powerful animal that can leap twenty feet and outrun deer over short distances, a healthy catamount can consume a white-tail a week in areas where deer populations are large. Gray wolves, also indigenous to the Blacks, were probably equally important in controlling the deer herds. Bobcats, still prevalent in the range, did and still do take young white-tails. As one longtime resident along the North Fork of the Swannanoa noted, "Wildcats . . . have been known to spring from elevated locations onto passing deer which were medium sized, as deer go, or even larger, [but they] are very destructive to fawns and yearling deer."[48]

These larger predators roamed the Blacks from top to bottom, using large sections of contiguous woodland in their search for food. But when it comes to long-distance foraging, few animals can match the omnivorous black bear. Feeding on anything within sight or smell, including fish, tree bark, berries, insects, and small mammals, a solitary bear may cover fifteen square miles during a single night's pillage. Because the same bear frequently turns up in disparate locales, estimating Black Mountain bear populations has always been difficult. They were, however, among the most visible animals at the beginning of the nineteenth century when naturalists began to investigate wildlife populations in the range. Telltale signs of bear habitat included clawed "bark and limbs of the balsam [Fraser fir] and black [red] spruce," where the beasts rose on their hind legs to "make long deep scratches in the bark with their forepaws." Early observers believed bears scratched for sport or "to let their companions know their whereabouts," an idea that retains credence among animal ecologists. Such bear trees, still visible in the backcountry, are also used for sharpening claws and shedding loose hair, especially in spring when the animals emerge from their lengthy semihibernation.[49]

Reptiles, too, can still surprise an inattentive hiker. The dry rocky outcrops of the lower south-facing slopes are notorious habitat for timber rattlesnakes. As the weather turns cooler, timber rattlers frequently congregate or "den up" in the rocks. There they spend the winter, often passing the cold months coiled together with venomous copperheads and nonvenomous species that crawl in to escape the ice and snow. The snakes tend to use the same sites year after year and, since timber rattlers have been known to live for three decades, a snake-infested cliff might (if unmolested by humans) remain prime reptile habitat for years.[50]

Lower-elevation hardwood forests have always been home to myriad birds. Along with blue jays, cardinals, wood thrushes, and other songbirds, many generations of ruffed grouse and wild turkeys have grown fat on the plentiful supply of insects, seeds, and mast. Crow-sized pileated woodpeckers drill nesting cavities high in dead trees, while robins frequent open areas in search of earthworms. Before passenger pigeons were exterminated by meat hunters in the nineteenth century, great flocks occasionally passed over the Blacks during their annual migration south, roosting among the oaks and chestnuts at the base of the mountains. Since the early 1800s bird watchers have paid particular attention to the upper peaks. Peregrine falcons and a variety of hawks (including Cooper's, sharp-shinned, and red-tailed) have nested in the Blacks. On rare occasions observers have seen bald and golden eagles, though neither bird makes its home in the region.[51]

Many of the fish, mammals, reptiles, and birds that inhabited the Black Mountains have been common elsewhere in the Appalachians or in the forests of the Piedmont and the coastal plain. But some that live in the highest elevations are rare elsewhere in the South. Mountaintop conifer forests, in fact, serve as an exotic animal preserve. Like Fraser fir, red spruce, and brook trout, various mammals and birds that now live among the evergreens stayed in the Blacks because temperatures remained low enough to sustain them, even after the glaciers retreated. In its appearance and ecological function the spruce-fir region is much like an island: a large patch of evergreens isolated on the summits with an ocean of hardwoods below. Animals that live up among the spruce and fir have difficulty surviving just a few hundred feet downslope.[52]

New England cottontail rabbits, northern flying squirrels, and rock voles are among the small mammals that inhabit the peaks, far removed from larger populations in higher latitudes. Another odd and rare resident is the northern water shrew, a strange little rodent that can trap air bubbles in the hair on its hind feet and, quite literally, run on water. Common ravens, which flourish in Canada, Greenland, and the American West, routinely navigate skies over the East's highest mountains. Red crossbills, cedar waxwings, Blackburnian warblers, and black-capped chickadees are but a few of the songbirds from northerly climes that have found at least a temporary home in the high-altitude South. Likewise, the diminutive northern saw-whet owl, which ranges as far north as southern Alaska, hunts shrews and mice in the upper reaches of the Blacks.[53]

The commingling of northern and southern animals in the highest Appalachians suggests that these are not simply northern habitats that have somehow been preserved intact. Instead, what has emerged on these elevated slopes is a unique, southern, high-mountain forest with its own quirks and idiosyncrasies. No animal epitomizes the peculiar qualities of Black Mountain life more than the spruce-fir moss spider, a minuscule member of the tarantula family only an eighth of an inch long. Discovered on the East's highest mountain in 1923, it is distantly related to some of the giant spiders of the tropical rain forests. In the Blacks this tiny arachnid feeds on springtails, infinitesimal insects that live in the thick, damp moss on the forest floor. Those insects require large patches of moss to breed and survive. Appropriately sized moss mats seem to occur only under Fraser firs. As the fir's range diminished when the glaciers retreated, the spider's habitat became increasingly restricted. Now the spruce-fir moss spider survives only in certain forests on the highest peaks of the southern Appalachians. It exists nowhere else in the world.[54]

The intricate association between spider and tree illustrates a fundamental ecological reality: in the turbulent world of the Blacks (or anywhere else in nature), that which affects one usually affects others. A prolonged blast of westerly wind that levels a fir also changes the growth of moss on the forest floor. Without the moss, the springtails disappear; if they go, spiders follow. Relationships between plants and animals are not only intricate and interconnected but also transient. Because of air currents and weather systems, what happens to tiny spiders in the Black Mountains is inevitably tied to events in distant places: the Atlantic, the Ohio Valley, the Gulf of Mexico, or the Arctic. During the earliest centuries of European settlement in the South, the Blacks lay on the far western periphery of North Carolina. Sheer distance from the more densely populated Piedmont and coastal plain made the mountains seem isolated and remote. Ecologically speaking, however, nothing was, or is, further from the truth.

That concept remains elusive for most individuals who think about nature in the high Appalachians. Scientists may debate the merits of patch dynamics and chaos theory, but Clements and Odum still have a lock on the popular imagination. Because the Blacks contain some old-growth timber and today are largely unsettled, many visitors to the region find it easy to imagine some (probably ancient) time when the mountains were inviolate, when spruce-fir, northern hardwoods, and Appalachian hardwoods formed perfect climax forests at appropriate elevations—a time when trees were thick, animals were plentiful, and nature was balanced and pristine. In part, this affinity for flawless original nature stems from an inability to comprehend the magnitude of the earth's geologic past. People have lived in North America only during the last part of the Holocene. They have been active in and around the Black Mountains for an even shorter period. Moreover, humans enjoy a life expectancy that is only a fraction of that of a red spruce or a white oak. As individuals and as a species, our sojourn in the uplands has been incredibly brief. Against a temporal backdrop that includes moving continents, changing climates, and emerging forests, the human presence in the East's high mountains emerges as a sudden and recent plot twist, a mere blip in the infinite trajectory of time. Yet we ignore our transience. For generations humans have defined pristine nature—in the Black Mountains and elsewhere—as an unblemished world that existed when people (especially literate white people who could pass on their observations) first appeared on the scene.[55]

Several writers have suggested that this problematic perception reflects one of the most powerful "core myths" of Western culture: the notion of nature as Eden. Stated in its simplest terms, the myth, which relies on popular perceptions of Clements and Odum as much as on biblical scripture, holds that if humans can avoid disrupting and corrupting it, nature is productive and perfect—in a word, Paradise. Indeed, much of American history has been written and read with that idea in mind. In the eyes of the first white settlers, the continent was wild and dangerous, a savage land. From that point those who embrace civilization and technology see a narrative of progress. Plowing fields, irrigating crops, and building cities and roads stand as heroic efforts to subdue an uncooperative earth and restore or "reinvent" the flawless world of original nature, of Eden. But others, especially environmentalists, read the narrative differently. For them, America was perfect in its untamed state. Its subsequent history is not progressive but, rather, a story of steady decline in which a once-productive and pristine continent has, at the hands of greedy and corrupt humans, "become a degraded desert."[56]

Nature as Eden inevitably looms large as we begin to consider people in the Black Mountains. Whether we choose to regard humans as a positive or a negative influence—whether we elect to write their story as a narrative of progress or declension or something in between—the temptation remains to think of nature in the Blacks as separate from human experience, an original world of stability and perfection. Virtually every facet of the nonhuman history of the region mitigates against that assumption. Which mountains are original? The gneiss and schist of a billion years ago? The original cross range? The weathered peaks encountered by humans? Which environment is pristine? The tundra and boreal forests of the Pleistocene? The relic spruce-fir? The recently arrived hardwoods? Which animals represent authentic nature? Giant beavers and ground sloths? The black bears, deer, and elk seen by white settlers? Such questions have no easy answers. As ecologist Daniel Botkin has written, the inescapable conclusion is that nature is not static, "like a single musical chord sounded forever." Even when "undisturbed by human influence," the East's highest peaks, like all of nature, more closely resemble "a symphony whose harmonies arise from variation and change over every interval of time." A changeless perfect landscape—the Black Mountain Eden—has "never existed except in our imagination."[57]

But even when (and maybe because) it occasionally conjures faulty images, human imagination is a powerful force and one to be reckoned with when we begin to consider how people have changed the Black Mountains. Wherever humans go, they seek to manipulate the nonhuman world to their advantage. Because people are cognitive beings, how they perceive nature and act within it usually hinges on their own experiences and ambitions. (Indeed, the decision to rely on modern theories such as patch dynamics and chaos to explain forest change is, in the end, based on my own set of observations and perceptions.) Some scholars who study environmental change have gone so far as to suggest that nature itself "is a profoundly human construction." This does not mean that the Black Mountains have no life of their own. As an uneasy night spent on Bear Wallow Ridge demonstrates, trees, plants, wildlife, and weather are real entities that exist apart from human experience. But how we think about nature and, consequently, about life in the Black Mountains is usually "so entangled with our own values and assumptions that the two can never fully be separated."[58]

The first people who ventured into the Blacks often had strikingly dissimilar ways of thinking about nature and their places in it. Some came to hunt and farm; some, to search for treasure and profit; others, to inspect and classify. But no matter how carefully they planned, how self-serving or altruistic their motives, people were only the newest agents of change in an already highly changeable world. Black Mountain plants and animals had long been linked to complex environments that extended far beyond the summits of the tallest peaks. With the arrival of *Homo sapiens*, the lives of moss spiders, brook trout, Fraser fir, red spruce, and a host of other organisms inevitably became entangled with the perceptions, hopes, and fears of human beings. The future of a unique southern landscape hung in the balance.

TWO

Footprints

SEPTEMBER

SWANNANOA VALLEY

An hour after dawn, fog is as thick as traffic in the westbound lanes of Interstate 40. Since daylight I have driven along the Catawba River, tracking its main prong upstream past the town of Old Fort. At the Eastern Continental Divide, where this superhighway crests, I leave the Atlantic-bound streams behind and descend 400 feet into the Swannanoa Valley. For eighteen miles straight ahead, elevations rarely exceed 2,500 feet. Yet just eight miles north, shrouded in September haze, stand the highest peaks in the eastern United States.

This amazing mountain corridor is, in part, the geologic handiwork of the Swannanoa River, a stream that rises just west of the divide. Gaining strength as it cascades from that slope, the river swings southwest, passing through a series of shoals and rapids until it meets its North Fork falling fast from the Blacks. I pull up beside the Swannanoa not far from that confluence, where it emerges into a series of broad, flat bottomlands. Out here, a few miles from the interstate, the landscape is a patchwork of pastures, woodlots, plowed fields, and gardens, clearly a place where humans have marked the land. That is, perhaps, as it should be. Thousands of years ago this magnificent valley was home to some of the first people to see and explore the Black Mountains.

Much of this fine land now belongs to Warren Wilson College, part of a 300-acre farm worked by faculty, staff, and students. But on one corner of the plot, hard by the river, next to grazing cattle and fields plowed clean by modern machinery, lies a remarkable patch of ground. It is a small tract covering roughly three acres. Fifty years ago local farmers found arrowheads and pottery shards scattered across its surface. Today archaeologists know it as the Warren Wilson site. Since 1965 they have carefully sifted its soil, searching for artifacts left by native people. Ancient tools and projectile points have surfaced in other valleys in and around the Blacks, notably along the upper Toe and Cane Rivers. But no site has been scrutinized as long and as carefully as Warren Wilson.

Today "the dig" (as the folks at Warren Wilson call it) is deserted. The summer's work has ended, and the excavations lie covered with black plastic. As the sun burns away the high fog, it is easy to see what drew the first people to this spot. The river is not large, but its flow is steady, even in this season of lingering high pressure and persistent drought. The dull hum of cicadas and steadily building morning heat remind me that summers are longer and winters less cruel than on the surrounding peaks. I sit by the sluggish stream, sipping the last of my morning coffee. Long shafts of sunlight push their way through sycamore, locust, poplar, and pine. A gray squirrel barks from the far bank. I picture others walking here, working here, establishing the first links between humans and the natural world.

People began trooping through the Swannanoa Valley about 8,000 years ago during the warmest millennia of the Holocene. They traveled in small, family-centered groups of fifty to seventy-five individuals usually led by an older male who had proven his prowess as a hunter. They were part of what anthropologists call the Archaic tradition, a seminomadic way of life that combined hunting and fishing with gathering wild foods. Those who came to the Swannanoa probably discovered the valley while hunting along the upper Holston and French Broad Rivers. The newcomers carried various tools and weapons, including powerful weighted spear-throwers known as atlatls. Archaic folk made temporary homes along the river, digging fire pits and setting up bark-covered huts. In spring they used traps and spears to take fish from Black Mountain streams. Women searched the woods for early fruit and edible plants. With the first frosts of autumn they ventured onto the surrounding mountain slopes to gather hickory nuts, acorns, and chestnuts from the high ridges to the north.[1]

Meanwhile men fanned out across the Blacks in search of deer and other game. Cloaked in deerskins (complete with head and antlers) for camouflage, they slipped quietly along the rivers, stalking small herds of white-tails. Stealth was crucial. Using an atlatl, a hunter could toss a spear that would travel the length of a modern football field, but the weapon was most deadly at a range of twenty to forty yards. Even when hurled with maximum force from a standing position, the projectile rarely dropped a deer in its tracks. Sometimes hunters tracked a wounded animal for miles, following bloody trails through thick vegetation, before administering the coup de grace with a spear driven through its heart. Around campfires at their Swannanoa base camp, men cleaned the meat and apportioned it to the group. Women scraped and tanned deerskins for clothing. Artisans fashioned antlers and hooves into tools, fishhooks, and ornaments. If game remained plentiful, the group continued to camp along the Swannanoa; if not, they moved on to another river, another valley. For the next 5,000 years such scenes unfolded again and again in the shadow of the Blacks.[2]

About 3,000 years ago, however, as the Holocene climate cooled, the Warren Wilson site began to attract a different clientele. The new arrivals were people of the Woodland tradition, a way of life that endured in the Appalachians until roughly A.D. 1000. Woodland people traveled in larger bands than their Archaic ancestors and probably identified themselves as members of social groups that extended beyond their immediate kin. They, too, came to the Swannanoa with spears and handheld projectiles. But in time Woodland folk began to employ a new weapon: the bow and arrow. A bow offered no more killing power than an atlatl, but arrows were easier to carry and could be released quicker and from greater distances. Moreover, archers could shoot from dense cover without having to stand and risk spooking game.[3]

Like their Archaic predecessors, Woodland hunters found much to like about the Swannanoa Valley. On the flat ground along the river, deciduous trees from the mountains mingled with vegetation common to the Piedmont and eastern lowlands, creating a rich and diverse arboreal environment. A typical ten-square-mile patch of forest might produce as many as 750,000 mature trees and another 786,000 seedlings. Those same ten square miles supported 40,000 gray and fox squirrels, 200 wild turkeys, 400 white-tailed deer, and a vigorous population of black bears, opossums, raccoons, and rabbits. Animal remains recovered at Warren Wilson suggest that Woodland people regularly feasted on all those species, with venison and bear meat the most common fare. The river itself furnished a number of potentially palatable critters, including beavers (valued for fur as well as flesh), various fish, frogs, and the hellbender, a giant salamander (commonly a foot or more in length) that inhabits lower-elevation streams. Bones of mountain lions, bobcats, gray foxes, and least weasels have also turned up at Warren Wilson, though they might have been killed for hides, not food.[4]

At certain seasons the trees themselves were a cornucopia of tasty produce. Grapes, cherries, and persimmons were warm-weather staples. Under optimal conditions a single hickory tree could produce three bushels of nuts, providing more than forty pounds of food rich in fat and essential amino acids. (Indeed, some estimates suggest that seven mature hickory trees might furnish enough nutrition to sustain one person for an entire year.) Acorns, which could be boiled to extract nutritious oil, were even more plentiful. And chestnuts parched over a slow fire provided a seasonal repast to chase the chill of the coldest mountain night.[5]

Yet even in such luxuriant forests, life was not always easy, for nature was not always cooperative. Capricious mountain weather, usually in the form of spring drought or a late hard frost, might restrict the yearly production of mast, driving away game and creating a lean year for humans and animals alike. Like Woodland people across the Southeast, those in the Swannanoa Valley found it necessary to store nuts and other seasonal delicacies for use in scarce times. Acorns laid by in a fat year were more than just a hedge against poor hunting or a hard winter. People do not stash food unless they intend to return to it. Woodland people were probably the first humans to live for extended periods at the Warren Wilson site, and they remained there for months at a time using stored foodstuffs to survive the worst that nature could dish out.[6]

In the turbulent world of the Blacks, bad weather could be a blessing as well as a curse. In spring, melting snow from the south rim of the range and torrential rains pelting Pinnacle and Graybeard frequently forced the Swannanoa out of its banks. When the river receded, it left rich black dirt that settled like thick powder along the floodplain. Using that soil, native people developed another highly effective technique for coping with food shortages. They became gardeners.

Along the Swannanoa, as in other populated valleys of eastern America, gardening originated with the casual, even accidental cultivation of weeds. Wherever Archaic and Woodland people went, they built shelters, dug fire pits, and buried garbage. Such activities always disturbed alluvial soil by the rivers. Discarded nutshells, fish and animal remains, and other organic material enriched the ground, creating irregular patches of broken, fertile earth around human habitations. In the ever-changing world of mountain forests, where plants incessantly fought one another for space, such prime real estate did not remain vacant for long. Weeds and grasses adapted to open areas quickly moved onto the bare ground. As Archaic and Woodland folk lingered along the Swannanoa, some weeds evolved into "camp followers," showing up again and again wherever humans stayed for more than a month or two. A few of those plants had edible seeds and leaves, easily harvested by people.[7]

Modern varieties of such flora still flourish at Warren Wilson. Among the most common are the gangly plants known as *chenopodium* (pronounced keen'-uh-pode-ee-um). Strains of *chenopodium* now growing near the Swannanoa include the leafy weeds called goosefoot or lamb's-quarters. Their leaves and seed clusters can be eaten raw and, with a bit of oil and vinegar, make a passable—if slightly bitter—salad.[8] Another favorite was a distant cousin of ragweed commonly called marsh elder or sumpweed. It produces tiny seeds as oily and nutritious as hickory nuts. In addition, early Swannanoa gardeners harvested sunflowers, small gourds, and squash, all descendants of wild plants. As useful greenery became established on disturbed ground, it was a quick and logical step for people to pull competing plants from around the favored species and scatter their seeds across a wider range of the Swannanoa floodplain.[9]

Exactly when weeds became crops remains a mystery. Archaic people in the valleys of Arkansas, Tennessee, and Illinois experimented with gardening about 5,000 years ago. In Appalachian valleys, however, the growing season was shorter and cultivation was more difficult than in the Midwest. Moreover, some of the most important cultigens seem to have followed humans into the mountains. Sunflowers came from the Great Plains and the American West; squash apparently descended from wild gourds native to the Ozarks. Marsh elder cultivated at Warren Wilson probably belongs to an extinct strain that once grew in eastern forests. Its eventual disappearance from the region is perhaps the best evidence of its dependence on people. Given the constraints of climate and the close association of crops with human migrations, agriculture must have been slow to make its way to the mountains. Gardening probably unfolded in fits and starts during late Archaic or early Woodland times and became well established early in the first millennium A.D.[10]

Serendipitous agriculture blended easily with hunting and gathering and, within the limits of local weather, provided Woodland people with an affluent life. If we could tour a Woodland village along the Swannanoa about A.D. 600 or so, we might find an acre or more of cleared ground worn completely smooth by human traffic. Temporary gardens pop up here and there—near the shelters, along the river, at the edge of the clearing. Well-worn paths lead from the habitations into the mountains beyond. Small saplings have been taken for weapons and building material.

Woodland people have also picked the forest clean of deadfalls and other usable firewood, leaving open woods that extend for several miles in every direction. This is no pristine landscape. Wherever we turn, the environment reflects the daily habits of the humans who live there.[11]

Between 800 and 1500 A.D. the Swannanoa and surrounding valleys underwent another transformation. This time the principal agent of change was an exotic, 7,000-year-old Mexican crop called maize, or in less precise, unscientific parlance, corn. It appeared on the fringes of Appalachia about 1,800 years ago but for eight or nine centuries remained an insignificant cultigen. Perhaps it took that long for the exotic grain to adapt to the cool, wet climate of the eastern woodlands. Or maybe well-fed Woodland people at first sensed no need to raise it in quantity. Whatever the reason, corn's status changed dramatically between 900 and 1000 A.D. In a little more than two generations it emerged as a dominant grain "from southern Georgia to southern Ontario and from the Atlantic coast to Minnesota."[12]

Easily cached for future use, maize offered superior insurance against lean seasons, and as corn culture took root, human populations boomed. But corn was not an easy crop. Seeds had to be stored, harvests had to be managed, and surpluses had to be distributed. Those tasks had to be integrated with hunting, cultivating other plants, and gathering wild foods. To cope with corn and the larger populations it nourished, native people apparently opted for a higher degree of social organization. Around A.D. 1000 cultures across the Southeast began to align themselves into new polities called chiefdoms, made up of towns governed by a leader who claimed descent from an important deity and set himself apart from common folk. With the emergence of chiefdoms, the Woodland tradition faded, and a new way of life, the one archaeologists call Mississippian, began.[13]

Because milder climates were more conducive to corn, the largest Mississippian chiefdoms developed far from the Black Mountains: in the foothills adjacent to the Blue Ridge, the Tennessee Valley, and south of the Great Smokies along the Little Tennessee River. In those locales Mississippian people farmed vast expanses of river bottomlands. In time they added beans (another domesticate from Mexico that helped replace nitrogen taken out of the soil by maize) to their agricultural repertoire. They built imposing towns and huge ceremonial mounds, often enclosed by wooden palisades. By the 1300s some of the larger, better-organized societies had developed into "paramount chiefdoms" dominated by a leader who exerted control over several towns across a large area.[14]

Mississippian culture came to the Swannanoa Valley, too, but in a slightly different guise. Growing seasons were shorter, bottomlands were less extensive, and valleys were more remote. Maize did not become an important crop until sometime after 1100 A.D.; beans arrived a century later. Organization into chiefdoms probably took 200 or 300 years longer than in the flatlands. Even with access to corn, Appalachian folk continued to rely heavily on acorns, hickory nuts, and other wild foods. Old cultigens also hung on. Marsh elder remained important in the mountains well after it had diminished in other areas. This mountain way of life is distinctive enough to have its own name. Archaeologists call it the Pisgah culture.[15]

Retention of old ways served Pisgah people well. Archaeological evidence from some of the bigger Mississippian towns, including those along the Little Tennessee River, suggests that natives there frequently battled nutritional deficiencies resulting from a diet too rich in corn. Erosion and soil exhaustion might also have plagued those who grew maize in quantity. Indeed, by the mid-1300s some larger Mississippian chiefdoms had gone into decline, partly because their populations had depleted local resources. Swannanoa natives apparently avoided those problems, but they did not leave the valley unchanged. Pisgah people lived year-round on the river. They built large villages on three to five acres of cleared land surrounded by elaborate wooden palisades. Within the enclosures they erected equally impressive houses that measured twenty to twenty-five feet across. Constructed with posts at every corner, the dwellings had roofs of poplar bark (or some equally pliable wood) and walls of woven saplings plastered with mud and clay. One hundred to 300 people resided in a typical town.[16]

Although they lacked metal technology, Pisgah people had stone axes capable of felling trees five inches in diameter. Within just a few years daily woodcutting might clear several acres of land, leaving only sprouting stumps and a few larger trees. Gathering enough firewood to support 100 or more people also required constant foraging. In time Pisgah women and children (the designated wood gatherers) ranged far beyond the towns and even retrieved deadfalls from forests on the lower mountain slopes.[17]

The ecological effects of such activities can be read in the dregs of ancient campfires. Charcoal recovered from Warren Wilson shows that people there usually stoked their fire pits with poplar, birch, pine, and

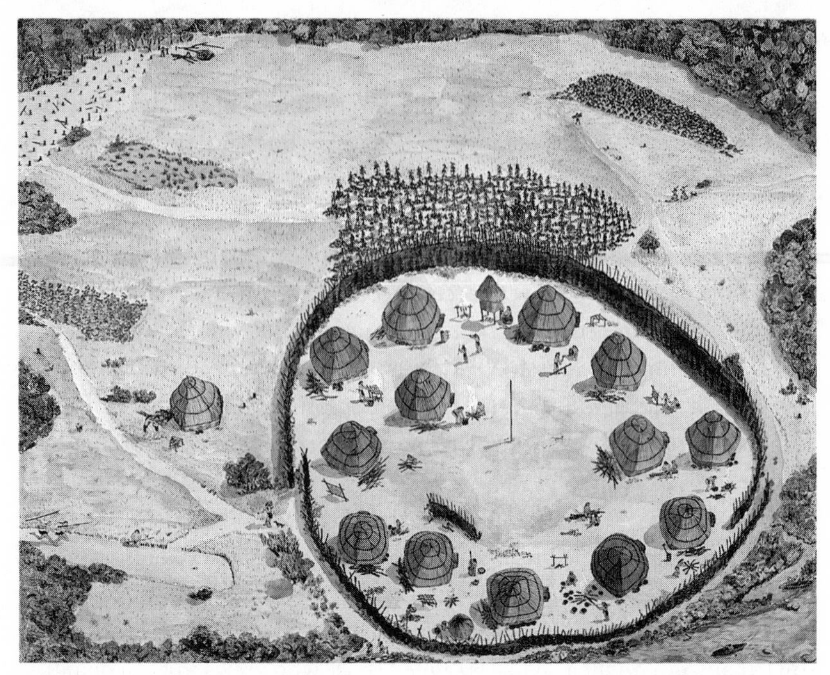

Artist's representation of a Pisgah village on the Warren Wilson site, ca. 1420. The Swannanoa River is in the foreground. Corn grows in garden plots outside the palisade, and the surrounding forests show the effects of native clearing practices. Pen and wash drawing by Gwen Diehn. Used with permission.

locust, all of which are typical of young forests.[18] Indeed, those species flourish along the Swannanoa today, especially where trees have begun to reclaim land cleared fifty or sixty years ago. Following several thousand years of on-again, off-again human occupation, the woods adjacent to Pisgah villages might have resembled the forests that still survive by the river.

Campfires were not the only blazes smoldering in the valley. Like native people across America, those around the Black Mountains probably used fire to alter the surrounding landscape. Much of what we know about fire in the mountains comes courtesy of those turbulent weather systems that swirled across the region thousands of years ago. Carried by smoke or washed from high ridges by hard rain, charcoal from Archaic, Woodland, and Mississippian fires settled to the ground near native towns. Most of it has since disappeared. But purely by chance some particles fell into mountain ponds and peat bogs. There the charcoal mixed with tiny pollen spores blown from nearby plants and trees, providing a handy, highly stratified archaeological record of the burning practices of the native inhabitants and the vegetation it encouraged. So far the best such samples have come from places fairly distant from the Blacks, namely the Little Tennessee River Valley and the Chattooga River watershed in southwestern North Carolina. From those sediments paleoecologists have learned that Archaic people set light ground fires on the slopes and ridges where they hunted and gathered nuts. Later, during Woodland and Mississippian times, natives kindled such blazes closer to where they lived, burning wooded floodplains and stream terraces.[19]

If people in Black Mountain valleys employed fire in similar ways (without local sediment samples we can only speculate), light fires set by Archaic and Woodland hunters probably spread through the hardwoods on the lower slopes. The largest Black Mountain trees—those that flourished in moist, sheltered coves—remained untouched, as did the northern hardwoods and spruce-fir of the highest peaks. But at elevations between 2,000 and 3,000 feet periodic burning encouraged the growth of various oaks and American chestnut, all of which have thick bark and easily survive light ground fires. Red and sugar maples (which otherwise might have replaced the oaks) are not nearly so resistant to fire and virtually disappeared from areas frequently burned. Fire also cleared the forest floor of woody plants and underbrush, so that the taller trees towered over a meadow of grasses and bracken fern. These parklike "oak orchards" were a huge boon to food gatherers. Acorns and chestnuts were easier to pluck from ferns and grasses than from brushy undergrowth, and in seasons of abundant mast such forests attracted deer and other wildlife, providing hunters with an accessible—if somewhat unreliable—game supply.[20]

On level ground, fire was the farmer's ally. Even when planted with nitrogen-fixing beans, corn is an exhausting crop; indeed, it depletes soil faster than any other grain. When Pisgah people needed fresh ground, they sought a small (usually one acre or less) forested plot within easy walking distance of the town. They probably felled large trees by girdling their trunks with stone axes and setting fire to the adjacent undergrowth. With their bark slashed and roots singed, the trees soon died and could be pushed over and burned again, creating suitable ground for a new field. It was hard work, and the natives must have used each field as long as possible, perhaps even allowing old tracts to lie fallow for a short time to recover some fertility and then planting them again. If the rains fell, cultivated fields soon became overgrown with various food plants. Beans climbed cornstalks; sunflowers bobbed above a dense carpet of squash and gourds. Any patch of dirt left uncovered invited hardy camp followers such as *chenopodium*. Wild strawberries and blackberries, both of which favor open ground, also invaded agricultural plots. If we could walk by the Swannanoa in the mid-1400s, we would likely find several acres of tangled gardens and weedy old fields punctuated by stumpy woodlots and old trees too tall to succumb to fire or stone axe.[21]

Due to the demand for wood, mast, game, and agricultural land, the Swannanoa Valley supported only one or two Pisgah towns at any given time. Prolonged drought, unceasing rains, a hard winter, or some other climatic quirk might destroy crops or reduce animal populations, forcing entire villages to relocate. But Pisgah culture endured in the region into the fifteenth century and beyond. Though anthropologists still have much to learn about native lineage, it seems likely that Pisgah hunters and farmers at Warren Wilson (and elsewhere on the French Broad and the Nolichucky) were the immediate ancestors of the people early European explorers generically misidentified as "Indians." White settlers later knew them as Cherokees.[22]

This Indian world—Appalachia before the white man—has always occupied a peculiar place in popular thinking about the Black Mountains. It has been easy to imagine natives as nature's caretakers living an idyllic life in old-growth forests, never at odds with their environment, and never stretching local resources beyond their limits. But those images, like visions of some lost Black Mountain Eden, are creations of the modern mind. By the end of the fifteenth century the tall mountains stood watch over valleys where people built palisaded villages, planted crops from foreign lands, and used fire to transform forests and manage game populations. To the west and south, beyond the Blue Ridge and the Smokies, the Indian world was one of populous towns and paramount chiefdoms, vast cornfields, and cleared land. It was not a static world of ecological perfection or continuous bounty but, rather, a world in which people understood the land and its ways, where natives enjoyed easy abundance when they could but had no qualms about taking steps to ensure sustenance when nature turned harsh and unpredictable. By the mid-sixteenth century it was also a world on the cusp of change.

FEBRUARY

THUNDER HILL OVERLOOK, BLUE RIDGE PARKWAY

Autumn's drought is a memory. We now face one of the wettest winters on record. Starting in November, storm after storm careened across North America into the

Carolinas. Late December brought a foot and a half of snow; early January, ten inches of rain and sleet. Ten days later another two feet of wet snow fell on the Appalachians. The frozen precipitation has taken its toll on the high-elevation forests. As I pull into this deserted parking area, I see broken timber everywhere: along the roadside, hanging precariously from high branches, and tangled in undergrowth. On every ridge the hardwoods have splintered tops, as if some manic giant had walked among them swinging a dull scythe. In late afternoon, even after a weeklong warm spell, snow lies in irregular patches on north- and west-facing slopes. Thirty-degree temperatures and a brisk wind make me glad I dressed warmly. I hear pops and pings from the truck engine as it cools in the chilly air.

Thunder Hill Overlook hangs on the eastern lip of the Blue Ridge not far from the resort town of Blowing Rock. The parking lot rests on a cliff 3,785 feet above sea level, overlooking the Yadkin and Catawba River Valleys. During the height of summer it is a favorite stop for tourists and photographers. But on this crystalline February evening I am alone. I walk the parkway south for a half-mile or so and find a seat on a rocky outcrop in an adjoining pasture. The distinct profile of Grandfather Mountain rises on my right. On the southwestern horizon, in the vermilion smudge of a winter sunset, the Blacks stand like a stone wall, snowy peaks outlined in sharp relief against a darkening sky. The air is so clear that I can see the dark side of a quarter-moon hanging in the east.

As daylight fades, the land below comes alive with light. From here I can see deep into the Piedmont, past the cities of Lenoir, Hickory, and Morganton. Far away, visible to the naked eye but best seen with binoculars, is the Charlotte skyline. Nights like this afford a rare opportunity to sense the sheer enormity of the Appalachians, the abrupt shift in elevation and increasingly rugged terrain as one ventures inland from the coast.

Walking back to the parking area, I think about how the Blue Ridge must have appeared to the first European explorers. Then as now, the long string of mountains towered over the Piedmont, looming as a formidable, though hardly insurmountable, obstacle to overland travel. To negotiate it the newcomers used routes charted by native people. Like modern highways those trails stuck close to major rivers and creeks, meandering through gaps and low places and providing passage at the safest points. Most early explorers took one look at the tallest peaks and decided that their business lay elsewhere, usually in some broad valley far removed from winter's tree-shattering wrath. But as the first Europeans moved along the rivers on the west side of the Blue Ridge, they occasionally came close to the Blacks, establishing a fleeting nexus between sixteenth-century Spain and the East's highest mountains. Sometimes, especially when they talked with Indians, the Spanish heard stories that were hard to ignore. Buried in the valleys were precious metals and gems. Whoever found them would be richer than they could ever imagine. Now and then—perhaps as the explorers sat around their campfires on clear nights like this—the high peaks beckoned, enticing the strangers with visions of treasure.

On May 25, 1540, 600 Spanish soldiers awoke twenty-five miles due south of Thunder Hill, in the powerful Indian chiefdom of Joara, not far from modern Morganton. Over the preceding ten months the Europeans had trekked overland from Florida, looting the land for their king. Along the way they had stolen corn from local storehouses, fought bloody battles with native warriors, and kidnapped Indians to act as slaves, translators, and emissaries. During the previous ten days the expedition had plowed across the Piedmont, moving up from South Carolina, passing just west of Charlotte, and pausing for a night or two near Hickory before arriving at Joara. Either out of fear or as an expression of their intended goodwill, Indians there provided food and safe shelter that allowed the soldiers some much needed rest. But the group's commander was in a hurry and ordered his men to march toward the line of blue mountains visible in the distance. The soldiers had little choice. They followed the iron-willed commander they knew as the "Captain." History remembers him as Hernando de Soto, the first European to traverse the Appalachians.

Cloaked in body armor and brandishing swords, lances, and crossbows, de Soto and some of his men rode horseback. Others walked. Large dogs, also dressed in armor and trained to attack on command, trotted beside the horsemen. Three hundred hogs, brought along as a mobile food supply, rooted their way through the adjacent forests. From Joara the odd assemblage rambled north, apparently along an ancient Indian trail that followed essentially the same route as N.C. 181. On the western horizon de Soto's men saw the sharp silhouettes of Shortoff Mountain, Hawksbill, and Table Rock before they made camp at the top of the mountain at Jonas Ridge. The next day, probably near Linville Falls, they took a hard left onto another path and descended a long incline to a shallow river, which they crossed in a swift, knee-deep current. In all likelihood the stream they forded was the upper North Toe, not far from the Avery County community of Ingalls. During the last days of May the entourage worked its way down the North Toe, along the modern path of U.S. Route 19E, and spent one night where the Toe and Cane join to form the Nolichucky. By the end of the month the soldiers were well downriver, near Erwin, Tennessee, at an Indian town called Guasili. For at least part of the trip across the high ridges—especially as he followed the North Toe—Hernando de Soto must have been within sight of the Black Mountains.[23]

Several accounts of de Soto's travels survive, but descriptions of land around the Blacks are spare. The trip took five days, during which the expedition crossed "over very rough and lofty mountains" and marched through "very good country." They camped in "a savannah" and an "oak grove," saw "plenty of pasturage for cattle," and passed "streams with little water, though they flowed rapidly." The odd mountain weather got some attention: an almost offhand remark that the group "endured great cold, although it was already the twenty-sixth of May." In a summary list of "Diversities and Peculiarities" encountered during the expedition, one chronicler commented on a prevalent upland tree by tersely noting that "wherever there are mountains, there are chestnuts." Another five days or so west of Guasili the expedition paused for two weeks at Chiaha, a palisaded Indian town on the French Broad River, well into Tennessee. They were ecstatic at the prospect of flat land, since "the horses were tired and thin and the Christians [de Soto's men] likewise fatigued."[24]

We yearn for more. Where are the majestic peaks green with spring? The verdant valleys? Cathedral-like forests? How could de Soto's men travel through such extraordinary country for the first time and remain so strangely laconic? The answer has as much to do with our perception of nature as with theirs. Modern Americans usually read the works of "first white men"—from de Soto to Daniel Boone to Lewis and Clark—in a peculiar and misleading way. We do not see the natural world as they saw it but, rather, as "we would have seen it in their place," as if we were hiking through the region with nothing to do but admire the landscape. No real first white man could afford that luxury.[25] Like other explorers, de Soto was at work, not play. He was in the mountains to vanquish the natives, reconnoiter the land, and set aside a portion for his private estate. What mattered most was his labor, manifested in the rigors of the journey: crossing high ridges, surveying prospective pasturage, enduring cold weather, and resting horses and men weary to the bone after ten days on treacherous terrain. He had neither time nor inclination to wax eloquent about a sylvan landscape that we might have found irresistible.

What the Captain and his employers did have time for was gold—or anything else that glittered and promised to line their pockets. Wherever they stopped, the Spaniards showed the natives jewelry and asked where rare stones and shiny yellow metal might be found. But by the time they

Hernando de Soto, Spanish explorer, whose quest for gold and precious gems took him near the Black Mountains. He was probably the first European to see the range, though he apparently did not venture into it.

got to the mountains, the soldiers had precious little to fill their coffers. (The big discovery was a cache of freshwater pearls that one of the company later lost.) But the Captain saw or heard something at Joara that piqued his curiosity. As one of his writers noted, "There was better disposition to look for gold mines than in all that [country] they had passed through and seen in that northern part [of their journey]."[26]

Maybe it was the Blue Ridge itself, visible in the distance, that prompted such optimism. Mountains are logical places for mines. Or perhaps de Soto noticed that the Indians at Joara used copper, a metal they acquired from natives to the north and west. The Captain might also have seen another substance, one for which the Black Mountains have since become famous. It had a luster like pearls and occurred in a wide variety of colors: green, light brown, ruby, lilac, black, and translucent white. Native craftsmen cut jewelry and ornaments from it. Native people in eastern America had traded it among themselves since the Woodland period. Today we know it as sheet mica, or muscovite.[27]

Small grains of mica can be found in almost any Appalachian soil, but sheet mica is rare. It forms in pegmatite, a coarse-grained granite, deep beneath the earth's surface. Pegmatite frequently occurs in larger crystalline rocks, including the ancient gneiss and schist of the Blacks. Along the region's major rivers millions of years of rain, snow, and sleet have steadily eaten away at the surface material, making age-old deposits available to miners. Today the area around Spruce Pine in Mitchell County, not far from where de Soto waded the North Toe, produces the finest sheet mica in the world.[28]

But if de Soto collected any Black Mountain minerals, he kept mum about it. Mica, after all, was a poor substitute for gold. Besides, while the expedition rested at Chiaha, the Captain learned of other Indians who lived somewhere on the upper Nolichucky and mined shiny metal. He immediately sent two of his men to check out the rumor and ordered the rest of the contingent to head down the Tennessee Valley. Lured by more promising lands to the west, the Spaniards left the Blacks and their mica-rich pegmatite behind as the entourage moved out of the high Appalachians.[29]

Three hundred years later, when nineteenth-century miners began to dig for mica in Mitchell County, they uncovered evidence that native people had, in fact, worked the best deposits long before. Washington Caruthers Kerr, a geologist who prepared reports for the North Carolina Geologic Survey in 1875, took careful note of such discoveries. At the Silvers (also known as the Sinkhole) Mine near the tiny hamlet of Bandana, workers found "open cuts and extensive dumps on the surface" and "small tunnels and shafts where the formations were soft." Much later, in the 1950s, when owners expanded operations at Sinkhole, they found another labyrinth of tunnels, "evidently the remains of work carried on by the aborigines." In the same general area, near Bakersville, Kerr described an old excavation with "open pits forty to fifty feet wide." Dirt had once been piled "in huge heaps about the margins of the pits." Giant oaks grew on the mounds among numerous old trees that had fallen and decayed, suggesting that the mine had been abandoned several hundred years earlier. Kerr could find "no clue to the object of these extensive works, unless it was to obtain the large plates of mica or crystals of kyanite [a bluish, quartzlike mineral], both of which abound in the coarse granite rock." Since Kerr's day modern mica and feldspar operations have obliterated the old tunnels and excavations, leaving archaeologists to wonder who worked them, when, and for how long. But if those North Toe mines existed in 1540, then de Soto, uninformed or, more likely, unimpressed, rode past them.[30]

After leaving the mountains, the Captain's men traveled southwest across Alabama, Mississippi, and Arkansas into Texas. It was a two-and-half-year journey to disaster. The company ran short of supplies and had its ranks depleted in several costly battles with natives in the Deep South.

In late May 1542, on the banks of the Mississippi River, de Soto died of some mysterious "fever." Shortly afterward the once-impressive entourage fell into disarray. Forced to butcher their remaining hogs and flee determined Indian warriors, roughly half the original company of 600 managed to stagger down the Mississippi to safety in Mexico. A search party found the ragtag contingent at Vera Cruz in 1543. They had no gold and no precious stones; instead they spoke only of brutal overland travel and the stubborn resistance of native people. The evidence was irrefutable. The Captain had failed miserably.[31]

But dreams of untold riches in uncharted mountains died hard. In their final reports at least two of de Soto's chroniclers hinted that his soldiers had found gold and copper while they parleyed with natives on the upper Nolichucky. That rumor was alive and well when a second Spanish explorer, Juan Pardo, returned to Joara in 1566. Pardo's primary objective was to find an overland route to connect his country's outposts on the south Atlantic coast with some of its silver mines in Mexico. In the meantime he was to explore the region, meet the Indians, convert them to Christianity, and make them allies of Spain.

Traveling some of the same Indian trails as de Soto, Pardo arrived at Joara shortly after Christmas, on the feast of St. John the Apostle, according to his Christian calendar. The weather was freezing, and the mountain peaks (including the Blacks) were blanketed with snow. Worried that crossing the Blue Ridge in the dead of winter would prove disastrous, Pardo stayed at Joara for two weeks before returning to the Carolina coast. He left behind a small, hastily constructed fort (christened Fort San Juan in honor of the feast day on which he arrived) and thirty heavily armed men under the command of Sergeant Hernando Moyano de Morales.

Charged with the awesome responsibilities of holding the land for Spain and saving Indian souls from eternal damnation, Moyano had plenty to occupy his time. But the Blue Ridge winter was long and the soldiers grew restless. Perhaps local people showed Moyano copper and mica; maybe they told him of that place not far away where Indians mined precious minerals. Whatever the story, it rekindled old dreams and lured the Spaniards out of the Piedmont. At some point early in the spring of 1567, with a handful of his men and some warriors from Joara, Moyano lit out for Indian towns in the mountains, four days' journey away. He might have been headed for the upper Nolichucky, the same region that had tempted de Soto's men with tales of gold. This time, however, mountain people wanted nothing to do with the Spanish and their newfound friends.

Moyano fought two memorable battles somewhere across the Blue Ridge, allegedly killing many Indians. But like de Soto before him, Moyano found no gold and no mines. And like the Captain, he eventually wound up at Chiaha on the French Broad, resting his men after a rigorous journey.

Meanwhile, on the coast, Pardo made plans to rendezvous with Moyano, who was thought to be surrounded by hostile natives. This time Pardo left earlier, in September 1567, and by the last week of the month he was back in Joara. From there he went up the Catawba River, crossed the Eastern Continental Divide into the Swannanoa Valley, made his way around Graybeard and the southern end of the Blacks, and stopped for several hours at an Indian town called Toecae, near modern Asheville. His route from there has been a matter of considerable debate and conjecture among historians and archaeologists, but somehow (probably via the Pigeon River and an overland trail to the French Broad) he crossed the mountains and met Moyano at Chiaha. With Moyano's contingent in tow, Pardo's group apparently turned south, picking their way around the Great Smokies and back to Toecae before returning through the Swannanoa Gap to Joara and the coast. They never found a route to the mines of Mexico.[32]

Surviving accounts of Pardo's travels are much slimmer than the volumes left by de Soto. Like the Captain, Pardo was ever heedful that he was in the mountains on business for Spain. His work, especially his speeches and instructions to the Indians, got most of the attention. Even so, we can occasionally glimpse the mountains as Pardo saw them. The Catawba near Joara was a "high-volume river" surrounded by "as beautiful a land" as "in all of Spain." A writer described the Swannanoa Valley and the Asheville region as "a place which is over the top of the ridge." Arriving in early October as forest mast matured, the Spaniards discovered "many grapes, many chestnut trees, many nuts, and quantities of other fruits." The alluvial floodplains impressed them as "flat lowlands" with soil "not very rough to the touch." Much like the native people who had preceded them, Pardo's men found it "very good land where harvests of all sorts can be made."[33]

Good land, indeed. But it was not enough to bring Pardo back. Over the last half of the sixteenth century the Spanish increasingly found themselves overextended in North America, unable to hold the southeastern coast—much less the interior—against other European powers. Undermanned and poorly supplied, Fort San Juan and other outposts established by Pardo were abandoned. Rumors of gold, silver, copper, and even diamonds persisted. A few hardy souls took up the search for treasure, but those efforts met with frustration. Eventually Spain gave up on the region, preferring the real riches of Mexico and Peru to the phantom wealth of the southern Appalachians.[34]

Throughout their travels around the Blacks, de Soto and Pardo almost never strayed from established trails. They cut few trees (except for Pardo's ephemeral forts) and rarely killed game. But even this tenuous and short-lived connection between Spain and the Black Mountains likely had profound ecological consequences for the local environment. Like all European explorers, de Soto, Pardo, Moyano, and company may have been walking repositories for Old World viruses and bacteria that killed native people in droves.

The native inhabitants of eastern America were no strangers to disease. For generations they had battled numerous illnesses, including rheumatism, arthritis, respiratory disorders, venereal syphilis, and, possibly, tuberculosis. However, for ten thousand years, since their earliest ancestors had crossed the Bering Strait land bridge, the natives had remained isolated from many of Europe's oldest afflictions. They knew nothing of smallpox, measles, and influenza, infections that slipped silently into their bodies when a stranger coughed, sneezed, or spoke. European animals, too, might have carried lethal contagions. Pigs, like those that accompanied de Soto, often harbor a variety of microbes that, over time, can mutate into disease organisms harmful to humans. In addition, the hogs may have introduced anthrax, brucellosis, trichinosis, and several other ailments common to domestic animals into southern forests, diseases that had the potential to infect both wildlife and native people.[35]

Serious epidemics might have been unleashed in the region even before the Spanish arrived. While in the Piedmont de Soto found several "large uninhabited towns, choked with vegetation." In typically matter-of-fact fashion, a writer noted that the area had been abandoned two years earlier due to "a plague in that land," pestilence that might have been carried overland by Indians trading with Spanish settlements in Florida or along the South Carolina coast. But if de Soto or Pardo saw abandoned towns along the North Toe, Swannanoa, or Nolichucky, they gave no notice, and there are few reports of sickness among the soldiers as they made their way across the Blue Ridge.[36] But the Spanish also found few Indians between Joara and Guasili. We are left to wonder: Had disease already invaded the region? Did de Soto and Pardo (or their animals) introduce it into the valleys surrounding the Black Mountains? How many Indians died as a result of the incursions? So many questions, such meager evidence.

What we do know is that the following century was a turbulent time in the southeastern interior. Native populations went into sharp decline. Some of the largest Mississippian chiefdoms, with their vast cornfields and elaborate towns, simply disappeared or broke up into smaller polities. In part the demographic disaster might have stemmed from prolonged droughts, crop failures, or other unforeseeable twists of nature. But epidemic disease seems a more likely and comprehensive explanation. If so, we can only imagine the terror that accompanied every episode of pestilence. Shamans and medicine men were powerless; conventional herbal remedies, ineffective. When an Indian town fell sick, fields went untended, hunting grounds were unburned, and acorns and chestnuts remained unharvested. Winter's chill winds inevitably brought famine or starvation. Natives who survived the calamity were often on the move, migrating to new lands and banding together in concentrated settlements more easily defended against enemies, both native and European.[37]

The Cherokees, descendants of the Pisgah people, may have been among those who closed ranks and relocated during the biological upheaval. They moved south and west out of the higher Appalachians into the hill country between the upper Tennessee and upper Savannah Rivers. Because those settlements remained relatively isolated from European outposts along the Atlantic coast, mountain natives had time to recover from epidemics introduced by Spanish explorers. By 1700 as many as 30,000 Cherokees probably lived in the region. In western North Carolina important Cherokee towns sprang up west of the Black Mountains along the Tuckasegee and Little Tennessee Rivers. The Blacks became part of an important buffer zone between the Cherokees and their Piedmont neighbors. Cherokees hunted along the North Toe River and its tributaries, taking deer, bear, and turkey from the oak and chestnut forests.[38] But by the mid-eighteenth century the valleys surrounding the East's highest mountains had no permanent Indian settlements. Just as the region had lured de Soto with rumors of gold, it would entice English and French explorers in pursuit of other riches. This time the treasure hunters would not be so quick to leave.

APRIL

CHERRY LOG RIDGE

Water droplets gather on my jacket as I shoulder a day pack and walk into the gray woods. Weather like this is common in early spring, when clouds hang on the Blacks for days at a time, dousing leafless hardwoods with intermittent rain. Ahead the trail—ill-defined under the best circumstances—disappears into white haze as it winds its way up the headwaters of the South Toe toward the northeast face of Pinnacle. I came to this small tributary in search of brook trout, but given the rain and April chill, I decide to leave my fly rod in its case and simply hike upstream. I move slowly for a half-mile and then head off-trail, hopping from rock to rock up one of several small creeks that feed the left prong of the river. According to a soggy topographic map, elevation here is about 4,200 feet. Steep slopes rise 200 feet or higher on my left and right.

In the fog the woods seem as patchy and disheveled as ever. I see red and striped maples, yellow birches, hickories, eastern hemlocks, yellow poplars, and an upland magnolia commonly called a cucumber tree. Shrubby rhododendrons and mountain laurel clog the forest understory. In the inevitably imperfect language of ecological classification, this qualifies as an Appalachian cove forest, so named because its vegetation is well adapted to the clammy confines of sheltered ravines like this.

It is also the closest thing the Blacks have to a rain forest, and it nurtures a stunning array of plants and animals. More than twenty species of trees and more than a thousand varieties of herbaceous plants might be found in this picturesque cove. For twenty yards on either side of the creek, green shoots of mayapple and the small white flowers of bloodroot sway just inches above autumn's leftover leaf litter. Thumbing through a field guide, I discover other showy spring foliage—waxy dark green galax, white trillium, jack-in-the-pulpit, and Solomon's seal—most of which are not yet in bloom. For an hour I rummage along the stream bank looking for another distinctive plant. It has a single stalk and five sawtooth leaves. By all accounts it should be here; it often grows among the species I have already identified. But today I do not find it. Perhaps it is still too early in the season for its leaves to emerge at this elevation. Eventually I give up and walk back downstream, consoling myself with the thought that I am not the first to search for wild ginseng and come away empty-handed.

What has been a steady drizzle becomes a pelting rain. Hoping to see some of the cove forest's wildlife, I begin the long trek back to the trailhead. The wet weather masks my movements through the woods, but I find no animals. As evening approaches and the downpour intensifies, I hear the distinctive gobble of a wild turkey and notice the small half-moon hoofprints of a white-tailed deer pressed into the mud along the stream. At dusk on a rainy April weekday the cove forest seems deserted, isolated from human activity, a place far removed from commerce and industry. But that is a faulty image. The commercial ties that first bound the Blacks to other, more populous locales were established by European traders and explorers who sought wildlife and vegetation in areas like this. White-tailed deer, ginseng, and other plants became valuable commodities traded at a profit across the Atlantic. Just as surely as faraway winds and cloud movements shaped Appalachian weather, distant economic currents began to dictate the fates of Black Mountain plants and animals.

For three-quarters of a century after English people settled at Jamestown, the Carolina upcountry remained uncharted territory. A number of would-be mapmakers speculated about the proximity of the Appalachians to the Pacific Ocean. Some thought the great salt sea lay just beyond the mountains' western slopes, only ten days' walk from the Atlantic. Others knew it was farther but still believed a long arm of the Pacific stretched down the western flank of the range to the Gulf of Mexico. A third popular theory held that just across the Blue Ridge several "great and Navigable rivers" flowed all the way to the western ocean. Despite the perambulations of de Soto and Pardo, even the Piedmont was terra incognita. John Lederer, a German who claimed to have reached its western edge in 1670, returned with fantastic tales of a giant brackish lake ten leagues wide and a sandy desert that required two weeks' travel to cross.[39]

Exploring such foreboding terrain figured to be dangerous business. Still, a few Englishmen were willing to take the risk, for like de Soto, they heard the siren song of mountain riches. This time the prize was not gold, but leather—beautiful, buff-colored, suedelike leather fashioned from the supple skins of white-tailed deer. Spanish explorers had seen such leather during their travels in the mountains. But it was the English who first recognized the economic potential of deerskins.[40] By 1700 explorers from both Virginia and South Carolina had ventured far inland seeking Indians who could supply the valuable hides. Among the natives most important to the deerskin trade were the Cherokees, who had access to lush mountain valleys said to be teeming with deer. Some of those valleys lay at the foot of the Black Mountains.

Efforts to contact the Cherokees probably brought English explorers close to the Blacks in 1673. That year Abraham Wood, a wealthy Virginia planter who controlled a backcountry fort near Petersburg, sent two agents into western Carolina. One, James Needham, already had a reputation as a courageous backcountry explorer; the other, Gabriel Arthur, was an illiterate indentured servant. Escorted by eight native guides, Needham and Arthur journeyed from Wood's fort into the upper Yadkin River Valley, where they paused briefly at an Indian village called Sitteree. It was among the westernmost strongholds of the Piedmont natives, the last stop on the road to Cherokee country.

After a brief rest the party struck out cross-country, pushing their way through tangled forests and rhododendron thickets, sometimes riding but usually walking and leading their horses. For four days Needham and Arthur trekked due west, looking "upon the sun setting all the way," until they arrived on the Eastern Continental Divide. There, at a gap "not above two hundred paces wide," they crossed the Blue Ridge. On the other side they found the going easier and took only "halfe a day" to descend the west slope. Like their Spanish counterparts, Needham and Arthur were preoccupied with business and left only spotty descriptions of mountain geography. But somewhere west of the Blue Ridge the group passed five mountain rivers running northwest. All were easily forded and had "sandy bottoms, with peble [*sic*] stones." Were some of these the North or South Toe, the Cane, or their tributaries? Possibly. A due westerly course from Sitteree would have landed Needham and Arthur in modern Avery County, where they might have turned southwest down the North Toe to the Nolichucky, along the same trail that de Soto followed.[41]

After what he deemed successful negotiations with the Indians, Needham left Arthur with the Cherokees to learn their language and returned to Virginia. In the fall of 1673 Needham again set out for the mountains, accompanied by native guides and leading horses loaded with goods to be exchanged for skins. En route the pack train was attacked by Occaneechis who lived along the Roanoke River. They killed Needham and took his merchandise, probably in an attempt to dissuade Virginians from trading directly with the mountain Indians. Meanwhile Arthur remained among the Cherokees until the summer of 1674, when a local chief finally escorted him back to Wood's fort.[42]

Word of Needham's fate spread quickly across the South, but traders from Virginia and South Carolina continued to woo the Cherokees, as did French merchants operating from bases along the Mississippi River. Working through native intermediaries and directly with the Cherokees, colonial traders soon created a lucrative market for a wide array of mountain products, many of which could be found in the forested coves around the Blacks. Venison and wild turkey became staples of a thriving trade in wild game. Chestnuts, gathered and preserved by native women, were highly prized seasonal delicacies in colonial households. Pelts from beavers, otters, minks, and black bears also fetched good prices in Carolina's coastal towns.[43]

For English merchants, however, the prime lure of the mountain trade was always deerskins. Leather working had long been one of England's foremost businesses, but between 1710 and 1714 a series of deadly bovine diseases struck European cattle. In an effort to stem the epidemics, England temporarily banned the sale of infected cowhides from the Continent. As imports from France and other suppliers declined, demand for American leather soared. Gloves, saddles, harnesses, bookbinding—anything made of cowhide could be fashioned from deerskins. By the mid-1700s yellow buckskin breeches, once the work clothing of common laborers, had become popular among all English social classes. The breeches were, as one historian has observed, "the eighteenth-century equivalent of modern denim jeans."[44]

Colonial merchants and their native partners did their best to sate Europe's appetite for American leather. Erratic records from colonial port towns tell part of the story: a million deerskins were shipped from Virginia and South Carolina between 1698 and 1715; another 2 million had been sent from South Carolina alone by 1740 and as many as 175,000 *annually* by 1762; a million hides were shipped from Savannah between 1764 and 1773; and 310,000 left French Louisiana in the late 1750s. Because most of the skins taken from western North Carolina first went overland to one or more such outlets, it is tough to gauge exactly how many hides came from the Appalachians, let alone the Black Mountains. Some estimates indicate that during the 1730s (when the mountain trade was at its height), a single skillful Cherokee hunter ranging to the far corners of his territory—to regions like the valleys of the South Toe and Cane—could collect and cure 300 pounds of leather per year.[45] If so, a goodly number of Black Mountain deerskins probably turned up in London haberdasheries furnishing chic apparel for city folk who never set foot in an Appalachian forest.

In exchange for leather the Cherokees received European cloth, blankets, knives, axes, metal tools, guns and ammunition, and other items deemed essential for forest life. They also bartered for mirrors, scissors, earrings, belt buckles, and other decorative items, the so-called prestige goods that were symbols of status within Cherokee communities. But the trade was not, as some colonial merchants and politicians argued, an enterprise of equal benefit to both cultures. Thousands of southern Indians, including some Cherokees, were sold as slaves to masters in New England and the Caribbean. Worse yet, the trading paths between the coast and the mountains were conduits for pestilence. Carried overland by traders or native middlemen, smallpox struck the uplands in 1697, 1738, 1760, and 1780, killing thousands of Cherokees during every outbreak.[46]

As they became more involved in colonial commerce, Cherokees had to defend hunting grounds and trade routes against encroachment by whites and by other natives. Disputes over trade and land became more common; an atmosphere of mutual suspicion and posturing by colonial and native leaders made negotiations increasingly difficult. Between 1759 and 1761 and again during the American Revolution, various factions of Cherokees fought long and costly wars against Carolina and Virginia colonists. Over the century the natives saw their numbers decline drastically, from a post–de Soto high of 30,000 to fewer than 8,000 by 1800.[47] Militarily and politically the Indians were still a force to be reckoned with and would remain so until their forced removal from the region in 1838–39. But with their numbers diminished, Cherokee men sought deer and other game closer to home. The Black Mountains—always distant and now remote hunting grounds—gradually slipped from the natives' grasp.

Given the sheer volume of deerskins leaving the Appalachians, one might well wonder how any white-tails survived, in the Blacks or elsewhere. But deer are remarkably fecund and therefore remarkably resilient animals. Modern experiments show that with adequate food and appropriate range, a typical herd of ten deer can increase more than twentyfold in only five years. When Cherokee men fought—against other natives, white settlers, or some temporary coalition of the two—hunting stopped, trading slowed, and deer used the lull to recover. Local white-tail populations ebbed, and Cherokee traders sometimes complained about poor hunting; but deer never disappeared from the mountains. The best estimates of historians and animal ecologists suggest that across North America, deer numbers probably declined 35 to 50 percent during the height of the trade.[48]

Whether those figures hold for the Black Mountains is anybody's guess. Well removed from the Cherokee towns, the high mountain valleys probably drew fewer native hunters than similar lands to the west and south, especially during the century's last decades. As native hunters focused their energies closer to home, no Indians showed up to burn the oak and chestnut woods on the Blacks' lower slopes. When brushy undergrowth replaced grass and browse, deer probably found those woods less attractive than in times past. Drought, an early frost, annual variations in the mast supply, a slight uptick or downturn in predators, a sudden outbreak of disease or parasites, a stormy winter—any of these could momentarily increase or diminish local herds. Moreover, prices for deerskins rose and fell, competition for deerskins increased or lessened as new trading companies moved in or out, and hunting pressure waxed and waned in accordance with the Cherokees' desire for European goods. With all those variables, we may never know the exact ecological impact of the trade. The only certainty is that white-tailed deer were the first Black Mountain animals to feel the effects of distant urban markets. They would not be the last.[49]

Deer brought Cherokee hunters to the Blacks and Englishmen into Indian country, but exotic plants first lured white people to the highest peaks. From the earliest days of New World exploration, Europeans avidly collected American greenery, especially if it possessed curative power. In the early 1700s colonists discovered wild American ginseng, that elusive, five-leafed, red-berried plant with the oblong root that grew along the headwaters of mountain rivers. A bit of investigation confirmed that the American variety was much like Oriental ginseng, highly esteemed in China as a treatment for a wide array of physical and mental disorders. By the 1750s dried ginseng roots from the Appalachians, gathered and preserved by Cherokee collectors, began to find their way to England, France, and China.[50]

Ginseng was not Appalachia's only botanical attraction. A protracted naval war between England and France (spurred in part by the American Revolution) fostered a boom in shipbuilding that consumed much timber and left Europe's forests seriously depleted. The vast woodlands of America seemed to offer a solution. Perhaps a fast-growing tree, maybe one that stood in some uncharted mountain cove, could miraculously transform stumpy woods into magnificent forests. Ornamental species were even more valuable. In the eighteenth century collecting exotic plants became a passion among Europe's upper classes. Across the Continent well-to-do gardeners paid handsome prices for azaleas, magnolias, wild cherries, rhododendrons, goldenrods, Virginia creepers, and other American flora. Decorative foliage even served national interests. At Versailles, the opulent residence and showplace of French kings, royal gardeners created lavish displays that were the envy of monarchs far and wide. The Versailles gardens spanned nearly 4,000 acres and contained rare specimens from around the world, including 1,200 orange trees that grew within a glass enclosure.[51]

European scientists also took an avid interest in American flora. Beginning in 1735 the renowned Swedish botanist Carolus Linneaus developed a new system of plant classification. Before Linneaus's monumental work, species had been grouped together based on such human-centered traits as edibility, beauty, or medicinal value. The Linnean scheme required that plants be studied for their own sake and that they be classified on the basis of common natural characteristics, including methods of reproduction. As Linneaus's ideas gained credence, scientists began a mad rush to discover, name, and organize the world's greenery based on the new design. What better place to begin that enormous undertaking than in the rich and diverse forests of eastern America? And if, while tramping about in the New World, a would-be classifier happened on a dazzling new ornamental or a potent pharmaceutical, so much the better.[52]

One person who dreamed of exotic discoveries was a well-to-do French farmer named André Michaux. He was born in 1746 not far from the fabulous gardens of Versailles. The Michaux family worked 500 acres of fine French farmland, and André figured to remain on the property when he came of age. But when his young wife died in 1770 shortly after giving birth to a son, a grieving Michaux quietly decided to devote his life to botany. He gave over management of the farm to his brother and moved to Paris to be near France's best scientists. Later he studied in England and spent two years collecting plants in Persia. In 1785, just after an American naturalist named Thomas Jefferson arrived in Paris, Louis XVI appointed André Michaux France's royal botanist. Much to Michaux's delight, the king decided to send him to the United States. Michaux's primary mission was to seek new plants and seeds that would enrich French horticulture and facilitate botanical exchanges between the two countries. Before he was through, he would forge an important link between the East's highest mountains and the world's scientific community.

Michaux got his first look at the Appalachians during the rainy summer of 1787. Accompanied by his son, François-André, and native guides, the botanist followed a well-established trading path northwest from his Charleston home toward the southern flank of the Blue Ridge. There the group went about the business of "botanizing," collecting a variety of trees and shrubs from the Carolina uplands. The Michauxs returned to the region late in the autumn of 1788 to gather more foliage amid freezing rain and snow. On those first westward excursions the Frenchmen saw rhododendron, mountain laurel, wild ginseng and violets growing in cove forests, and white pines standing on dry, exposed slopes.[53]

By the time the weather had warmed in the spring of 1789, André Michaux was ready to try a new route. On the last day of May, with his son and a slave, he set out on horseback up the Catawba River. For much of the trip he retraced de Soto's route, and on June 13 the botanist rested at the base of the Blue Ridge, near Morganton, the old site of Joara. Two days'

Artist's rendition of André Michaux, botanist extraordinaire, who collected plants in the Black Mountains in the 1780s and 1790s. Painting by Richard Evans Younger; used by permission of the artist.

travel from Morganton put the party on the headwaters of the Catawba at a place called Turkey Cove, located north of Marion, North Carolina, near the community of Sevier.

Much to the chagrin of those who would reconstruct his route from there, surviving records of André Michaux's travels that spring are sketchy. Somewhere an Indian guide joined the party, and on June 17 the Frenchman left Turkey Cove to spend five days on what he knew as "Black Mountain." According to his notes he discovered an unknown species of wild azalea and other new plants before he returned to Turkey Cove on June 22. From there Michaux's group used tomahawks to hack their way through dense thickets of rhododendron and laurel en route to Yellow Mountain, located northeast of the Blacks in Avery County. In those days before Lewis and Clark, local people touted Yellow Mountain as the "highest in all North America." After visiting that peak Michaux intended to go on to Kentucky but was dissuaded by rumors of Indian trouble. Instead he headed east into Virginia and north to Philadelphia and New York before returning to Charleston in late September.[54]

The Michauxs hardly had time to unpack before André began planning another mountain excursion. In early November he and François-André again went up the Catawba to Turkey Cove. In the drizzly, biting cold of late autumn they crossed the Blue Ridge and passed a South Toe waterfall ("Cataractes meridi. de Taw river") on the east side of the Blacks. There, during the last week of November, André found a new species of arrowwood and pulled it from ground already white with ice and snow. By December 5, after more collecting around Turkey Cove, father and son headed back to Charleston laden with some 2,500 trees, shrubs, and smaller specimens. Many of those plants probably came from the lower slopes of the Blacks. André's incomplete notations of his discoveries listed acorns from mountain oaks, various types of wild azaleas, mountain laurel, and rhododendron, all packed in seven chests for shipment to Paris.[55]

While the Michauxs harvested greenery in the East's highest mountains, their native country was in turmoil. The outbreak of the French Revolution followed by the arrest and eventual execution of Louis XVI made it increasingly difficult for André to secure financial backing for his excursions in America. In 1790 François-André returned to France, ostensibly for health reasons, but also to look after the family lands during the worsening crisis. Yet even in the face of such difficulties, the elder Michaux still managed to range far from his Charleston home, exploring north to Quebec and west into the Ohio country during the next five years. By midsummer, 1794 he was ready for a return to Turkey Cove and "la Montagne noire."[56]

After exploring "Linville's high mountains" and collecting plants from "a place called Crabtree" (perhaps Crabtree Mountain near Spruce Pine or a settlement called Crab Orchard along the North Toe), on August 10 Michaux arrived at "the foot of Black Mountain," where he spent two days prowling the forests. Judging from the specimens he assembled, he climbed higher this time. Along with the ubiquitous laurel and rhododendron, the list from his journal includes striped maple, American beech, red spruce, and mountain ash, all of which flourish above 4,000 feet. By August 16 he was back on the Toe, where he contracted with a local resident and hunter named Martin Davenport to take him farther into the mountains.

For two glorious weeks that summer Michaux and Davenport roamed across the high ridges visible in the distance. Resting only when it rained, they scaled Roan Mountain (on the North Carolina–Tennessee border); they scrambled to the peak of Yellow Mountain and, on August 27, headed east to climb Grandfather. Taken with Grandfather's distinct profile and rocky escarpments, Michaux became convinced that it, not Yellow Mountain, was "the highest of all North America." In his exuberance at reaching the summit (and in keeping with the political temper of the times) the ebullient explorer belted out a chorus of the "Marseillaise" and shouted, "Long live America and the Republic of France; long live Liberty!"[57]

It was perhaps the most exhilarating moment of his mountain travels and, in part, a farewell. After making his way back to Charleston in late autumn, Michaux did not return to the region until the following spring, when he visited Davenport in early May. He spent two weeks botanizing nearby before embarking on a marathon trip across Tennessee to the Illinois Territory and the Mississippi River. In March 1796 the Frenchman finally got back to the Toe River, where he spent another week recuperating from his travels and gathering mountain flora. This brief sojourn along the Toe was, as Michaux's primary biographers have written, "his last journey into the region he loved."[58]

Hoping to secure more funds for further work in America, Michaux sailed for France in mid-August 1796. A shipwreck off the Holland coast nearly killed him and dumped forty boxes of precious seeds and plants into the Atlantic. But with the help of some local folk the intrepid botanist recovered his health and most of his cargo before heading overland to Paris. There he quietly resumed his life in the scientific community and set about writing two lengthy books. The first, a treatise on American oaks, went to press in 1801. The second, a multivolume opus of American flora, was in progress when Michaux left on an expedition to the South Seas, where he hoped to see the botanical wonders of New Guinea and the northern reaches of Australia. He made it as far as Madagascar, where he contracted malaria and died in November 1802.[59]

Not counting his time at Davenport's in 1795 and 1796, Michaux spent less than three weeks in the Black Mountains. Yet his reputation as an explorer remains strong; in popular literature he is often cited as the first white person to visit the range. De Soto, Pardo, Needham and Arthur, and even Michaux's guide, Martin Davenport (who as a hunter must have been well acquainted with the Blacks), probably have better claims to that distinction. Indeed, Michaux was not even the only botanist at work in the region in the late eighteenth century. John Fraser, a Scot who collected plants at the behest of Russia's Catherine the Great, explored the highest peaks of the Smokies (and might have ventured into the Blacks, though it is impossible to say for sure), where he discovered the southern fir that now bears his name. For a time he traveled with Michaux in the Carolina Piedmont. (The Frenchman apparently regarded Fraser as a rank amateur and was happy when the garrulous Scot went his own way.) But Michaux was first to write of a trip to La Montagne Noire, and in the public imagination he survives as the "first white man" of the high peaks. Professional and amateur botanists alike delight in pointing out the many trees he discovered and named, including the cucumber tree and the Catawba, or purple, rhododendron.[60]

Michaux's travels and journal influenced Black Mountain history for years to come. Decades later his collections, housed in the Museum of Natural History at Paris's Jardin des Plants, would inspire other botanists to venture into "les hautes montaignes de Caroline." Even more important was a brief observation Michaux made during his final visit with Davenport in March 1796. Finding a sprout of flowering hazelwood, the botanist wrote that the plant grew "only in the highest mountains or in Canada." Those very words, "the highest mountains," would one day help spark one of the most sensational scientific and political controversies in North Carolina history.[61] But all of that lay in the future. At the turn of the century, as word of Michaux's death filtered back to the United States, most North Carolinians were less concerned with wild plants than with another, potentially more valuable Black Mountain commodity: land.

JUNE

ESKOTA, CANE RIVER VALLEY

Is anything more perfectly pastoral than an unmown hayfield in the last week of June? Splashed with morning sun, this simple patch of waist-high grass stretches like an emerald carpet from a pot-holed gravel road to the Cane River. A light breeze fans the field, rippling the tasseled hay like water jostled in a shallow bowl. Two tiger swallowtail butterflies, momentarily rousted from blossoms of pink clover, glide past. A half-dozen red-winged blackbirds wheel up from the river and light near an old barn at the edge of the road. I lean on a locust fence post, momentarily mesmerized, absorbing the sweet, leafy, fresh-dirt smell of agrarian life.

The vista is so quaintly placid that it is easy to forget the hayfield's purpose. In another week or so, mowing machines will move in. The grass will be gathered, bound in giant rolls, and stored in the barn. If it rains, farmers will mow again in two months. Come December, summer hay will be supplemental feed for cattle, keeping them fat and healthy through the worst of the Black Mountain winter. From all indications this is not a dairy farm. The owners may keep a cow or two for breeding stock. But most likely, cattle raised here will find their way to local slaughterhouses or the livestock markets of Asheville. Difficult as it is to grasp on this postcard-perfect country day, this hayfield is part of a business. Like deerskins and ginseng, the domestic animals it nurtures are products that move between country and city. The barns that store the hay, the fences that enclose this field, and the roads that connect Eskota with other towns all result, in part, from the sale of livestock. If urban people did not demand beef, this picturesque Cane River landscape with its butterflies, blackbirds, and bucolic charm might disappear.

On the return trip to my town I drive slowly for a while, pondering the irony and taking in the surrounding terrain. The danger of killing frost has long since passed; household gardens, planted with beans, corn, squash, and lettuce—food for families living along the Cane—are coming on fast. Larger plots nearby have been given over to cash crops such as string beans, cabbage, and burley tobacco. The countryside is the hayfield writ large: a patchwork of woods, clipped lawns, and recently plowed earth that reflects a curious blend of agrarian custom and urban needs. Reading rural mailboxes along the road, I find names synonymous with early settlement of the valley: Wilson, Smith, Griffith, Young, Bailey, and Boone. These are old appellations that roll easily off the tongue and reflect the heritage of Northern Europe and the British Isles. Fitting, perhaps. By the time white farmers found their way to the Cane and other Black Mountain rivers, they had been well schooled in the ways of commerce, and from the first they used the rural landscape to attract city money.

According to a 1763 proclamation by the British government, lands beyond the crest of the Blue Ridge were off-limits to American settlers. But as with so many of its edicts, the crown could not enforce the injunction, and colonists simply ignored it. The Toe River Valley drew a smattering of permanent residents before the American Revolution; in succeeding years, many others passed through the region on their way to settlements along Tennessee's Watauga, Holston, and Nolichucky Rivers. As for the Blacks themselves, treaties with the Cherokees signed in 1777 and 1785 finalized the de facto transfer of ownership begun during the deerskin trade. Those agreements officially evicted native hunters from the East's tallest mountains and made the range part of Burke County, North Carolina.[62]

In theory the Blacks were now open to any North Carolinian who could claim a tract, get it surveyed, and pay the appropriate fees: fifty shillings to one pound per hundred acres. But most of those who first laid claims to the mountains were not settlers but wealthy speculators. Planters and merchants from the coastal plain, government officeholders, out-of-state investors—anyone savvy enough to hire a surveyor or bribe a county official—snatched up the best parcels. In 1785 Waightstill Avery, North Carolina's first attorney general (and a good friend of André Michaux's), claimed virtually all of the North and South Toe Valleys. John Gray Blount, from faraway Beaufort County, staked out another 320,640 acres, a tract that encompassed the entire Black Mountain range. In the dead of winter in 1795, John Brown, agent for a Philadelphia-based syndicate, braved driving snow and numbing cold as he zigzagged from the Toe River south to Salisbury. In only three weeks he secured a half-million acres for his investors; almost all of his claims were falsely registered under the names of North Carolina citizens. Some veterans of the American Revolution received land grants along the Toe and Cane, and a few squatters managed to hang on to isolated parcels. But like the trade in deerskins, traffic in Black Mountain real estate was driven by the appetites and actions of businessmen who lived far from the high peaks.[63]

Most would-be immigrants to the Blacks first had to purchase or lease acreage from one of these absentee owners. As a result the white population grew slowly at first. When the federal government completed its first census in 1790, eighty families lived along the Toe. Perhaps as few as 500 people resided between Grandfather Mountain and the eastern slopes of the Blacks. Indeed, most settlers seemed to prefer the warmer Swannanoa Valley. Shortly after the census the Swannanoa was populous enough to become part of Buncombe County, which was carved from Burke in 1791. The advent of county government drove up real estate prices, and speculators eventually relinquished their hold on other lands near the Blacks. During the first decades of the nineteenth century, people migrated toward the high peaks from all directions: east from the French Broad, west from the Piedmont, southeast from Tennessee, and southwest from Pennsylvania and Maryland. In 1833 North Carolina deemed it necessary to create yet another new western county, called Yancey. It spanned a vast territory from the western side of the Blue Ridge south and west across the North Toe River Valley. It took in virtually all of the Black Mountains.

Most of Yancey's earliest residents farmed the same river bottomlands that had once sustained Pisgah people. Indeed, the pattern of white settlement was so distinct that certain family names quickly became synonymous with individual valleys and streams. The North Fork of the Swannanoa was home to the Burnetts; the Ivy River, to Ogles and Brigmans; the North Toe, to Brights, McCrackens, Wisemans, Davenports, and Youngs; the lower Cane, to Edwardses, Hensleys, Higginses, and Rays. They fed themselves with corn, the same grain (though by now a different variety) that had revolutionized Indian society. Beans, squash, potatoes, and pumpkins—all domesticated in America—still flourished in valley fields. But the new plots differed from native gardens in one crucial respect: white farmers hoped to do more than simply survive nature's seasonal whims; they were in the mountains to turn a profit.[64]

In the early nineteenth century the same mountain corridors that had brought explorers overland from the Atlantic coast now linked the Blacks with towns such as Morganton, Lenoir, and Asheville, where merchants were quick to buy and barter for mountain products. From the foothill towns, goods went overland to Charlotte, Salisbury, Charleston, and Savannah. Initially, white settlers were just the newest suppliers of items that had once been staples of the Indian trade. Stopping at Martin Davenport's household on the North Toe River, John Brown, the Philadelphia speculator, saw more than a hundred deerskins, cured and ready for market. By the 1830s two Yancey County residents had built a ginseng "factory," where they processed some 86,000 pounds of the root. Over time, however, mountain farmers developed new specialties. Valley climates proved ideal for tobacco, oats, buckwheat, and rye. Corn and apples were popular, too, especially in liquid form. In 1840 Yancey County had thirty-two distilleries that annually turned out 5,790 gallons of whiskey, cider, and brandy.[65]

But when it came to profits, neither grains nor spirits could rival the most important commodities produced on mountain farms: livestock. Hogs and cattle arrived along Black Mountain rivers with the first white settlers. In the cool upland valleys, where cattle escaped the warm weather diseases that plagued them in the lowland South, livestock populations boomed. Indeed, by 1850 in Yancey County domestic animals outnumbered people roughly ten to one.[66]

Yancey farmers also had the good fortune to live near Asheville, one of the Southeast's premiere cow towns. Each year, beginning in August, skilled drovers from Ohio, Kentucky, and Tennessee brought hogs, cattle, sheep, horses, mules, and even ducks and turkeys to Buncombe County. From there livestock moved south to markets in Spartanburg and Charleston. During the 1840s hundreds of thousands of animals passed below the southern rim of the Blacks every year.[67] Farmers on the Cane and Toe simply took their herds overland to Asheville stockyards, where locally reared animals joined the annual parade to market.

The crops and livestock that kept farmers fed and financed did not raise themselves. In thickly forested bottomlands settlers practiced what is commonly called slash-and-burn agriculture. To prepare new ground they first girdled the tallest hardwoods and then burned off the smaller trees and undergrowth, creating semicleared patches locally known as deaden-

A deadening, or partially cleared agricultural plot, in Mitchell County, North Carolina. The standing trees were girdled two years before, and this is the second corn crop planted on the site. Though taken in 1901, this photograph shows agricultural techniques used by some of the earliest white settlers in the Black Mountains. U.S. Forest Service.

ings. Those plots could be tilled with the simplest farm implements, and the rich, humic soil initially proved ideal for corn, grains, and vegetables. As summer waned, circulation of warm and cool air between deadenings and surrounding forests gave rise to thick fog and heavy dews that helped stave off killing frosts. During winter half-dead trees shaded the ground so that it retained moisture from seasonal rain and snow. Root systems from the dying timber reduced erosion, an important consideration on sloping terrain. In spring the plot could be burned again and covered with nitrogen-rich ash.[68]

Black Mountain agriculture was a eclectic amalgam of American and European habits. Clearing methods bore a striking resemblance to Pisgah and Cherokee farming. Other practices reflected the Celtic and Germanic heritage of the region's first white settlers. For generations their ancestors in Ireland or the German Rhineland had raised livestock on open range. Irish farmers also used various systems of field rotation, most of which depended on fire for initial clearing and allowed exhausted cropland to grow up in brush and forest.[69]

Peculiarities of the local landscape also dictated what Black Mountain farmers could do. Girdling and burning were simply the quickest and easiest techniques for getting crops into the ground. But as with native farming, practicality did not make for ecological perfection. Within three to four years seasonal burning and removal of the standing trees left fields completely exposed to intense mountain rains. Erosion gradually increased, nutrients were leached from the soil, and inevitably the tract became infertile. When that happened, a farmer had no alternative but to create another deadening, leaving the old ground to revert to hardwood forest, a long, complicated, and uncertain process subject to all the nuances of nature. In the Blacks natural reforestation of a depleted tract might require twenty-five years or longer.[70] During the first decades of the nineteenth century the farms along the Toe and Cane must have been a patchwork of new deadenings, cleared plots, and old fields reverting to woodland. As long as farmers had enough land to allow exhausted ground to lie fallow, the system could be sustained, no matter what the demands of the household or the urban markets.

Hardwood forests were important to livestock, too. When the first settlers came to the Blacks, North Carolina had no laws that required cattle owners to fence their herds; indeed, for much of the nineteenth century the state made it *illegal* "to interfere with livestock grazing on any land not under cultivation . . . *no matter who owned the lands or the animals.*" The vast Black Mountain forests, held by absentee owners, became common grazing lands open to all settlers. Cattle and hogs lived off the woods, their diets supplemented only with a bit of salt and corn provided by their owners.[71]

Set loose to fend for themselves in the forests along the Toe and Cane, cattle immediately developed preferences for certain Black Mountain flora. Native grasses and herbaceous plants, neither of which were plentiful at lower elevations, were bovine favorites. When that forage disappeared, cattle browsed the trees. Yellow poplar was probably first to suffer. Modern studies show that when turned into forests like those along the Cane and Toe, cattle quickly defoliate and kill 75 percent of yellow poplars less than five feet tall. By biting and pulling at larger trees (twelve to fifteen feet high and up to an inch and a half in diameter) the animals can "ride them down" and take the leaves. Once they consume all the yellow poplars within reach, cattle move on to black locust, ash, birch, and eventually chestnut. In addition the animals trample and compact soil among the hardwoods, making it difficult for the trees to regenerate. Over time the most palatable species completely disappear from heavily grazed woodlands and are replaced by red maples, hickories, and eastern hemlock, all of which cattle find unsavory.[72]

Although much grazing took place in the river valleys, livestock also made their way to higher ground. Of particular interest to Black Mountain cattle owners were the random, open patches of grasses and shrubs in the region's high-elevation forests. In the nineteenth century local people called them mountain meadows or fields; today ecologists know the scruffy grasslands as balds. Despite years of study by scientists the exact origins of mountain balds remain shrouded in mystery. Clearing by native or white farmers, cataclysmic wildfire, variations in soil, and changes in local wind and weather patterns have all been cited as possible explanations. Whatever the cause, such open areas were common in and around the Blacks by the early 1800s. On the Ivy River, John Ogle, one of the region's earliest settlers, apparently raised cattle on grassy land located above 5,000 feet, a site later known as Ogle Meadows. Livestock also grazed the summits of the Great Craggies and Roan Mountain.[73]

Driving a herd of cattle 5,000 feet up into the Blacks was no easy task, but many farmers thought it worth the trouble. For one thing, the farther the animals ranged from river bottomlands, the easier it was to keep the beasts from eating or trampling unfenced crops. In addition many nineteenth-century scientists believed that the soils of high ridges were more fertile and better suited to grasses than those of the valleys. In fact the opposite is true; generally speaking, the more elevated the ground, the thinner and more acidic the soil. But the dark dirt under the evergreens resembled that of Europe's finest pastures, and some settlers believed that if cleared of trees, high-elevation sites would immediately sprout lush carpets of grass. Herds that summered at high elevation also seemed less susceptible to milk sickness, a dread malady that not only killed cattle but also poisoned humans who consumed milk from afflicted cows. Cattle usually contract the disease by consuming toxic plants and mushrooms common in wet habitats. Nineteenth-century people never understood the exact cause, but they did associate milk sickness with the damp forests of the bottomlands, often attributing the affliction to the mists, dews, and fogs that settled along the rivers on warm evenings. Any Black Mountain farmer who followed conventional wisdom knew that in summer, cattle should move to higher and drier ground.[74]

Like the forest that surrounds it, a mountain bald is not static. Depending on moisture and soil conditions, the meadows are subject to invasion by a wide variety of plants and trees. Blueberries, thornless blackberries, rhododendron, laurel, pin cherries, beech, and birch are only some of the species that frequently take root among the grasses. In the Blacks and other settled regions, cattle kept that from happening. Every summer when herders took their stock to the high meadows, the animals grazed selectively on herbs and newly sprouting broad-leafed trees, in effect pruning extraneous vegetation from the bald. The plants that did gain a toehold tended to be shrubby species like wild azalea, rhododendron, and mountain laurel, none of which cattle favored. Like wind and fire, livestock became agents of disturbance, helping preserve and enlarge grassy environments in regions that otherwise might have been covered with trees. Indeed, cattle are so efficient at keeping down invading vegetation that some scholars have suggested that places such as Ogle Meadows "owe their treelessness to deliberate clearing for the purpose of grazing."[75]

As the weather cooled and livestock moved back into the valleys, the highest Black Mountain balds took on characteristics of "the arctic tundra and wind-swept beaches of [America's] northeastern coast," habitats not seen in the Blacks since the last days of the Pleistocene. Snow buntings and other birds from the far north congregated on the balds in late winter. A host of shrews and woods mice also found the grass attractive, as did New England cottontail rabbits. During winter's heaviest snowfalls bobcats and least weasels prowled the meadows in search of smaller mammals. In what ranks as one of nature's most ironic gambits, a domestic environment maintained by humans and farm animals became home to some of the Blacks' most wondrous wild creatures.[76]

Whether they grazed high meadows or munched yellow poplar and birch along the Toe and Cane, livestock were fair game for the region's largest carnivores. Any mountain lion that could bring down a deer could easily make a meal of a calf. Black bears sometimes raided valley farms in search of half-feral pigs. But in the minds of those who settled along the Toe and Cane, no predators were more dangerous than gray wolves. Across eastern America settlers believed that wolves were savage beasts that attacked farm animals for sport and usually killed more than they could eat. Many modern studies give the lie to that image, suggesting that when both livestock and wild foods are plentiful, wolves much prefer the latter. But the wolf's reputation as a killer sealed its fate. From 1836 until 1857 Yancey County collected a special tax used to pay any hunter who turned in a wolf's head to local officials. Throughout the nineteenth century and into the twentieth, Black Mountain farmers routinely shot any wolf, mountain lion, bobcat, or black bear that crossed their paths. As livestock populations rose, native carnivores declined. It was the beginning of an ecological disruption from which the Blacks never recovered.[77]

While they converted crops into cash and made the woods safe for cattle and hogs, settlers also went about the uniquely human chore of naming places. Accurate maps were in short supply, but when local people spoke of the land, they used terms that anyone familiar with Black Mountain geography would understand. Two major rivers retained vestiges of names given them by native people. "Swannanoa" probably came from the Cherokee word *suwalinunahi*, also the native designation for a trail that led out of Joara, up the Catawba River (along the route of Interstate 40), and across the Eastern Continental Divide. "Toe" is apparently an English corruption of *Estatoe* (pronounced Est-a-to-wee), the name of a Cherokee town and, legend has it, of an Indian woman who drowned herself in the river after her lover's death. Spanish explorers moved too fast to leave much of their language behind, but their explorations also helped preserve native nomenclature. "Appalachian" likely derived from *Appalachee*, the name of a powerful Indian chiefdom de Soto had explored in Florida. *Tocae*, the name of the native town near Asheville that Pardo visited, perhaps survives as "Toecane," a local community and general shorthand for the region around the Blacks.[78]

European and American immigrants added their own appellations, christening Graybeard, Pinnacle, the Great Craggies, and the Blacks themselves during the first years of settlement. Over time other peaks acquired European surnames. Gibbs Mountain, near the northern end of the range, was named for an itinerant Methodist minister. Sometimes topography dictated descriptions, as in the case of Deep Gap, a low place south of Gibbs Mountain, or Potato Knob, an oval-shaped summit near the bend in the fishhook. Wildlife and vegetation also figured in the naming process. Cattail Peak might have been named for native mountain lions. An adjacent mountain, Balsam Cone, was covered with red spruce and Fraser fir, trees locally known as he-balsam and she-balsam.[79]

Those names bespoke a long history of human activity in the Black Mountains. From the time people first walked along the Swannanoa until cattle and sheep tramped across mountain balds, *Homo sapiens* had repeatedly reshaped the landscape to suit their needs. Connections between people and land were more than physical. Much of what had transpired in the region resulted directly from human imagination, from how people thought about the mountains and what they might provide, be it security, subsistence, instant wealth, cheap land, or steady profits.

Yet in the early nineteenth century the Blacks were hardly unique among the Appalachians. Subtle environmental changes associated with commercial agriculture and livestock production were occurring in river valleys throughout the range. As André Michaux and others could attest, most knowledgeable folk did not yet believe that the East's highest summit stood between the Cane and Toe. Plenty of other mountains (Grandfather, Yellow, and Roan among them) were more celebrated for altitude and scenic beauty. That was about to change. Thanks to a professor, a politician, and a mountain man, the Blacks would soon become the most famous peaks in North Carolina.

THREE

Mitchell's Mountain

MAY

CHAPEL HILL

According to a relief map that hangs near my writing desk, the University of North Carolina stands roughly 180 miles from the East's highest peak. But today, as I battle morning commuters in a frantic search for downtown parking, the mountains seem a world away. At 8:00 A.M. the temperature on Chapel Hill's main street hovers near eighty degrees. Deciduous trees are already in full leaf, providing a powerful visual antidote to the still-gray oaks and budding maples I left behind in the Blacks two days ago. The contrast between the regions is more than climatic. In recent years the explosive growth and urbanization of the North Carolina Piedmont have transformed this college town into a bustling city. Shopping plazas and upscale subdivisions spread out from Chapel Hill into the surrounding countryside, turning what was once prime farmland into an asphalt maze of residential streets, cul-de-sacs, and access roads.

Walking toward the university, though, I feel the rush-hour pace slacken a bit. Spring semester classes ended two weeks ago; most of the students have long since departed. I pause briefly at a two-foot-high stone wall that marks the boundary between town and campus. "The Wall" (as the students call it) is a nondescript but well-known UNC landmark first erected nearly a century and a half ago. Today it serves notice that even in this urban atmosphere, the Black Mountains are closer than they seem. The geologist who commissioned the original stonework was none other than Elisha Mitchell, university professor of science, ordained Presbyterian minister, and legendary explorer of the East's highest peaks.

To step across the Wall is to step into Mitchell's world. Even now, at the modern university, evidence of his presence abounds. A science building bears his name. Books from his personal collection gather dust on library shelves; a few volumes still bear his trademark signature, "E. Mitchell." Those same libraries hold collections of his letters, memorandum books, course syllabi, and other papers. Eventually I will see them all. But first I seek out the most celebrated piece of Mitchell memorabilia: a large, silver-plated pocket watch housed in the Gallery of the North Carolina Collection, the university's matchless array of documents and artifacts pertaining to state history.

Mitchell carried the watch in the summer of 1857 when he ventured into the Blacks for the last time, a trip that he hoped might settle a long and rancorous debate with a former student named Thomas Clingman. For two years the two had sniped at each other—most recently in the Asheville newspapers—arguing fiercely about who had been first to measure the highest mountain in the East. But on the evening of June 27 Elisha Mitchell fell to his death while walking alone through an isolated ravine on the headwaters of the Cane River. When searchers recovered his broken timepiece, its hands were stuck at 8:19:56 P.M., presumably the exact hour, minute, and second of the fatal accident. The watch remains frozen at that instant in time, a macabre reminder of the professor's death and the feud that preceded it.

The saga of Elisha Mitchell and Thomas Clingman is an old and oft-told story full of pathos and irony that still ranks as a central event in North Carolina's antebellum history. But even now few who repeat the tragic tale know all its nuances or recognize how it altered public perceptions of the Black Mountains. To tell that story—to understand how an argument between two men lent new meaning to a mountain landscape—we must begin here, on this campus, when Chapel Hill was a rustic village in the middle of a Piedmont forest, when a young professor, recently arrived from New England, began to read and wonder about the rugged country to the west.

Elisha Mitchell was born in Washington, Connecticut, in 1793, about a year before André Michaux declared Grandfather Mountain the highest peak in America. The Mitchells came from old Yankee stock, a lineage that extended to the 1630s and included the Puritan missionary John Eliot. By the time Elisha came of age, his family had the necessary means and social standing to send him to Yale. There he studied science with Benjamin Silliman, eminent professor of chemistry and natural history and founder of the *American Journal of Science and Arts*. Popularly called *Silliman's Journal*, the prestigious periodical was perhaps "the greatest single influence in the development of the American scientific community" during the nineteenth century.[1]

Mitchell entered that community at an opportune moment. Fifty years earlier American science had been dominated by "gentleman amateurs," men with independent income who devoted their abundant leisure time to botany, astronomy, or other personal interests. When Silliman began teaching at Yale in 1802, he was one of fewer than twenty scientists working full time in the United States. But less than a decade later the academic climate had changed. The nation had a new assortment of state-funded colleges, most of which needed university-trained specialists to fill faculty rosters. In 1816 William Gaston, University of North Carolina trustee and U.S. congressman, learned of two young professionals fresh from Silliman's tutelage who sought academic appointments. One was Elisha Mitchell; the other was a Yale classmate named Denison Olmsted. Gaston eventually brought both men to Chapel Hill. Olmsted taught geology, mineralogy, and chemistry; Mitchell, who arrived on the last day of January 1818, took charge of mathematics and natural philosophy, a course similar to physics.[2]

Moving south was a shock for the young professor. Thick forests of white oak and loblolly pine still covered much of Piedmont North Carolina. Chapel Hill and the other villages in the region were accessible only by way of crude horse paths or deeply rutted wagon roads. Mitchell's new wife, Maria North, traveled some of those routes when she joined her husband at the university in late December 1819. Leaving Tarboro in the northeastern coastal plain, she rode seventy miles "without passing any village" before arriving in Raleigh. There she and her traveling companions "left all civilization behind us, and went on through woods and woods" for another whole day until the stage stopped at the university. After touring the half-dozen university buildings and seeing the forty-odd ramshackle houses that made up Chapel Hill, Maria noted, with the abiding optimism of a newlywed, that the provincial college town was "much more pleasant than I expected."[3]

While Maria adapted to life in the wilds of Chapel Hill, her husband taught mathematics and science on a campus populated by 100 rowdy boys. In keeping with his New England upbringing Mitchell was a devout and introspective man steeped in the traditions of Calvinism and constantly concerned about the state of his soul. In 1816, immediately after he left Yale, he studied at the Theological Seminary in Andover, Massachusetts, and for a short time preached as a licensed Congregational minister in Connecticut. During that period he kept a diary in which, like his Puritan forebears, he brooded incessantly about his "inability to reject pleasure, pride, humor," and other enticements of the secular world. Though he discontinued the journal shortly after he moved to Chapel Hill, his Congregational faith never wavered. Ordained by the Orange County Presbytery in nearby Hillsborough, he led daily prayers at the university and often preached in local churches. Mitchell was no religious firebrand. Indeed, his tranquil demeanor and infinite patience (especially with rambunctious, irreverent students) were the stuff of campus legend. But as a colleague later wrote, the professor never passed up a chance to "declare to his fellow men the will of God for their salvation."[4]

Mitchell also had a Calvinist's disdain for trivial pastimes. During his first years in Chapel Hill he became an avid collector of Piedmont plants, often spending five afternoons a week in fields and forests near the university. But botany was no mere hobby. Quickly frustrated by his poor knowledge of southern flora, Mitchell enlisted the help of Lewis David Schweinitz, a Moravian theologian and expert on local greenery who lived in Salem, North Carolina. For several years the two men carried on regular correspondence and traded specimens collected during forays into the Carolina countryside. Mitchell admitted that his expertise never approached that of his Moravian mentor. But thanks to Schweinitz the professor became acquainted with some of America's most renowned naturalists, including John Torrey and Asa Gray.[5]

Mitchell was also a passionate collector of books, for both himself and the university. (Indeed, by the time of his death his library contained over 2,000 volumes and easily ranked as one of the best scientific collections in the South.) Many of his early purchases reflected his newfound interest in botany. Among his first acquisitions were André Michaux's volume on American oaks, his *Flora of North America*, and *North American Sylva*, a multivolume work authored by François-André Michaux and based on his and his father's travels. Somewhere in those books Mitchell must have read about the unique environment of North Carolina's high mountains and the exotic foliage that grew there. At some point during those early years at the university he resolved to explore the western part of the state.[6]

His chance came sooner than he expected. In 1825 Denison Olmsted left North Carolina to return to Yale. A year or so earlier Olmsted had begun North Carolina's first Geological and Mineralogical Survey. Funded with $250 from the state legislature, the survey was designed to provide the first scientific assessment of North Carolina's geography and natural resources. When Olmsted departed, responsibility for the survey fell to

Elisha Mitchell, geologist, botanist, university professor, Presbyterian minister, and Black Mountain explorer. North Carolina Collection, University of North Carolina Library, Chapel Hill.

Mitchell. He began to travel extensively across the state—mostly on horseback but occasionally on foot—during the summer, when the university was not in session. In 1827 and 1828, after touring the Coastal Plain and the Piedmont, he headed west, intent on having a look at the Blue Ridge and the peaks beyond.[7]

Although taken with the region's scenic splendor, Mitchell found the mountains an intolerably primitive place. During his early journeys he repeatedly chastised Appalachian folk for what he regarded as laziness. While traveling in Ashe County he noted that some of the lands he saw were "as fine as the good parts of N[ew] England." But, he lamented, the local "people lacked industry." They neglected their livestock and lived in "unsightly log hovels" when they might easily have enjoyed "vast herds of cattle," great flocks of sheep, and "painted frame houses."[8]

Such comments likely sprang from the professor's Calvinistic theology. Two hundred years earlier, when Puritan colonists stepped ashore in New England, they had often decried the uncultivated state of the land. They called it a wilderness and cowered in fear of the wild animals and wild men who lurked in its dark forests. The Devil was there, too. Set down in desolate country, cut off from the divine influence of church and civilization, even the most devout Christian might be lured into an easy life of sin and savagery. Indeed, Mitchell's distant relative John Eliot often expressed his concerns about "wilderness-temptations" that encouraged well-meaning folk to abandon all discipline and live carefree in the woods. If God's people were to maintain proper deportment, the wilderness had to be tamed and the savage natives driven away. Settlers had to cut trees, plant fields, and create a productive, pastoral landscape where good might triumph over evil.[9]

Two generations removed from Eliot, Elisha Mitchell never saw the wild New England that had threatened his ancestors, and the professor's religion was in many ways different from that of his Puritan forebears. But the basic notion that Christians should not live too close to nature echoed incessantly through his early descriptions of western North Carolina. For him the mountain wilderness, like that of early New England, was a dangerously decadent place. Learning of one western resident who kept two wives and had fathered children with several other women, Mitchell complained that the mountains were "a terrible place for such irregularities." Corrupted by the forests around them, white people lived like savages. Men spent most of their time hunting and roaming the woods, while women became "schquaws, very pretty ones [the professor admitted] but squaws notwithstanding."[10] Mitchell even seems to have worried that he might be led astray in the wilderness. Though he spent much time outdoors, he rarely slept overnight in the forest. Instead he sought lodging wherever he could find it and usually stayed with local people (many of whom he did not know) along the route. When invited, he preached in nearby churches, no doubt urging his listeners to stay on the disciplined path that led to salvation.

That sort of zeal might seem odd for a scientist, but it was in keeping with the temper of the times. By the late 1820s many of the lawyers, merchants, and professional men who served in the North Carolina legislature shared the professor's belief that wilderness stood in the way of civility and progress, especially in a state where the largest towns and even the university could be reached only after days of travel through heavily forested terrain. The politicians, however, worried more about money than morality. In 1790 North Carolina had been the third most populous state in the Union. But a steady stream of emigration, largely owing to a stagnant economy, dropped it to fifth by 1820. Land values declined, and the state's tax base diminished dramatically. Reform-minded legislators saw the geological survey as a first step toward finding better routes for roads, canals, and river traffic—improvements that might bolster commerce, stimulate economic development, and keep ambitious citizens (and their tax dollars) at home.[11]

In outlining their plan for the mountains, Mitchell and his legislative cohorts embraced the popular notion that the region's soils were well suited to grasses and livestock. Given good roads and access to markets, mountain folk could provide the state with wool and dairy products, the same commodities that had revolutionized New England. Since Mitchell was head of the geological survey, his role in the transformation was crucial. If the mountain landscape could be charted, it could be civilized, and once civilized, its people might be delivered out of the wilderness into prosperity. In 1828, funded by the legislature and driven by divine directive, the Presbyterian professor began the work that made him famous. He started measuring mountains.[12]

Having read of Michaux's triumphant excursion to Grandfather Mountain, Mitchell was eager to see what passed for the highest peak. When he finally reached Grandfather's summit in mid-July 1828, "the day was fine" and "the prospect . . . all but infinite." From his vantage point he could make out "the endless ridges of Tennessee," Yellow Mountain, Roan Mountain, and—on the far southwestern horizon, still slightly obscured by "a few flying clouds"—the massive, sprawling "Black Mountain of Buncombe." In keeping with his assignment he immediately tried to bring order to the scene. "It was a question with us," he later recalled, "whether the Black and Roan Mountains were not higher than the Grandfather." He carefully noted that he had gotten the same impression a year earlier while observing the range from Morganton, and he concluded that if he ever spent "another summer in these parts," he might head for "Toe River and investigate the district lying between and around these high mountains." That evening, after descending Grandfather, Mitchell camped under a protruding rock ledge with several companions, one of the few occasions (and perhaps the only time) he slept outdoors in 1828. Despite the company and the comfort of a roaring fire, Mitchell remained ill at ease in the wild, tossing restlessly beneath his blanket and looking at his watch "a good many times to see if it was not nearly morning."[13]

Not long after the professor settled back into his duties at the university, the search for the state's highest mountain took on new urgency. Following Michaux's discoveries, several amateur naturalists (including John C. Calhoun of South Carolina) had suggested that one of the southern peaks might be taller than New Hampshire's Mount Washington, long touted by New Englanders as the highest of the Appalachians. By the early 1830s many prominent North Carolinians thought it time to test those theories. South of the Blacks, towns such as Asheville and Flat Rock had already begun to attract tourists, most of whom were wealthy socialites from South Carolina and Georgia who sought respite from the lowcountry's sweltering summers. Across the state, politicians and entrepreneurs dreamed of turnpikes, hotels, mineral springs, spas, and other amenities that might draw sightseers and bolster the stagnant economy. What better way to lure well-heeled visitors than to promise them a visit to the highest ground in eastern America?[14]

Among the most active boosters of western North Carolina was David Lowry Swain of Asheville, who after serving as the state's governor became president of its university in 1835. That same year, perhaps at Swain's behest, Elisha Mitchell left Chapel Hill to take the first scientific measurements of Roan, Grandfather, and the Black Mountains. No doubt the expedition also appealed to the professor's academic ego. By the 1830s America was well into its "Second Great Age of Discovery" and a bevy of explorers, scientists, and artists had begun to catalog the continent's natural wonders. In a world where Europeans routinely boasted of sophisticated culture and aristocratic traditions, Americans bragged about nature, pointing to the majesty of the Hudson River Valley, Niagara Falls, the Great Lakes, the Mississippi River, and the Rocky Mountains. Granted, a short trip to search out the East's highest peak hardly figured to put Mitchell in the same league with Lewis and Clark. But in a nation increasingly fascinated with "the natural, the big, the distinctive," Mitchell knew that measuring the eastern summit could cement his reputation within the scientific community. He also knew that he would have to work fast. He had only six weeks, the length of the university's summer break, and though he was familiar with Grandfather, he had never been to Roan or the Blacks.[15]

Like many of those who studied mountains in the early nineteenth century, Mitchell calculated elevation by measuring differences in barometric pressure. Like other Black Mountain explorers before him, he first went to Morganton, that age-old jumping-off place for western expeditions. Once he had established Morganton's elevation, an assistant stayed behind to watch a barometer there while Mitchell carried a second barometer into the mountains and made his own observations. Using a complicated mathematical formula, he planned to compare the readings and derive elevations for various sites. The technique was hardly foolproof. Variations of only 0.1 inch in barometric pressure might cause calculations to be off by more than 100 feet, which was no small consideration in a region subject to sudden summer storms and fast-moving low-pressure systems.[16]

Mitchell said little about conditions on Grandfather and Roan in 1835, but apparently those peaks proved relatively easy to measure. As the professor made his way into the Blacks, a large high-pressure system (probably a typical summertime Bermuda high) seems to have anchored off the Atlantic coast, bringing days of scattered clouds but little rain to western North Carolina. Though blessed with good weather, Mitchell soon encountered other problems. The so-called Black Mountain, which he had seen only from a distance, was actually fifteen miles long and had so many remote cones, domes, peaks, and pinnacles that the professor had trouble deciding where to place his barometer. After securing lodging in a local household, he first took a reading on Celo Knob at the north end of the range. Looking south from there, however, he quickly concluded that some other spot might be even higher. Hoping for a better vista, he headed for the Cane River Valley, where he asked two local farmers, Samuel Austin and William Wilson, to show him the highest mountain. They took him to Yeates Knob, a prominent peak near the point of the fishhook, today known as Big Butt.[17]

Gazing across the Cane River Valley at the north-south shank of the range, Mitchell saw several mountains that appeared more elevated than the peak on which he stood, but even his trained eye could not discern the tallest. To stand near Yeates Knob on a summer day is to know the professor's quandary. Although an observation tower now marks the true summit, several peaks on either side might easily be mistaken for its equal. In fact, the seven tallest mountains in the Blacks, all of which can be seen from Yeates Knob, vary in elevation by only 140 feet, differences easily discernible by Mitchell's barometers but not by casual observation.

However, some recent experiments conducted in the region suggest that the professor probably had both the mind and the means to solve his dilemma. In the summer of 1991 Perrin Wright, a mathematician from Florida State University with a keen interest in Black Mountain history, hauled fifty pounds of homemade survey equipment to a site not far from Yeates Knob. Without attempting to establish elevation at any point, Wright was able to rank, in perfect order from highest to lowest, ten of the eleven tallest peaks on the far side of the Cane. He used instruments comparable to those available in 1835 and based his work on the most fundamental laws of trigonometry and range finding, concepts well known in the early nineteenth century. Wright's research indicates that had Mitchell been so inclined, he might have remained on Yeates Knob for several days, done some basic trigonometric calculations, and compiled a similar ranking *without discerning exact elevations*. He could then have gone to the highest mountain (or any other) and deployed his barometer for several days. Once he had an accurate measurement for that peak, he might then have used his Yeates Knob survey to compute elevations for the rest.[18]

Perhaps Mitchell considered a similar strategy. Having taught courses in trigonometry and calculus, he surely understood the principles of long-distance surveying. Indeed, as he prepared for the 1835 trip, he noted that he intended to investigate "Positions by Trigonometry." But once into the Blacks, the professor found himself pressed for time. And as always, he was reluctant to camp in the woods. Moreover, his primary objective was not necessarily to find the highest peak in the range but, rather, to show that some southern Appalachian summit—be it Grandfather, Roan, or the enigmatic Black Mountain of Buncombe—outranked Mount Washington. As he stood on Yeates Knob in 1835, he probably found it more expedient (and less threatening) to locate some obviously high place and go there with a barometer. If he could record an elevation higher than Mount Washington's, his work was done.[19]

After much deliberation Mitchell fixed on a particular peak that lay between the "North and Middle forks of Caney River." The next morning, July 28, he set out to climb it. Accompanied by William Wilson and a new guide, Adoniram Allen, the professor made his way up a laurel- and rhododendron-choked "bear trail" that ended on a prominent summit. Maybe he intended to take extensive notes as he surveyed his surroundings, to somehow assure himself that he stood at the pinnacle of the range. He never got the chance. By the time the party had struggled up the slope, light afternoon clouds (spawned by the Bermuda high) hung on the high peak, enveloping it in a wet filmy haze and making it impossible for the men to get their bearings. Shivering in the dampness, Mitchell spent just two hours taking barometric readings before heading back to the Cane River settlements. There, he happily noted, the weather was clear "and the thermometer at 80." The trip to the mountaintop was brief, but at the time it seemed sufficient. Figures from the professor's barometer and the one at Morganton established the summit's elevation at 6,476 feet, more than enough to eclipse Mount Washington, then listed at 6,234.[20] A week or so later he returned to Chapel Hill to spread the good news.

In November Mitchell recounted his work in the *Raleigh Register*, one of North Carolina's premier newspapers. Explaining that "some gentlemen in the West" (his name for the region's publicists) had been "promised an account of the results," he provided new heights for various North Carolina mountains. "For the sake of comparison" he listed the measurements with elevations from several New England peaks, including Mount Washington. With careful understatement Mitchell wondered if the "highest peak of Black" might someday "attract an occasional visitor" from other parts of the state.[21] The boosters, however, were less restrained. A state long ridiculed as a regressive backwater now had a landmark of national significance, and as the *Register*'s editor exulted a week later, North Carolinians could justifiably "LOOK DOWN" on anyone who might "insolently venture to taunt us with inferiority."[22]

While promoters reveled in Mitchell's discovery, the professor returned to his classroom and began an article for *Silliman's Journal*. Hoping to gather more information and shore up his research, he returned to the Blacks in 1838. This time, however, he first went to Asheville (a town whose elevation was well established) and, while lodging with farmers along the Swannanoa's North Fork, spent most of his time exploring the southern end of the range. Based on barometric readings taken there and in Asheville, he calculated that one of the peaks on the southern rim measured

6,581 feet, or 105 feet higher than the foggy summit he had visited three years earlier.[23]

When he returned to visit the Blacks a third time in 1844, nearly a decade had passed since he first stood on Yeates Knob. Burnsville, the Yancey County seat, was now a respectable village. Thanks to the region's booming livestock business, mountain people seemed more interested in herding than hunting, a trend the professor regarded as an important step toward civility and salvation. Moreover, by the 1840s the Blacks and even Mitchell himself were finally on the map—or at least on *a* map. Amid the hoopla surrounding the 1835 discovery, Swain had somehow persuaded Roswell C. Smith, a prominent geographer, to list "Mount Mitchell" as "the highest point of land in the United States, east of the Rocky Mountains." The catchy appellation appeared in Smith's 1839 *Geography and Atlas* and soon gained wide acceptance among local residents. As he approached Burnsville in 1844, Mitchell innocently asked a Yancey County youth if the Black Mountain might be glimpsed from a nearby hill. The professor could scarcely contain his excitement when the lad "replied that he could show me Mount Mitchell from there."[24]

But where exactly was Mount Mitchell? As he embarked on his third mountain expedition, the man for whom it had been named could not be sure. In the six years since his last visit, the professor had discovered that he had seriously underestimated Morganton's elevation in 1835, a miscalculation that meant his initial measurement might be 150 feet too low. He also knew that the barometers he used in 1835 were much inferior to those available in 1844 and likely had given him faulty readings. And he still had to account for the new elevation of 6,581 feet recorded on the south rim in 1838.[25]

Hoping to settle those issues, Mitchell left a barometer in Asheville and went back to the Cane, where he hired William Riddle and his son, Marvil, as guides. On July 8, 1844, the three set out on foot toward the southern end of the Blacks, hoping to locate and measure the elusive summit. They followed the Cane's headwaters as far as they could, crossing rough ridges lush with summer vegetation. As Mitchell explained it, he climbed "all the way through laurels," with "the Whole Black Mountain before me." Early that afternoon, as he crawled through a rhododendron thicket on his hands and knees, he noted (in a famous phrase now rife with irony) that he "could not help thinking . . . what a comfortable place it would be to die in." Around 4:00 P.M. the group reached a prominent peak. But as the professor began his work, the weather turned ugly. With thunder rumbling in the distance and a bank of thick clouds moving across the surrounding mountains, Mitchell deployed his barometer and got a reading of 6,672 feet. By sundown he and his guides were back on the upper Cane, hopping from rock to rock downstream. At dusk the storm that had been brewing on the high ridges broke across the valley, forcing the men to take cover. Much to his dismay Mitchell spent a long, wet night huddled around a smoky fire before returning to the settlements the next day. All told, it was a two-day, twenty-mile trek, a "dreadful journey" that the professor hoped would be his last to the Black Mountain.[26]

Mitchell's obvious anxiety in 1844 suggests other ideas typical of his day and perhaps a subtle shift in his thinking about nature and wilderness. For more than a hundred years, as part of the intellectual movement known as Romanticism, artists, philosophers, and writers had celebrated the world's "sublime landscapes," those few breathtakingly beautiful places where the splendor of the natural world so overwhelmed humans that they could not help but acknowledge a divine presence. The wilderness remained a dangerous place, and one still had to be on guard against its evil influence. But for a few wary souls who went there alone and on foot, the majestic terrain, undisturbed by people, also offered a rare opportunity to glimpse the Creator's handiwork and marvel at his arresting power.[27] As an educated man of the nineteenth century Mitchell was well acquainted with such notions, and they undoubtedly began to influence his ideas about the Blacks. But try as he might, he could never escape the Calvinist notion that meeting his maker in a wild place might not be an altogether pleasant experience. Small wonder that the professor—as he crawled like a supplicant before the "Whole Black Mountain"—not only came face-to-face with God but also with his own mortality.

Though he stood (and occasionally knelt) in awe of the supernatural qualities of the Blacks, Mitchell never wavered from his progressive plans for the region. Like many of his contemporaries he still believed that the sooner such wild landscapes became safe for human habitation, the better. He returned to Chapel Hill and drafted two letters (both intended for public scrutiny) to the U.S. congressman who served the mountain district. The long missives again invoked a New Englandesque vision of the region's future. As Mitchell explained, he looked forward to a day when even the highest peak of the Blacks might be covered with lush meadows and vast herds of cattle and sheep.

Yet still the question lingered: Where was that highest peak? For a man who had thought he might die getting there, the professor spent precious little time describing the summit. In the 1844 letters he listed it only as the "Top of Black," added the new elevation of 6,672 feet, and suggested that it stood somewhere in the southern part of the range.[28] Perhaps in Mitchell's mind the mountain's exact location made little difference. He had several barometric readings, taken over ten years, that proved that the Blacks outranked Mount Washington and gave North Carolina clear title to the highest peak in the East. He was sure that the congressman to whom he addressed the letters would be pleased. Mitchell had known the mountain representative for nearly fifteen years, ever since his student days at the university. His name was Thomas Clingman.

AUGUST

MOUNT GIBBES/CLINGMAN'S PEAK

After a night of ferocious rain, morning fog hangs heavy on Fraser firs and mountain ash. As I pull into a small parking area at Stepp's Gap, I can see twenty yards ahead and no more. With an elevation of 6,100 feet or so, the gap is actually a saddle, or low spot, between much higher peaks. Named for Jesse Stepp, an early settler along the North Fork of the Swannanoa, the break in terrain also serves as an informal boundary between the northern and southern reaches of the Blacks. Hiking north, one crosses the highest mountain in the East and the long string of lesser peaks that culminate at Celo Knob. To the south lies the bend in the fishhook, the short half-circle of mountains that connects the Blacks with the Blue Ridge. Today I head south, walking an unpaved road around the far side of the range. With any luck the prevailing southwesterly wind and a climbing sun will clear the air as I ascend. Until then I pull my jacket close and trudge headlong into the cloud, listening to the rhythmic crunch of boots on wet gravel.

It takes less than a half-hour to affirm my faith in mountain weather. By mid-morning the fog lifts. After a brief hike through a clump of dripping rhododendron, I stand in breezy sunshine atop Mount Gibbes, a knobby, fir-covered, 6,500-foot cone that affords stunning views of the Cane and South Toe River Valleys. From the peak's narrow crest I can also look down on N.C. Route 128, the asphalt thoroughfare that connects the Blue Ridge Parkway with the East's tallest mountain. Now, at the height of summer, dozens of cars and recreational vehicles stream past, eager to reach the more famous northern summit. I return to the access road and follow it south. A quarter-mile farther down, where the gravel ends and the ridge swells into another 6,500-foot-plus promontory, I listen as the wind whistles through steel towers used by a distant radio station to bounce its signal across western North Carolina. Like

Mount Gibbes, this is a mountain that now draws few sightseers, a prosaic place that serves a pragmatic end.

A century and a half ago, however, these two peaks were as well known as any in the Blacks. Though the historical record remains sketchy, Mount Gibbes apparently got its name in the early 1850s just after a South Carolina physician, Robert Wilson Gibbes, calculated its elevation. A few years later Mount Gibbes gained even more notoriety when reports surfaced that Elisha Mitchell had visited the mountain during his early explorations of the Blacks. As the professor clashed with Thomas Clingman, detailed descriptions of the peak, including its topography and vegetation, circulated in Asheville newspapers.

The neighboring pinnacle, now so mundanely adorned with radio towers, played an even bigger role in the controversy. For over a decade it masqueraded as the highest mountain in the East. Local people called it Mount Mitchell, in honor of the professor. But now, having been exposed as an impostor, it has acquired another name. As anyone familiar with the region's current geography knows, the radio towers stand on Clingman's Peak. As if that were not confusing enough, the true summit of the Blacks—the mountain 2.35 miles north of here that lures modern tourists up Route 128, the one currently called Mount Mitchell—was also once known as Clingman's Peak.

I linger a few minutes on this southern summit, listening to the August wind and puzzling over the muddled nomenclature and the complicated argument that produced it. It was a long, bitter, and very public fight, one that eventually brought the Black Mountains into the national spotlight. It involved politics, public relations, and the economic future of North Carolina. But more than anything else it was a clash of egos, a churlish conflict between two of the state's most learned men, men who shared many of the same ideas and who had once been friends.

In 1829 at age seventeen Thomas Lanier Clingman left the Piedmont village of Huntsville to enroll at the University of North Carolina. A slight, wiry lad who had grown up hunting and fishing along the Yadkin River, Clingman cut an odd figure on campus. He had the peculiar habit of talking to himself as he wandered from class to class, and when his gawky demeanor and simple clothing drew taunts from upperclassmen, the newcomer occasionally answered with his fists, early evidence, perhaps, of a pugnacious personality. But during his first days in Chapel Hill the backwoods youth found a mentor and confidant in Elisha Mitchell. In Mitchell's classroom Clingman imbibed a healthy dose of the professor's zeal for progress and reveled in his discussions of the geological survey and the mountain wilderness. When Thomas graduated in 1832, he immediately set out on his own summer tour of the Blue Ridge, including Grandfather Mountain. From there, like his teacher, he caught his first glimpse of the majestic peaks that lay to the south.

Though he excelled in science, Clingman left Chapel Hill determined to be a lawyer. In the 1830s law provided the quickest and surest path into public service, and all who knew young Thomas agreed that he was a man of considerable ambition who aspired to national political office. After an apprenticeship with William A. Graham (a prominent Hillsborough attorney who later became governor of North Carolina), Clingman passed his bar examinations and won election to the state general assembly in 1835. He went to Raleigh as a member of the Whig Party, a national coalition that favored federal funding for internal improvements and had ties to David Lowry Swain and others from the university. But now, as a resident of Surry County, where the opposing Jacksonian Democrats traditionally ran strong, Clingman served just one legislative term before he lost his seat in 1836.

Convinced that he might never fulfill his destiny in his home county, the budding politician moved to Asheville. As the commercial center of western North Carolina, Asheville afforded decent opportunities for a fledgling lawyer. Moreover, it had a robust Whig contingent, and the nearby mountains provided a magnificent laboratory for geology and natural history. In those comfortable surroundings Clingman quickly resurrected his career. A skilled orator with a passionate commitment to improving transportation and trade facilities in western North Carolina, he won election to the state senate in 1840. Three years later he became the mountain district's representative in the U.S. Congress.

His politics reflected many of the same ideas about nature that influenced Mitchell. Lacking his mentor's theological bent, Clingman spent comparatively little time contemplating the relationship between wilderness and immorality. But he still believed that the mountains should be thoroughly civilized and made productive. For that to happen the region needed cash, preferably from outside investors with enough financial clout to bring immediate change. During the 1840s the Asheville politician became a "one-man Chamber of Commerce" constantly promoting "the beauty, climate, and natural resources of the North Carolina Mountains."[29]

Like Mitchell, Clingman considered his region ideal for livestock and dairy farming, a point he drove home in an 1844 essay written for a New York agricultural journal. The Blacks and other high mountains, Cling-

Thomas Lanier Clingman, politician, geologist, regional booster, and Black Mountain explorer. North Carolina Collection, University of North Carolina Library, Chapel Hill.

man explained, were "particularly fitted for pasture grounds," and their proximity to Asheville guaranteed easy access to markets in "the planting States south of us." The congressman also courted Yankee capitalists with tales of buried treasure. Beneath Appalachia's good soil, the congressman insisted, lay abundant supplies of gold, lead, copper, and iron ore, most of which had yet to be discovered. Boasting that his district had sufficient water power to "move more machinery than human labor can ever place there," Clingman urged northern entrepreneurs to invest in mining. As he described it, "opening valuable mines" with "capital from abroad" would create a new group of laborers who, in turn, might "furnish a good home market to the farmer."[30]

Clingman's fascination with mining and manufacturing grew out of his work in geology, another interest cultivated in Mitchell's classroom. The congressman, however, was no mere rock collector. He was, in fact, North Carolina's "foremost citizen geologist," one of those gentleman amateurs still prevalent in the world of nineteenth-century science. He read widely in the professional literature, corresponded regularly with the nation's most eminent geologists, and often went afield in search of rare specimens. Among his more notable finds were a large diamond and some flakes of platinum, both unearthed in Rutherford County gold mines and reported in *Silliman's Journal*.[31]

As a geologist and as a politician Clingman paid close attention to Mitchell's explorations of the Blacks and eagerly awaited news of the professor's latest discoveries. In early 1845 the congressman read with interest the letters describing the location of a new high peak near the south rim. But if anything in Mitchell's work disturbed Clingman, he gave no sign. What mattered most was publicity for his district, and in that cause his mentor was a powerful ally. For another decade professor and politician remained on good terms as they continued to encourage development of the Blacks and the surrounding area.

Then suddenly in 1855 the two partners in mountain promotion had a dramatic falling-out. For the next two years, while North Carolina and the nation looked on, teacher and student bickered like schoolboys over the region they loved. Like so many quarrels among friends, this one started with a misunderstanding. From 1839, when "Mount Mitchell" first appeared in Roswell Smith's atlas, people living near the Blacks had attached the name to various summits. By the early 1850s most local folk knew Mount Mitchell, which they believed to be the highest point in the East, as the mountain in the southern reaches of the Blacks that stood just south of Mount Gibbes (today's Clingman's Peak). Reading Mitchell's accounts and especially his last vague description of the "Top of Black," Clingman surmised that this was the same southern summit the professor had scaled in 1844. From Mitchell's letters Clingman also knew that the professor had embarked on the 1844 excursion "well satisfied" that he had not previously been on the range's highest mountain.[32]

Operating on those two assumptions and thoroughly familiar with Black Mountain geography, Clingman became convinced that a peak roughly three miles north, beyond Stepp's Gap (the current Mount Mitchell), was higher than the one then named for the professor. In the late summer of 1855 the congressman climbed the northern summit, took new barometric readings, and estimated its elevation at 6,941 feet. In October he wrote his own public letter, to Joseph Henry, secretary of the Smithsonian Institution and one of the most prominent men in American science. The congressman credited his teacher with proving that the "portion of the Black Mountain since called Mitchell's Peak, or Mount Mitchell, was higher than Mount Washington." But, Clingman noted, it had been cloudy in 1835 and 1844 when Mitchell had visited the mountains. When it came to identifying the summit, the professor "did not appear to feel at all confident on the subject." Copies of the congressman's letter, with Henry's notations referring to the highest mountain as "Clingman's Peak," appeared in newspapers in both Washington and Asheville.[33]

Although Henry released the news to the press, he was not about to put the Smithsonian's good name behind Clingman's claim without subjecting it to scientific scrutiny. For an expert opinion the secretary turned to Arnold Guyot, renowned professor of geography and geology at Princeton University. Born in Switzerland and educated in Germany, Guyot had been surveying the Appalachians since 1849 and was, at that very moment, hard at work establishing reliable elevations for the Blacks. Guyot's methods were time consuming, physically demanding, and far more meticulous than those employed by either Mitchell or Clingman. Remaining in the mountains for days at a stretch, Guyot took many barometric readings for each peak he climbed. Using a device called a water level (a surveyor's instrument capable of determining relative elevations of distant mountains) and mathematical calculations, he also estimated heights of summits he could not visit. His measurements were widely regarded as the best available, and much of his work in the Blacks went unchallenged well into the twentieth century. Checking Guyot's findings, Henry discovered that the mountain Clingman had measured in 1855 was, in fact, taller than the one called Mount Mitchell, though the congressman had overestimated the high peak's elevation by approximately 200 feet. Backed by Guyot's numbers, a Smithsonian report published in 1856 described "the highest point of the Black Mountain, now called Clingman's Peak," as "the most elevated spot on our continent, east of the Rocky Mountains."[34]

Mitchell's initial response to Clingman's discovery set the parameters for their long-running argument. In typically mild-mannered fashion the professor wrote Henry at the Smithsonian and carefully explained that local folk had simply attached his name to the wrong mountain. He noted that he had spotted the highest peak from Yeates Knob in 1835 and that it lay between the north and middle forks of the Cane, the general location of the new Clingman's Peak. But instead of claiming that he had climbed the mountain on that foggy July day in 1835, Mitchell—for reasons still unclear—recalled that his guides, William Wilson and Adoniram Allen, had mistakenly led him to a peak that lay "too far north." Not until 1844, he noted, did he return and ascend the true summit, which, he assured Henry, was the same one Clingman had visited in 1855. As several historians familiar with the dispute have noted, the professor never questioned whether the mountain Clingman measured in 1855 (today's Mount Mitchell) was tallest in the range. Instead, the two men argued over individual achievement: "The sole point of contention was whether Mitchell had preceded Clingman to the highest peak during his 1844 visit."[35]

Clingman had plenty of evidence to the contrary. Mitchell's own vague descriptions of the arduous 1844 journey up the Cane indicated that the professor had climbed a peak somewhere south of Stepp's Gap. Clingman also had a compelling piece of physical evidence, something he regarded as a smoking gun. In 1845, only a year after Mitchell's purported trip to the summit, Nehemiah Blackstock, a surveyor working in the Cane River Valley, had made his way into the southern reaches of the Blacks. While on Mount Gibbes, Blackstock had found "a piece of timber with a groove in it." To Blackstock it looked like part of a surveyor's water level. The wood had hardly decayed at all, indicating that someone with more than a casual interest in measuring mountains had been there not long before. That someone, Blackstock concluded, was Elisha Mitchell. The professor had simply misidentified Mount Gibbes or possibly the adjacent mountain (the modern Clingman's Peak) as the "Top of Black."[36]

Instead of immediately firing off another letter to Henry, Clingman first wrote Mitchell, suggesting that the professor return to the Blacks, refresh his memory, and reconsider his claims. It was a gracious gesture but not entirely altruistic. In 1856 Clingman did not need another controversy. He was too busy trying to negotiate the political minefield that surrounded the most explosive public questions of the day: slavery and southern secession. In some way those issues touched every facet of life in the antebellum South, and indirectly they became entangled with the developing squabble over the region's highest peak.[37]

Like many of his mountain constituents, Thomas Clingman owned no slaves. Nevertheless, as the national debate over slavery intensified, the congressman began to wonder if the South might be better served by leaving the Union. Some historians believe such views were completely consistent with his earlier pleas for progress and development. For years farmers from his district, including those who lived in and around the Blacks, had sold crops and livestock to buyers in the Carolina lowlands. The demise of slavery in those regions might ruin a lucrative market for mountain produce. Besides, in an independent South free from restrictive tariffs and competition from northern states, western North Carolina might finally establish the thriving industrial economy Clingman envisioned. Although the congressman remained popular with mountain voters, those ideas alienated him from fellow Whigs who, while supportive of slavery, rarely raised the issue of secession. By the early 1850s, as his views became more extreme, Clingman had made plans to abandon his old party and join the proslavery, pro-secession Democrats.[38]

The congressman also knew that a change in political affiliation might help his career. Although well aware of Clingman's desire for higher office, North Carolina's Whig leadership had repeatedly denied him a nomination for the U.S. Senate. In 1856, as he openly embraced the Democrats for the first time, he hoped his new friends might be more accommodating. He certainly did not wish to alienate them by getting into a protracted verbal battle with Elisha Mitchell, one of the most respected men in the state. As Clingman later noted, "I had political contests enough from time to time to occupy all the moments I could spare from other duties and was extremely averse to a discussion with . . . a former preceptor [and] a friend I had much valued."[39]

But Clingman had publicly questioned Mitchell's work, and with his discoveries and scientific reputation on the line, the professor could not allow the allegations to go unchallenged. In April 1856 Mitchell announced his intention to defend himself and his discoveries in the *Asheville Spectator*, a Whig newspaper that had been highly critical of Clingman and his decision to join the Democrats. Exactly why Mitchell chose the *Spectator* remains a mystery. The professor had friends among the Whigs but apparently took little active role in state politics. And though he hailed from New England, he had long since grown comfortable with southern ways. In Chapel Hill, where he taught the sons of southern planters, Mitchell not only owned several slaves but also defended the practice at home and abroad, once going so far as to preach a proslavery sermon to a congregation in his native Connecticut. Whatever his reasoning, when Mitchell enlisted the *Spectator* in his cause, his argument with Clingman became a highly charged political debate. With a crucial national election looming in the fall, Clingman planned to campaign vigorously for the Democrats. He could ill afford bad publicity, and he warned Mitchell that, if attacked in the *Spectator*, he would respond in kind.[40]

Their now-infamous war of words began in June and raged for six months, with the professor writing to the *Spectator* and the congressman responding in the *Asheville News*, a paper that had traditionally supported his views. Generally speaking, their positions did not change. Mitchell insisted that he had identified the highest peak in 1835 and climbed it in 1844. Using the professor's own records and pointing to Blackstock's discovery of the water level, Clingman argued that Mitchell's route up the Cane in 1844 inevitably led south toward Mount Gibbes, not north toward the summit.[41]

As the debate intensified, Clingman's expertise in examining the written record, a skill honed during years of courtroom practice and congressional debate, quickly put Mitchell on the defensive. By July 1856 the professor realized that he could no longer trust his twenty-year-old recollections. He decided to return to the Blacks, hoping to locate William and Marvil Riddle, his guides in 1844. In his effort to find the Riddles, Mitchell enlisted the aid of Zebulon Baird Vance, a family friend and state legislator from Asheville. He had known Vance since 1851 when Zeb (as his friends called him) had spent a year at the university. Vance quickly found the Riddles, but when he heard what they had to say, he elected not to take a statement until Mitchell could be present. When ill health forced the professor to return to Chapel Hill before he could speak with the guides, Clingman visited the Riddles and conducted his own interview. In a signed affidavit notarized by the Burnsville postmaster, William Riddle carefully laid out the path the professor had followed in 1844. It led straight to Mount Gibbes. As if that were not enough ammunition for Clingman, Marvil suddenly remembered that the professor had left a water level at the top of the mountain.[42]

By the time the Riddle testimony became public, what had begun as an academic debate over routes and measurements had already degenerated into a tawdry tantrum of verbal barbs and personal insults. In a dispatch filed early in the debate, Clingman caustically noted that the peaks of the Blacks stood "boldly and stubbornly" in their proper places and would never "change their outlines" to conform to the professor's "shifting representations." Given Mitchell's faulty memory, the congressman later wrote, it should come as no surprise that he had forgotten about the water level. Frustrated at every turn by an impudent student, the man once known for his unflappable ministerial demeanor flew into a rage. "You do not know what friendship is," he railed at Clingman in an August letter, and "whatever you may claim to feel of that kind, is hollow and pretended, or if real, is unreliable and worthless."[43]

Anger was a poor substitute for evidence, however, and despite his tirades the professor still produced nothing new to support his cause. From Chapel Hill in November he reiterated his belief that he had been first to identify the highest mountain from Yeates Knob in 1835. Then, in what amounted to a last-ditch effort to save face, he suggested that his failure to reach the summit in 1844 might be blamed on William Riddle, who had somehow led him to the wrong mountain. Clingman declined to respond, cleverly allowing his silence to suggest that Mitchell had conceded. Before the year was over, several geographers had listed the highest mountain as Clingman's Peak and attached Mitchell's name to the lesser mountain three miles south.[44]

When December slipped by without further exchange of letters, the ugly fracas appeared to be over. No one could have guessed that before another year passed, the conflict would flare anew. This time mountain people would be asked to take sides, and before the argument ended, a talkative local woodsman would become nearly as famous as the professor and the congressman.

NOVEMBER

HAIRY BEAR

In the sharp, clear autumn air, the snapping twig sounds like a rifle shot. Reflexively I start at the sound, anxiously scanning the woods. At this time of year I do not expect company. Most of the tourists have disappeared with the fallen leaves. Since midmorning I have walked alone on the Crest Trail, a heavily traveled path that runs north from the East's highest mountain along the very spine of the range. Now, with my lunch spread around me, I sit on a small slab of rock, roughly 6,593 feet above sea level, encircled by a thick stand of Fraser fir, perfectly situated for an encounter with a wandering animal. It does not help to know that in Elisha Mitchell's day this heavily forested peak was called Hairy Bear.

I hear another rustling in the trees to my right, but this time I relax. Judging from the sound, I decide the creature is small: a squirrel, I figure, or maybe a grouse or groundhog. I am wrong on all counts. With a whimper a small brown dog crawls timidly from the edge of the woods. She is emaciated, her skin stretched taut over protruding ribs. Briars, burs, and brown fir needles cling to her coat. Attached to her collar is a small radio transmitter, an unmistakable sign that this is a hunting dog, an animal trained to track black bears. I give her most of my food and a pint of water, sustenance for the hike back to the truck. After a quick phone call to a number embossed on the radio collar, the dog is en route to her Burnsville home. As it turns out, she has been alone in the woods for two weeks after straying from her owners during an October hunt in the South Toe Valley.

Similar incidents involving lost dogs are common in the Blacks during bear season. While large packs of purebred hounds scour woods and rocky outcrops in search of bears, hunters prowl nearby dirt roads in all-terrain vehicles. They communicate with two-way radios and monitor their dogs via signals transmitted from the electronic collars. When the dogs locate their quarry and give chase, the hunters shadow the pack in their vehicles until a bear is treed or crosses a road, where it can be killed. To a nonhunter like me it seems a mechanistic and impersonal exercise, a sport that requires little physical exertion and one in which the odds are always stacked solidly against the bear.

It was not always so. In the nineteenth century, bear hunters took to the mountains on foot or on horseback. For days at a time they trailed dogs across the high ridges, sleeping on the ground and taking their chances with mountain weather while they waited for the pack to do its work. It was a demanding and dangerous venture, one that required extraordinary stamina, expert marksmanship, and a keen sense of local geography. Bear hunting was also an integral part of male culture, a way in which men measured themselves against their peers and gained standing in their local communities.

During the 1850s, while Elisha Mitchell and Thomas Clingman trumpeted their scientific accomplishments, a rangy Cane River farmer living at the foot of Hairy Bear quietly established himself as the region's best bear hunter and one of its leading citizens. Before the decade was out, his outdoor expertise and skill in negotiating rugged terrain placed him at the center of the ongoing argument over the highest mountain. The peak once known as Hairy Bear is now simply called Big Tom. All who hike the Crest Trail know that the new name honors Big Tom Wilson, renowned bear hunter, tracker, finder of missing persons, and western North Carolina's most celebrated mountain man.

Born in 1825, Thomas David Wilson grew up the son of Scots-Irish farmers in the South Toe River Valley. At some point early in his life his family began to call him Big Tom, but exactly when and how he earned the nickname is uncertain. The title apparently had little to do with size. He was six feet, two inches tall, but spare and spindly. Probably the Wilsons used the informal title to distinguish Thomas David from younger family members who shared his first name.[45]

Had Elisha Mitchell known Big Tom in those years, the professor surely would have regarded him as one of those unfortunate folk hopelessly corrupted by nature. The lanky youth spent much of his free time exploring the forests near his home, "chasing the bear and deer," or "seeking to allure the wary trout from their haunts in the cool depths of the pools." According to family friends this early "life in the wilds developed him into a figure tall, straight, lithe, rawboned and sinewy, [with] a rugged constitution" and "a knowledge of woodcraft such as few men ever possess."[46]

In the early 1850s Big Tom moved across the Blacks to marry Niagara Ray, who lived on the upper Cane. The marriage might have been a step up for Wilson. In terms of land the Rays were well off. Niagara's father, Amos, held title to more than 13,000 acres along the river's headwaters. During Big Tom's first years on the Ray property his prestige as a woodsman continued to grow. Highly skilled at robbing bee gums, he sold some of the best honey to be had anywhere in the mountains and he continued to take his share of white-tailed deer and brook trout from the forests and streams on family lands. By the time he reached his late twenties, though, his real forte was stalking black bears.[47]

Like any predator, a worrisome bear—one that uprooted crops or killed pigs—might be shot or trapped at any season. But the most pleasurable and sporting way to take a bear was during the annual fall hunt. Preparations began early, during the last days of summer. While crops matured and cattle fattened themselves in the high meadows, farmers along the Cane and Toe took short trips into the mountains to look for bear sign. They searched for overturned rocks, shredded logs, and paw marks left by the animals as they feasted on grub worms, berries, early chestnuts, and other summer fare. An especially observant woodsman might even find a spot where a curious bruin had excavated a nest of yellow jackets or rousted a hapless groundhog from its burrow. Long before the first frosts, Big Tom and his Cane River neighbors had determined exactly where they would hunt that fall.

From late September through the first days of winter small groups of men left their farms and camped deep in the mountains. Using specially bred hounds (including Plotts, Blue Ticks, and Black and Tans) they hunted for a week or longer, hoping to lay in a supply of bear meat and collect a few skins to trade with the area's itinerant merchants. Once into the woods, a "driver" led the dogs, while other men fanned out ahead and took up positions in trees, on high bluffs, or on crudely constructed wooden stands. When the dogs located a bear, the hunters listened intently from their stations, ever alert to the direction of chase. A skilled marksman might fire at a bruin as it ran, but usually the men waited until the exhausted animal sought refuge in a tree. Alerted by the telltale change in canine voice indicating a bear had been cornered, the hunters then moved in and shot the animal from its perch. Due to the communal nature of the hunt, men often had the chance to show off their outdoor skills in front of their peers. When the expedition ended, the best marksmen and dog trainers left the woods with something far more precious than bearskins. They returned home basking in the admiration and envy of their friends.[48]

Many hunters must have been jealous of Big Tom Wilson. He had an amazing ability to negotiate steep terrain and relentlessly followed his dogs through the thickest stands of rhododendron and laurel. Unlike most of his friends, he often ventured into the mountains alone, carrying nothing more than a pocketful of cornmeal, always confident that he could kill enough game to feed himself and his hounds. Usually he took up a post on some isolated bluff and set the dogs loose to find a bear and chase it toward his stand. But if that plan failed, he simply followed the pack (sometimes for days), working as both driver and hunter, until a bear could be treed and shot. On such solitary excursions Big Tom hunted on foot, storing any bears he shot in a safe cache and returning later on horseback to retrieve them. Family tradition holds that he killed between 113 and 117 bears (an average of 2 to 3 per year) during his days as a hunter.[49]

In the Black Mountain settlements, where residents relied on storytelling to preserve their collective memories, talking about bears was almost as important as killing them. At church, around campfires, and across fences, men regaled neighbors and visitors alike with tales of their forest exploits. But shameless boasting rarely impressed local folk. A hunter's reputation hinged on his ability to recount every detail of a given hunt—the fierceness of the bear, the exact route of the chase, the courage of the dogs, and the killing shot—all without giving the appearance of bragging. When it came to that sort of descriptive, overly modest talk, Big Tom Wilson had no equal. Even today his descendants fondly recall his remarkable ability to "paint a picture with words."[50]

By the mid-1850s Big Tom had found a way to market his skills as a hunter and storyteller. His knowledge of the Blacks and his friendly manner made him the perfect mountain guide. Since 1835, when Elisha Mitchell had initially inquired about the tallest mountain, Big Tom's relatives had been helping strangers find their way around North Carolina's high peaks. Indeed, William Wilson, who served as one of Mitchell's first escorts, was Big Tom's cousin. Over the next twenty years, as other scientists followed in Mitchell's footsteps, guiding became something of a cottage industry in the Blacks. Moses Ashley Curtis, a botanist from Wilmington who visited the range in 1839, noted that in the company of his guides, he traveled "on foot or horse back along intricate cattle paths" and put up with primitive accommodations "such as you never dreamed of."[51]

By the time Big Tom began guiding, however, the business had changed. In the mid-1850s most mountain travelers were not scientists but seasonal tourists. Better stage roads and the new Western North Carolina Railroad (which, amid ongoing political arguments about its exact route, was inching its way toward the Blue Ridge) brought a slow but steady stream of summer sightseers from the Piedmont and the Carolina lowcountry. For well-to-do folk the favored venue for visiting the Blacks was Asheville's Eagle Hotel, operated by James Patton and his family. From the Eagle the Pattons' guides escorted guests up a winding horse trail that followed the North Fork of the Swannanoa. En route the travelers might stay at one of three mountain residences. The newest and most famous perhaps was the Mountain House, a lodge built by William Patton, another family member who lived in Charleston. Two stories high with roughly 2,000 square feet of floor space, it stood near Potato Knob, 5,200 feet above sea level. Other visitors opted to stay with Jesse Stepp, who also offered horses, guides, and cabins for rent, first in the North Fork Valley and later on the disputed summits. At Stepp's the quarters were cramped and the beds uncomfortable, but guests at least enjoyed some shelter from the elements as they toured the nearby peaks.[52]

Big Tom Wilson's service was an even more humble enterprise. Prospective clients simply reported to his house on the upper Cane. If the bear hunter happened to be home and unoccupied with other chores, he took the guests up the river's headwaters toward Mount Mitchell and Clingman's Peak. A well-maintained horse trail made it possible to visit the summits and return in one day, though most tourists preferred to spend the night in the mountains and watch the sunrise from the highest point in the East. When they went with Big Tom, overnight guests camped in open air or appropriated one of the crude shelters high in the mountains. His wife, Niagara, prepared hot meals for the travelers while they remained at the cabin, and she packed dried meat and other nonperishables for the trip.[53]

Despite the Spartan nature of his operation, Big Tom soon became one

The young Tom Wilson (*above*) and his Cane River cabin (*below*), as depicted in *Harper's New Monthly Magazine*, November 1857. Cabin reproduction from North Carolina Collection, University of North Carolina Library, Chapel Hill.

of the area's most sought-after guides. Urban visitors admired his prowess with a rifle, his quiet confidence, and his easy affinity for the land. Those unaccustomed to mountain cuisine also found much to like about Niagara's cooking. After a meal of "rich juicy Yancey beef," biscuits, and apple pie, one client wondered whether "Queen Victoria or President [James] Buchanan ever ate so glorious a meal." But for most sightseers the biggest attraction was Big Tom's remarkable gift for telling tales. As one of his sons later remarked, the gregarious mountain man could leave home in the middle of a story "and talk right over Mount Mitchell and all the way back, and never make a break."[54]

Most of Big Tom's clients arrived in summer or early autumn. But in 1857 in the dead of winter (two months or so after the initial controversy between Mitchell and Clingman had died down), a famous visitor from the north called at the Cane River cabin. The out-of-season tourist's name was David Hunter Strother, but thousands of American readers knew him as Porte Crayon, talented writer and illustrator for *Harper's New Monthly Magazine*. Crayon was among the earliest practitioners of that peculiarly American literary craft known as local color writing. He traveled widely in the United States—especially in the South—collecting stories of intriguing people and places that he then serialized for *Harper's*. During the 1850s the magazine's 200,000 mostly urban subscribers counted on Crayon for the latest pictures and descriptions of American life. Stories of so-called pioneers, those rough-and-tumble folk who lived on the margins of an increasingly urbanized eastern America, were particularly popular with *Harper's* readers, and Porte Crayon could not wait to introduce them to Big Tom Wilson.[55]

Crayon probably arrived in the Blacks accompanied by his wife and daughter. But his fictionalized account, published as "A Winter in the South," tells the story of several Virginians on a journey to New Orleans. After making their way into North Carolina from Tennessee, the group seeks out Big Tom for a trip into the high mountains. Against the bear hunter's advice, the men decide to make the trip in a single day and take no food. By the time they reach the summit, the greenhorns are so famished that they briefly contemplate eating one of Big Tom's dogs. But the guide's ingenuity saves both the hound and the trip. Telling his guests that "an ideer jist struck me," Big Tom decides to break into a summer tourist cabin, perhaps one owned by the Pattons or Jesse Stepp. There he finds dried corn and bran, which provide sufficient sustenance to see the group safely down the mountain. Arriving at the Wilson cabin "shrouded in the gloom of night," the men revel in one of Niagara's famous meals, stuffing themselves with pork, chicken, pumpkins, and cabbage "until they [are] nearly blind."

Even with its embellishments, Crayon's characterization of Big Tom reflects many of the qualities that endeared the bear hunter to his clients. Throughout the daylong adventure he proves a knowledgeable and patient companion. He takes time to point out various Black Mountain landmarks and shows his inexperienced guests where to find water and how to parch corn over an open fire. Big Tom also offers the travelers constant encouragement. Once, during an extended rest stop, he rousts the disconsolate party by exclaiming, "Men! if we stop here ciphering and grumbling, we'll never see the top of the Black." It was, the *Harper's* journalist noted, a well-timed remark that "stirred up the dying embers of that enthusiasm which had hitherto sustained them."[56]

Before he left the mountains, Crayon made detailed drawings of Wilson's cabin, his family, and the bear hunter himself, a rare (and perhaps the only) depiction of Big Tom as a young man. In the drawing the guide stands with his dogs in a rhododendron thicket, leaning casually on his rifle, the perfect mountain man. In any other year that sketch and the accompanying essay would have given Big Tom the kind of free advertising that any guide might envy. But when *Harper's* finally ran the article in November 1857, the affable woodsman scarcely needed more publicity. By then anyone who picked up a North Carolina newspaper already knew about Big Tom Wilson. He was the man who found the body of Elisha Mitchell.

JANUARY

BIG BUTT TRAIL

At 4,800 feet I find snow. It is light for this time of year, a half-inch or so, the result of yesterday's sudden blast of Arctic air. For nearly a month before, warm Gulf winds buffeted the Blacks, bringing plenty of rain—even thunderstorms—but little frozen precipitation. The snow is somehow reassuring, confirmation that things are returning to normal. Resting against a fallen maple, I take in the contrast. Above, against the white backdrop, moss and lichens growing on the light bark of birch and beech give the woods a strangely metallic, silver-green look. Below, staring down a long valley toward the town of Barnardsville, I can still detect the drab hues of late autumn, a collage of dull yellow, burnt orange, and brown.

The change in scenery is also proof of the hiker's adage that in the mountains all trails run in the same direction: straight up. This one is no exception. Starting high on a ridge that separates the Ivy and Cane River basins, it climbs more than a thousand feet into the Blacks. Under a pack filled with heavy winter gear, I feel every inch of elevation gain. Finally, at 5,600 feet, the tortuous path crests, and I make camp at Flat Spring Gap, a wonderfully level patch of ground covered with wind-twisted hardwoods. Here the snow is thicker, laced with the tracks and urine stains of black bears kept from hibernation by unseasonably mild weather. My footprints mingle with theirs as I head toward the trail's namesake, that 6,000-foot peak called Big Butt, once known as Yeates Knob. It is the site from which Elisha Mitchell attempted to locate the East's highest mountain in 1835, and memories of that view brought him back in 1857.

I cannot duplicate Mitchell's experience. My ideas about nature are different from his. But free from crowds of summer hikers and autumn hunters, with only insomniac bears for company, I sense some of the awe (and indeed some of the uneasiness) that the professor knew during those visits. With the leaves down and the valley spread before me like a giant stage, I can identify all the places that figured prominently in the drama between Mitchell and Clingman: Celo Knob, Mount Gibbes, and the two peaks that alternately bore the names of the antagonists. Aided by the clear weather and historic hindsight, I can also pinpoint a deep chasm carved by an upper prong of Sugar Camp Creek as it drops from the summit en route to the Cane. On that headwater stream once known as Sugar Camp Fork the professor lost his life.

As the winter daylight fades, I stir from my place near Yeates Knob and begin the descent back to Flat Spring Gap. Despite yesterday's snow, temperatures remain surprisingly warm, the wind strangely calm. After supper I bank the fire and take advantage of a rare opportunity to sit outside and watch a January evening settle on the Blacks. At dusk, through the forest cover I can still see the dim outlines of the mountains across the Cane. Red lights from the radio tower on Clingman's Peak blink in the distance. Like all winter nights, this one is long. In the inky silence I have time to think once more about the professor's accident and its aftermath. The sensational event captured headlines across the state and again thrust the Black Mountains into the national spotlight. But it was also an episode well suited to creative embellishment, and in the years that followed, the stories told about Mitchell's death—more so than the tragedy itself—helped give the high peaks a new place in the public imagination. As a result the Black Mountains became not just a physical but a cultural landscape, a place shaped as much by stories of human experience as by the whims of nature.

During the early months of 1857, while Big Tom Wilson entertained Porte Crayon, Elisha Mitchell had time to reconsider his early travels in the Blacks. As he reviewed his notes and the public exchanges with Clingman, the professor became convinced that he still had a legitimate claim to the tallest mountain because he had seen it from Yeates Knob in 1835. With the end of the spring term at the university, Mitchell decided that the best way to press his claim was to return to the Blacks, take some new measurements, and talk with William Wilson, his original guide. Accompanied by his son Charles, daughter Margaret, and a slave named William, the professor left Chapel Hill in June 1857. It had been exactly twenty-two years since his first excursion to the Black Mountains. Much had changed in the interim. Mitchell was now sixty-four years old, and though still agile enough to climb mountains, he had lost some of the vigor he enjoyed as a young man. Fortunately it was much easier to get to the summit now. The professor and his family simply went to Asheville and then took the tourist route up the North Fork of the Swannanoa to one of Jesse Stepp's cabins. From there the professor could easily explore nearby peaks for days without ever having to sleep outdoors.[57]

While Margaret remained at Stepp's, Mitchell and his son began calculating elevations for several peaks in the southern reaches of the Blacks. The professor worked diligently with new barometers and a water level, taking what were probably his most meticulous mountain measurements. (According to some accounts he even placed a barometer in Savannah, Georgia, so that colleagues there could make sea-level observations of barometric pressure for comparison with readings taken in the mountains.) By late June, Mitchell seemed in good spirits. From Patton's Mountain House, just up the valley from the Stepp cabin, he wrote his eldest daughter, Mary, reporting that he and Charles were happily "keeping the batchelor's hall, getting on very well—in excellent health—living mostly on bread and bacon." It was his last letter.[58]

On Saturday, June 27, Mitchell decided to put his experiments aside and visit the Cane River settlements. Though vague about his plans, he probably intended to preach the next day at one of the local churches and then conduct his long-anticipated interview with William Wilson. Early Saturday afternoon Mitchell sent Charles to stay with his sister at the Stepp cabin and hiked out across the trails that led to the highest peak. As afternoon became evening, a slow-moving thunderstorm rolled across the Blacks, drenching the forests in a steady downpour. Even as the rains fell, Mitchell's children apparently saw no need for concern. Their father was traveling alone (maybe for the first time ever in the Blacks), but he knew the terrain well and had clear directions to a path that led from the summit, down the Cane River tributaries, to Big Tom Wilson's cabin, a local landmark. As usual he planned to find lodging there or at some other residence before nightfall.

Mitchell was to meet his son back at the Mountain House the following Monday. When his father failed to show by Wednesday, Charles began a frantic search that, two days later, took him to Big Tom's door. Upon hearing the story, the bear hunter immediately feared the worst. As Big Tom later remembered, "I said then, 'If he hasn't been here, and did not return to where he was to have met you, he is dead on that mountain.'"[59]

As word of Mitchell's disappearance spread, search parties swarmed over the Blacks. Zeb Vance and sixty men from the Swannanoa Valley combed the Buncombe County side of the range. On the Cane, Big Tom put together his own contingent and probed the Yancey side. For several days neither group found anything significant. Finally, with enthusiasm for the effort flagging, Big Tom decided to retrace the professor's likely route from the top of the Blacks. Just below the disputed summit, on the Cane River side, he and several companions discovered human footprints. They followed the tracks through moss, across rotten logs, and into a laurel thicket before the prints disappeared on an embankment high above Sugar Camp Fork. After noticing some roots that looked as if they had been dislodged by a man's foot, Big Tom clambered over a rock ledge down to the stream. There, in a small pool at the base of a forty-foot waterfall, he found Elisha Mitchell. As Big Tom described it, "Underneath a pine log . . . I saw his body, and called to the boys, 'Here he is, Poor old fellow!'" After recovering Mitchell's broken watch, Big Tom concluded that the professor had been delayed by the storm as he left the Mountain House. In the gathering darkness he had somehow strayed from the trail, lost his footing on the rain-soaked precipice, and tumbled into the stream. He had been dead eleven days.[60]

Mitchell's body remained in Sugar Camp Fork for another forty-eight hours until the Burnsville coroner inspected the site and held an inquest. After the coroner officially ruled the death an accident, Big Tom and nine of his Black Mountain neighbors carried the body back to the high peak. At the request of Mitchell's family his remains were taken to Asheville and interred in a Presbyterian churchyard. As a colleague later noted, the professor's "untimely end" left his work to be "completed by the pious hands of others."[61]

In the ensuing weeks, while the family mourned, Mitchell's friends took to that task with vigor. The professor was scarcely dead and buried in Asheville before Zeb Vance began a massive campaign to reassert Mitch-

Photograph of an 1857 lithograph showing the waterfall on Sugar Camp Fork (now called Mitchell Falls) where Elisha Mitchell lost his life (*above*). Mitchell's broken pocket watch (*below*), recovered from his body, marked the time of death at 8:19:56 P.M. on June 27. North Carolina Collection, Pack Memorial Library, Asheville, and North Carolina Collection, University of North Carolina Library, Chapel Hill.

ell's claim to the high peak. An ambitious (but as yet little-known) Whig with ownership interest in the *Asheville Spectator*, Vance viewed Thomas Clingman as a conniving turncoat who had abandoned his party simply to further his own career. By convincing North Carolinians that Mitchell had been first to the summit, Vance hoped not only to rescue a friend's reputation but also to discredit a man he and many other Whigs regarded as a traitor. For help Vance turned to Charles Phillips, one of Mitchell's colleagues at the university. A professor of mathematics and engineering, Phillips understood the various techniques used in measuring mountains. More important, he was a fine writer who had the literary skills to persuade the public and, if necessary, battle Clingman in print.[62]

In late July 1857, barely a month after the professor's death, Vance set out for Yancey County to gather evidence. With sympathy for Mitchell's grieving family running high, Vance found local people eager to cooperate. Samuel Austin, who had been with the professor on his first trek to Yeates Knob, swore that the professor had clearly identified the highest peak from there in 1835. A few days later William Wilson signed an affidavit stating that he had taken the professor to the top of the tallest mountain the following day. Wilson also remembered that the summit had been "called by the whole Country Mount Mitchell" until 1844 when "some persons [had] transferred his [Mitchell's] name to another peak." Nathaniel Allen, son of Adoniram Allen, Mitchell's other guide for the first trip, added that he had often heard his father talk about the 1835 route and knew "from his description that he was on what is now called the highest peak."

Vance then called on Big Tom Wilson. Big Tom had not known Mitchell well and perhaps had never met the professor, except in death. But Big Tom's reputation as a "bold, active, sagacious and whole souled mountaineer," coupled with his new notoriety as the finder of Mitchell's body, made him an especially useful cohort. Recalling conversations that had taken place years earlier, the bear hunter testified that he had often heard William Wilson and Adoniram Allen speak of taking the professor to the highest peak, a mountain that had "been universally called Mt. Mitchell by the Caney River people." In early September, at Vance's encouragement, Big Tom joined several local hunters as they escorted William Wilson back to the peak he had visited with Mitchell in 1835. Allegedly working solely from memory, the elderly guide led the group straight to the summit.[63]

Mitchell's partisans still had to contend with Blackstock's discovery of the water level and William Riddle's notarized statement that he had taken the professor to Mount Gibbes, not the summit, in 1844. But by late October, Vance had those matters well in hand. He found two local residents willing to testify that they had since heard Riddle say that "he was mistaken" about the 1844 expedition and now believed "Dr. Mitchell's proposition was correct." As for the much-ballyhooed water level, Vance secured statements from two other members of Blackstock's survey party, both of whom—incredibly—were now ready to swear that they had found the instrument on the highest peak, not Mount Gibbes. Jesse Stepp also weighed in as a Mitchell supporter, claiming that he had found the letter "M" gouged in a tree trunk when he had cleared some timber from the summit to allow for a better view. According to Stepp the carving looked to be twenty years old, confirming—to his mind anyway—that the professor had carved his initial there at least as early as 1844 and, quite possibly, in 1835. (With the tree now conveniently gone, those who might have doubted the tale had no way to check Stepp's story.)[64]

Buoyed by the new "evidence" and the outpouring of sympathy for Mitchell, Charles Phillips took the case to the newspapers, jousting with Clingman in a second public debate that lasted from October 1857 through February 1858. Forced to refight a battle he thought he had already won, Clingman tried the same arguments he had used successfully just a year earlier. Always careful to express his sorrow at the professor's death, the congressman reminded readers that Mitchell himself had repeatedly said that he had not reached the summit until 1844. Clingman also observed that he had talked with William Wilson in April 1857, two months before Mitchell's death. In that conversation the guide had seemed less certain about where he had taken the professor in 1835 and had noted that he and Mitchell had argued about which peak was higher. On the subject of traditional mountain names, Clingman again pointed out that for years prior to the professor's accident the lesser peak south of Stepp's Gap had been called Mount Mitchell.[65]

In the end, however, lawyerly logic proved no match for emotional eloquence. In a long article for the *North Carolina University Magazine* published in March 1858, Phillips portrayed Mitchell as a brave but cautious scientist, a man too modest to make outrageous claims even if they might enhance his academic reputation. Working under terribly harsh conditions, Phillips contended, Mitchell had identified the summit from Yeates Knob and climbed it the following day, proving to everyone's satisfaction that the Blacks were the highest mountains in eastern America. Citing the recent testimony about William Riddle and the water level, Phillips also dismissed the "gainsayers" who had criticized the professor's hazy recollections of his 1844 trip. The new evidence spoke for itself. Elisha Mitchell had laid claim to the high peak in 1835 and might well have gone back in 1844. No fair-minded observer could conclude otherwise.[66]

Vance and Phillips could scarcely have chosen a better way to make their case. The signed affidavits and sworn testimony gave the essay the ring of authority. The article's quiet intellectual tone suggested that it was the product of careful research by reasonable men interested only in the truth. Publication in the university magazine ensured that it would be widely read by sympathetic alumni and other influential people across the state. But Mitchell's supporters did not stop there. In early 1858, shortly after the essay appeared, they began to lobby Mitchell's children to move their father's body to the high peak. David Lowry Swain, now part-owner of a tract near the summit, secured the burial plot and offered it "to the Trustees of the University on condition that it shall be called Mt. Mitchell." In an elaborate ceremony on June 16 a large gathering of family and friends (including Vance, Swain, and William Wilson) saw the professor's remains reinterred on the mountain that had cost him his life. Realizing that he could not buck public sentiment, Clingman never renewed his claim. Instead he turned his attention to measuring several peaks in the Great Smokies (which he mistakenly thought might be taller than the Blacks) and left his mentor to rest in peace on the highest mountain in the East.[67]

Was Mitchell finally buried on a mountain that he discovered? The question inevitably haunts anyone familiar with the details of the dispute. During his early years in the Blacks the professor was hardly the careful, methodical scientist depicted by Vance and Phillips. But neither was he the disoriented novice portrayed in Clingman's searing newspaper attacks. When Mitchell first ventured into the Blacks in 1835, even local residents had only the vaguest sense of the region's overall topography. The professor's initial description of the summit—that prominent peak between the north and middle forks of the Cane—is equally nebulous (the river has numerous forks and tributaries that drain the west flank of the Blacks), but given his vantage point on Yeates Knob, it does suggest that he had fixed on a peak somewhere near the range's midpoint, the approximate location of the modern Mount Mitchell.

Whether he climbed it the next day is a more vexing question, not least because of the ambiguous recollections of William Wilson. Wilson's 1857 testimony and his ability to find the original route twenty-two years after the fact might provide compelling evidence in Mitchell's favor. But Wilson offered his statement at a time when sympathy for the professor ran high,

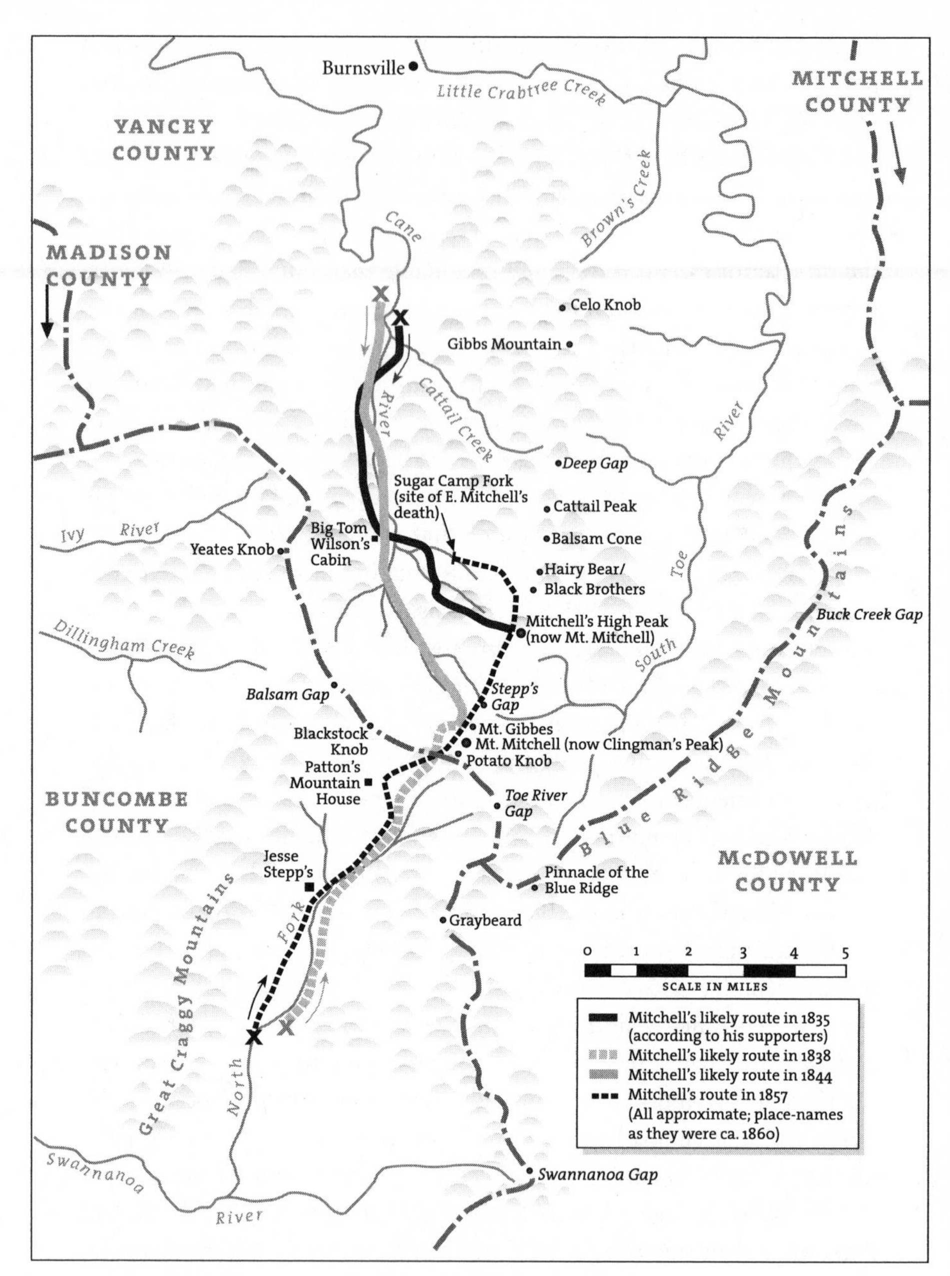

Important Sites and Routes in the Mitchell-Clingman Controversy

and Clingman steadfastly maintained that the aging guide had changed his story. Moreover, the celebrated re-creation of the first expedition took place at Vance's instigation and in the company of the professor's supporters. Some of them, including Big Tom Wilson, knew the summit well and might easily have pointed Mitchell's erstwhile companion in the right direction.[68]

With all the equivocal testimony and overblown political rhetoric that clouds the Mitchell-Clingman controversy, perhaps it might be worthwhile to look again at the professor's own observations, especially his estimates of elevation in the Blacks. For the most part, modern historians have shied away from using those numbers to prove that Mitchell climbed the highest peak. Given the professor's miscalculation of Morganton's elevation, the inferior quality of the barometers used in 1835, and the nearly identical elevations of nearby mountains, he might have gotten similar readings elsewhere in the range. Moreover, the professor's early assessments of other mountains, including Roan and Grandfather, were off by several hundred feet, discrepancies that cannot be completely explained by his faulty measurement at Morganton.[69]

All caveats notwithstanding, however, most of Mitchell's estimates for the Black Mountains *were* uncannily accurate. In the late 1850s a new survey done with a water level and coordinated with Arnold Guyot's work formally established the summit's height at 6,711 feet. If one allows for Mitchell's initial miscalculation of Morganton's elevation, then his 1835 measurement—taken at a time when lingering high pressure allowed for reasonably good barometric readings—was off by only 21 feet. Indeed, that first measurement falls within 6 feet of the mountain's currently accepted elevation of 6,684, a mark set by the U.S. Geological Survey. The 1835 numbers, recorded in the professor's own hand twenty years before the dispute, may provide some of the most convincing (and least tainted) evidence that he did, in fact, ascend the mountain that now bears his name.[70]

Why, then, did he so adamantly insist that he did not find the high peak until years later? Again his numbers provide the basis for some interesting speculation. In 1838 Mitchell recorded an elevation of 6,581 feet in the southern reaches of the Blacks. That measurement, also taken in good weather and coordinated with reliable information from Asheville, is within 20 feet of the currently accepted elevation for Mount Gibbes and within 40 feet of that for the modern Clingman's Peak, suggesting that the professor might well have explored those mountains during his second expedition. Due to his earlier miscalculation of Morganton's elevation, the new reading was significantly higher than that recorded in 1835. In all likelihood the 1838 trip convinced the professor that the tallest peak would eventually be found south of Stepp's Gap. For reasons still unclear (perhaps because he remained reluctant to trust his earliest work), Mitchell seems to have clung to that notion even after he became aware of the Morganton mistake.

If that is the case, then Clingman was almost certainly correct in his assertion that Mitchell went back to the Mount Gibbes area in 1844. Such a trip would have made perfect sense in light of the professor's most recent findings. Even if one ignores the conflicting testimony about the water level, Mitchell's own description of the route, written just days after the fact, also suggests that he and the Riddles ended up somewhere south of the true summit. The professor's 1844 measurement of 6,672 is more than 100 feet too high for either Mount Gibbes or the modern Clingman's Peak. But he read his barometer with a thunderstorm approaching. The accompanying drop in pressure (which might have given a falsely high reading) might well account for that discrepancy.

Considering the confusion that dogged Mitchell during his early work in the Blacks, it seems all the more remarkable that he never abandoned the barometer in favor of a visual trigonometric survey like that conducted by Perrin Wright in 1991. By the late 1840s (when Mitchell began to have doubts about his early work) such surveys were in vogue and well publicized throughout the scientific community. Indeed, George Everest, British surveyor general of India, used range-finding and trigonometric calculations during his early work in the Himalayas. Those techniques, which eventually helped pinpoint the highest mountain in the world, might easily have located the East's tallest peak long before Clingman got involved. But a trigonometric survey required several days of intensive field work. For all his interest in Black Mountain elevations, Elisha Mitchell—always uncomfortable with life in the wilderness—could never bring himself to spend a night on a mountaintop.[71]

Without an accurate trigonometric survey, it remained for Clingman, with an assist from Arnold Guyot, to point out his mentor's mistakes. Though few recognized it at the time, Clingman's insistence that Mitchell had been on Mount Gibbes in 1844 probably turned scientific attention away from that part of the range and eventually helped fix the summit's exact location. Guyot, in fact, found reason to praise both professor and congressman. Mitchell was the "daring pioneer" who first "proved the superior height" of the southern mountains. But Clingman deserved equal credit for the "first clear, accurate, and most graphic description of the Black Mountain." Unwilling to get caught up in the naming controversy, Guyot repeatedly referred to the disputed summit as Black Dome, a name used by some of the region's early settlers.[72]

While plausible enough in retrospect, such a compromise proved impossible during the late 1850s. With Mitchell safely entombed on the summit and Clingman effectively cowed in the press, the new story of the professor's investigations—the one told so well by Vance and Phillips—carried the day. In the public mind Elisha Mitchell became a hero and martyr. Eulogists lauded him as a New Englander who had learned to love the South, an intellectual who mingled easily with common folk, and a man who had risked his life to establish one of his adopted state's best-known landmarks. In contrast Clingman acquired a reputation as a backbiting scoundrel. A few hard-line Whigs even publicly branded the congressman a murderer. Thanks to a gubernatorial appointment, Clingman finally received his coveted seat in the Senate in 1858 and won reelection in 1860. That same year, after another long and heated argument in the press, Guyot got the senator's name attached to the highest point in the Great Smoky Mountains, a peak today known as Clingman's Dome. But southern secession took Clingman out of Washington, and in the years after the Civil War, lingering memories of his bitter fight with Mitchell sullied the once-powerful congressman's public image. Though he remained active in state politics, he never again held elected office.[73]

Meanwhile along the Cane River, Big Tom Wilson became a celebrity. The *Harper's* article contributed to his fame, but most of those who visited his cabin now wanted to hear about the discovery of Mitchell's body. Born talker that he was, the mountain man always obliged. Whenever he told the tale, Big Tom insisted that when he found the professor, he intended to bury him "on top of the mountain he had lost his life exploring" but had respectfully yielded to the family's desire to take the body to Asheville. Just as Vance and Phillips had hoped, Big Tom's status in the local community and his gregarious nature helped perpetuate the carefully crafted public images of Mitchell as hero and Clingman as villain. Big Tom's business also profited. In the years immediately after he found Mitchell's body, the amiable guide rarely lacked for summer clients. Like others involved in the dispute, the bear hunter discovered that his life would never be the same.[74]

The mountain landscape had changed, too. At the time of Mitchell's death Yancey County had 8,000 residents who worked roughly 1,000 farms. Most of their holdings were small, ranging between 20 and 50 acres, but 110 Yancey families worked tracts of 100 or more acres. Of the 900-odd farms in neighboring Buncombe County, well over half encompassed more than 50 acres. Livestock reared there and elsewhere in western North Carolina provided nearly half the cheese, a third of the butter, and a fifth of the wool produced in the state. Though agriculture dominated the region's economy, Yancey County also had over thirty "manufacturing concerns," including several small sawmills, two tanneries, and eighteen mills for grinding flour and meal.[75] Mitchell and Clingman ended up enemies, but their long campaign to civilize the mountain wilderness and to make it productive had already begun to pay big economic dividends.

Careful observers noted other changes, too. When William and Big Tom Wilson tried to re-create Mitchell's original trip to the summit, they discovered that a wildfire, probably kindled by hunters or campers, had burned a large section of the mountain's western face, leaving only "charred and fallen trees." (Indeed, Big Tom later suggested that this "fire scald," which looked like "someone's field," might have lured Mitchell off the trail—and toward Sugar Camp Fork—on that fateful evening in 1857.) A bit higher up William Wilson noticed that the grass and weeds were now thicker than in times past. He attributed the lush growth to the absence of deer that, though plentiful in 1835, had since been hunted out or driven into other, more isolated areas.[76]

But the most dramatic transformation resulted from the region's burgeoning tourist trade. By the end of the 1850s high peaks that had once seemed wild and remote were drawing hundreds of visitors each year. Horse trails and footpaths snaked up from the Cane, Toe, and Swannanoa. Summer cabins and campsites dotted the upper ridges and—after Stepp and others had cleared the summit of trees—a wooden observation tower stood near the professor's grave. (While the controversy raged, local folk moved the structure from one mountain to another until the tower, like Mitchell's body, finally rested on the true summit.) Much like Elisha Mitchell in 1844, sightseers found vistas that evoked that odd sense of forbidding isolation and ethereal power common in nineteenth-century notions of nature and wilderness. As one tourist from Raleigh observed in late 1857, "The entire view was magnificent, the grand, the stern, the terrible predominating, but softened here and there by the beautiful, the pleasing, the romantic." Upon leaving the mountaintop he found himself overwhelmed "with a wild sense of strife, an impression of the sublime and rugged, the tremendous reality of life."[77]

It was the story, however—the carefully concocted tale of a courageous explorer, a scheming politician, and a stalwart mountain man—that quickly became the main attraction. Never mind that the narrative occasionally strayed from the truth or that Mitchell's partisans had manufactured evidence to support the professor's claims. Travelers from across the South flocked to view the precipitous terrain on which the tragic plot had unfolded. Indeed, Mitchell's death only seemed to enhance the mountains' ethereal qualities. The same Raleigh tourist who stood awestruck on the summit also viewed "with saddened feelings" the fateful waterfall on Sugar Camp Fork. "It is a wild and gloomy place," he declared, "but God was there, and there the christian and the scholar met him."[78]

New place-names added to the ambience. The professor's supporters had asked only that he be honored on the highest mountain. They got more. By the 1860s, using language sanctioned by the state legislature, mapmakers had stripped Thomas Clingman's name from the summit and, for the moment, christened it Mitchell's High Peak. Farther south the mountain that local people had once believed to be highest still retained its popular designation as Mount Mitchell, a name transferred to the summit sometime in the 1870s. Decades would pass before any mountain in the Blacks again bore the name of the professor's rival.[79]

The name changes only confirmed what many people across the state already knew. The tallest mountain in the East—that fir-covered dome of granitic rock that, purely by chance, protruded into the atmosphere a few feet farther than the surrounding pinnacles—was no longer just a natural and geologic curiosity. It had been transformed into a martyr's mausoleum, a place for pilgrimage and reflection, and a source of state pride. Mapmakers called it Mount Mitchell, but it was really Mitchell's Mountain, a landmark that would always be linked to stories of the professor's life and death. Rightly or wrongly, from this point on its ecological fate would be forever bound up with its new status as a North Carolina icon.

FOUR

Modernity

JULY

MOUNT MITCHELL

It is the season of fog and tourists. A few minutes before noon, thick clouds blow in from the southeast, shrouding the Blacks in a translucent haze and obliterating the view from the summit. The sightseers have come anyway. In the asphalt parking lot on the mountain's eastern face I find cars from New Jersey, Florida, Illinois, Ohio, and Alabama, just a few of the myriad vehicles that annually carry more than 300,000 visitors to this spot. Today the tourists spill onto the pavement, donning sweatshirts and half-joking, half-complaining about the July cold and dampness. Within minutes they begin the short walk to the crest of the East's highest mountain and Elisha Mitchell's grave.

I dutifully pull on a sweater and take my place amid the throng. But as the group moves toward the summit, I veer down a side path in search of another, older visitor attraction. These days it is mostly a geologic curiosity: a giant slab of granite—some fifteen yards long and perhaps twenty feet wide—that overhangs an old trail several hundred feet below the mountaintop. In the 1850s, though, it was a well-known landmark. Travelers called it the Cave, the Shelving Rock, or the Sleeping Rock, and it provided natural shelter for those who stayed overnight in the high mountains. A light rain begins as I step beneath the cliff.

For the better part of a half-century now the site has been closed to overnight guests, and nature has reclaimed much of the surrounding terrain. Fraser fir seedlings cling to the thin soil on top of the shelf. Native rhododendrons, blooming hot pink in this summer season, protrude at odd angles from the sides of the enclo-

sure. Near one end of the ledge, where rainwater drips steadily from the granite roof, grasses and ferns cluster in thick, feathery clumps. But farther back, well under the ledge, the soil is bare and powder-dry. Tracks from mice, voles, and other crevice-loving rodents cover the floor. In one corner, at the very rear of the structure, sooty gray walls recall campfires kindled decades ago. I sit for a moment watching the rain and listening to the patter of tourists returning from the summit.

At a site like this, where the past so obviously intersects with the present, it is tempting to suggest that little has changed since Mitchell's day. But in truth both mountain and people are different now. For the Blacks, as for much of America, the late nineteenth century was a time of transition during which the region became part of the modern world. In a curious way the Sleeping Rock serves as an index of the change. In the years after the professor's accident the shelter attracted the usual assortment of campers, hunters, scientists, tourists, and writers. But they were of a generation different from Mitchell and Clingman's, with new notions about nature and the human place in it. Their story, like that of the modern mountain, begins where much of recent southern history begins: with the Civil War.

The war, or the "late unpleasantness" as some southerners call it, came slowly to the Blacks. No famous general, Union or Confederate, marched troops up the Cane or the South Toe. No great armies squared off at Ogle's Meadow or Flat Spring Gap. Instead the bloody conflict closed in bit by bit, day by day, disrupting long-established agricultural routines, siphoning off resources, and finally trapping mountain people in a vise grip of poverty and violence. Like much of western North Carolina, Yancey County was a place of divided allegiances. In its northern reaches Union sentiment ran so strong that citizens there finally demanded separation from their neighbors to the south. The state obliged by creating Mitchell County (ironically named for the slaveholding professor) in 1861. In southern Yancey, which encompassed the East's highest peaks, most residents sided with the Confederacy and were at least willing—if not always eager—to defend the slaveholding planters who bought livestock and produce from mountain farms. Big Tom Wilson served as fifer and head musician for a North Carolina brigade, while Nathaniel Allen and Samuel Austin, sons of Elisha Mitchell's first guides, joined a local contingent of volunteers known as the Black Mountain Boys. In 1862 they marched to Richmond to defend the city against Union general George B. McClellan during his Peninsula Campaign.[1]

In those first heady days of secession and war, some of the Black Mountain Boys signed on for the duration, figuring to win a quick victory and come home within a year. Others enlisted for a twelve-month hitch. They made few provisions for their families, leaving wives and children to run their farms until they returned. Their trust was probably well placed. Across southern Appalachia many women skillfully managed their husbands' lands. They bartered for seed and implements, supervised farmhands, acquired new real estate, and even bought and sold slaves. Throughout the region women also toiled behind the plow, tilling older tracts along river bottomlands.[2]

When it came to the heavy work of burning and clearing new fields, however, they sorely missed their husbands' labor. By 1862 they also had trouble finding hired help. That year the Confederate Congress passed the Conscription Act, which made all men aged eighteen to thirty-five years eligible for military service. The act offered exceptions to citizens working in certain industries and allowed those of sufficient means to hire substitutes to fight in their places. But for most men of the Blacks, the new law meant that they must either take up arms or evade the draft. Either way they left home. As more and more able-bodied workers disappeared, farm families along the Cane and South Toe faced a full-blown ecological and economic crisis.[3]

With labor in critically short supply and clearing of new land at a standstill, the system of field rotation that had sustained Black Mountain agriculture for a half-century broke down. Traditionally farmers had been able to store surplus grain both for their use and to barter or sell to neighbors. Now, however, older, depleted fields barely yielded enough corn and wheat to feed a single family. Amid the chaos of war farmers still had to cope with the chaos of nature. Three bone-dry summers between 1862 and 1864 left local corn crops withering on the stalk. Each autumn, as the droughts relented, unseasonably cold weather set in and eliminated any chance for late-season recovery. Desperate for relief, local folk turned to an old friend. Zeb Vance, the Asheville Whig and staunch defender of Elisha Mitchell, was now governor of North Carolina, and mountain people bombarded him with letters and petitions asking for help. A citizens' group from Mitchell County complained that the "hard freezes that visited this country much earlier this season than heretofore" had left local residents with less than "half enough grain to bread their own families" and none "to let their neighbors." Things were no better in Yancey. A Burnsville man informed the governor that "the absence of the labouring class, together with the early frost in the Fall has cut us short . . . of at least five thousand bushels of corn."[4]

In times past when crops failed, local people had relied on livestock to fend off famine. Because the animals fed on wild forage, cattle and hogs could survive all but the worst summer droughts. Slaughtered in winter, even a few head of poorly nourished livestock might easily see a family through the year. However, to preserve beef and bacon for storage, mountain people needed salt, a commodity now in critically short supply. Before the war Black Mountain folk had acquired their salt from various mines in Virginia. But by 1862 encroaching Union troops threatened those sources. In addition many mountain roads, which were barely passable in peacetime, had fallen into disrepair, and regional suppliers found it risky to transport salt into southern Appalachia. To cope with the shortage, some farmers kept their animals alive through the winter and slaughtered them one at a time, preserving what meat they could and selling the rest. Still, without salt the vast herds that grazed the forests and mountain balds provided little long-term insurance against starvation.[5]

To make matters worse, the Confederacy had no qualms about taking local produce to feed hungry soldiers. Starting in 1861 the government instituted a policy of impressment, which meant that southern generals could legally appropriate "needed food and materiel" to supply their men. Two years later Confederate legislators followed up with the "tax-in-kind," a measure that allowed the government to collect one-tenth of various crops, including such Black Mountain staples as "wheat, corn, oats, rye, and buckwheat." According to the new laws farmers were to receive cash for impressed livestock or "certificates of credit" redeemable for the fair value of their animals. But with food stores depleted and inflation rampant, merchants and speculators frequently offered better prices than the government, creating a thriving black market that also pulled resources from Black Mountain farms.[6]

The constant search for sustenance occasionally pitted military leaders against one another. By the fall of 1863 beef was so scarce in the Cane River Valley that an officer of the Burnsville home guard (a loosely organized militia entrusted with keeping order in the community) pointedly told Governor Vance that no matter who asked, those charged with local defense "would suffer no [more] cattle drives out of [Yancey and Mitchell] counties." To make sure his order stood, the officer began confiscating all livestock in both counties for use by the home guard. "It would," he explained, "take all the provisions that wane here to support [his] troops." He neglected to say how the rest of the citizenry might survive.[7]

Eventually Vance's government did offer help. From 1862 on, the state provided food to needy families and began a massive program of poor relief throughout North Carolina. In Yancey County women with husbands in active service got one share of the local allotment; those widowed by the war received two shares. But such measures usually proved inadequate, and many families discovered that even the state's unprecedented largesse did little to ameliorate the ruinous effects of conscription, drought, autumn freezes, and impressment. Every winter until the war ended, the grim threat of starvation hung over the valleys like late-morning fog.[8]

As chronic hunger became a fact of life along the Cane and South Toe, sizable groups of former Confederate soldiers, including some of the Black Mountain Boys, drifted back into the region. In the eyes of the government they were deserters, some of the 24 percent of western North Carolina troops who, for whatever reason, grew tired of the war and came home. A few of the outlaws hid out by themselves. More commonly, though, fugitives banded together in small, semiorganized units and sought refuge at the Sleeping Rock and other secluded sites among the high peaks. Armed with weapons brought from the front, they waged a sporadic but relentless war against the home guard, regular army soldiers, and anyone else who might take them into custody. They had plenty of company. Confederate draft dodgers, locally known as "out-liers," and small groups of Union sympathizers ("Tories" in southern parlance) also took refuge in the Blacks, bushwhacking Confederate troops and skirmishing with the home guard.[9]

The roving bands of armed men wreaked havoc on the landscape. Sweeping out of the mountains at night, they regularly ransacked farms along the Cane and South Toe, carrying off crops and livestock and terrorizing local residents. When they could not fill their bellies by pillaging, the renegades "lived off the forests." Hunting year-round, they took large numbers of deer, bear, and smaller animals. For centuries before the war wildlife populations had waxed and waned in response to weather, disease, and a host of other factors. But now for the first time since white settlement, human hunters took a noticeable toll. According to one ardent Confederate who lived on the Swannanoa's North Fork, "By the time of the surrender almost all of the game, even to the squirrels, had been shot, snared, or trapped." Blaming the carnage on "bushwhackers, deserters, and raiding Yankees," he noted that "the few deer and bear left in the area had taken refuge 'over on the Cane River in Yancey,'" an area so rugged (and not coincidentally, so Confederate) that roving Unionist hunters "usually did not follow game when it reached there."[10]

As the fighting drew to a close, Yankee sympathizers were not the only ones who avoided the upper Cane. The war put a huge crimp in the region's developing tourist trade. With slaves freed and the southern economy in shambles, many lowcountry planters could no longer afford a summer in the mountains. Moreover, it became increasingly difficult to get there. The Western North Carolina Railroad, which had once extended nearly to Morganton, had been ripped apart by Union general George M. Stoneman during his 1865 march across the state.[11] Even if would-be tourists somehow found their way to Asheville or some other mountain resort, the general climate of lawlessness in Yancey County made an overnight trip to Mitchell's High Peak a dangerous undertaking. Nearly a decade would pass before the sightseers returned en masse to the East's highest mountain.

As the war abated and tourist traffic dwindled, new shockwaves—this time from the world of science—rippled through the region's forests. For much of the nineteenth century scientists had been debating certain elements of the concept now called evolution. Indeed, Elisha Mitchell was probably well informed on the subject when he explored the Blacks. His library included works by Jean Baptiste Lamarck, a French naturalist who believed that environmental influences produced gradual change in plants and animals. Mitchell also read Charles Lyell, the eminent British geologist who argued that the earth was much older than Genesis (or most scientists) allowed and that its surface had changed markedly over geologic time. Mitchell apparently had no trouble reconciling such ideas with his Presbyterian theology. As he noted in 1842, God instructed people "in their duties to himself and each other," leaving them free "to make advances in the sciences by means of the faculties he has given them."[12]

But the professor did not live to witness the real firestorm over evolution. That began two years after his death, in 1859, with the publication of Charles Darwin's *Origin of Species*. Almost immediately the book created deep rifts within the scientific community. The sticking point was not evolution itself (that was old news) but the accompanying hypothesis of natural selection, the seemingly random process by which widespread change occurred. To many nineteenth-century scientists the theory (announced simultaneously by Darwin and a younger biologist named Alfred Russell Wallace) seemed speculative and incomplete. Others were simply uncomfortable with happenstance as an explanation for the diversity of life. Among those who spent much time pondering the mechanism of evolution was Edward Drinker Cope, an avid collector of fossils in the American West and one of the nation's best-known paleontologists. In September 1869, ten years after the appearance of *Origin of Species*, Cope's interest in evolution brought him to the Black Mountains.[13]

Part and parcel of Darwinian theory was the idea that species distribution changes over geographic distances, or in layman's terms, animals and plants in one region differ from those in another because they evolve in dissimilar surroundings. That notion prompted both Darwin and Wallace to conduct much of their research on tropical islands where isolation and the demands of a distinct environment sped up the evolutionary process. Cope had little trouble with that concept. While in the Blacks, he spent much of his time describing various birds and mammals that inhabited the islandlike spruce-fir region, concluding that animals near "the crest of the Alleghany Mountains" (his name for the Appalachians) resembled those of more northern latitudes. His observations not only provided fresh evidence regarding climate and species distribution but also helped establish the popular perception of the Black Mountains as Canada in Carolina.[14]

Cope was willing to follow Darwin and Wallace only so far, however. A man of strong religious convictions, the paleontologist could never accept the randomness of natural selection and spent much of his life looking for an alternative. During his stay in the Blacks he paid particular attention to mountain salamanders, constantly examining them for tentacles and other traits that suggested earlier, less-advanced stages of development. Such characteristics, he thought, might indicate that evolution progressed in a regular pattern that reflected some master design, an idea implicit in the work of Lamarck and other pre-Darwinian theorists. As Cope noted, the "irregular preservation of a larval character," observed in the amphibians of the East's highest peaks, might one day be "of interest in connection with the theory of evolution." To most scientists, though, a few anomalies in Black Mountain salamanders made little difference. By the late nineteenth century Lamarck's ideas, even as refined by Cope and his contemporaries, fell out of favor as researchers began to reconcile natural selection with the emerging field of genetics. The Blacks were once again center stage in a scientific controversy. But instead of arguments about discovery and the heights of mountains, the issue was now the very origin and order of life itself.[15]

While scientists debated Darwinism, the general public developed new and more practical interests in the natural world. In the late nineteenth century, middle-class Americans, especially those living in eastern cities, became utterly obsessed with the outdoors. In nature they saw the chance to escape the frantic pace of urban life, to recall the nation's frontier past, and to pit themselves against the environment in their own Darwinian struggle for survival. In wilderness they found a soothing antidote to civilization, a place of recreation and recovery that brought out the best in humanity. Increasingly, eastern urbanites began to draw distinct geographic divisions between labor and leisure. The city, with its factories and regimented daily routines, was the place for work. The woods, where no time clocks beckoned and no irate supervisor railed about productivity, were the place for play. In terms of sheer physical exertion, those who trekked or camped in rugged terrain might, in fact, work harder than any city dweller. But it was noble labor, closer to nature, and in keeping with the tradition that had once made America strong.[16] The notion of wilderness as cure for the nation's ills would have been anathema to Elisha Mitchell. But as calm returned to western North Carolina and the region's economy recovered from the devastation of the Civil War, this emerging outdoor ethic brought visitors back to the mountain that bore his name.

Hikers were among the first to rediscover the Blacks. In 1876 a group of scientists and college professors convened at the Massachusetts Institute of Technology to form the Appalachian Mountain Club (AMC), an organization dedicated to exploring the peaks "of New England and adjacent regions." The first of many such clubs formed across the nation, the AMC helped foster a national craze for hiking and mountain climbing. At the same time farther south, construction of the Western North Carolina Railroad resumed in earnest. By 1880 trains had begun chugging across the Eastern Continental Divide, from Morganton to Asheville, providing easy access to the North Fork of the Swannanoa and the trails into the Blacks. Shortly thereafter hikers from the AMC began testing their mettle against the East's tallest mountain.[17]

For many who had scaled the peaks of New England, Mount Mitchell proved a giant disappointment. The fifteen-mile trek from the train station was long but not especially rigorous, and one hiker marveled that the southern mountain, though "much higher than [New England's Mount] Adams or Washington," could be climbed "free from difficulty or fatigue." Besides, even after two decades of war and neglect, one could still find the all-too-familiar trappings of civilization. Patton's Mountain House was easily identifiable, though its roof and walls had collapsed into rubble. Near the summit an old weather observatory, manned for a few years in the early 1870s, lay in ruins. Around the Sleeping Rock overnight campers found stumps and scraggly trees, the inevitable result of renegade soldiers,

Hikers near the summit of Mount Mitchell, ca. 1916. By the early twentieth century the Black Mountains were a popular destination for wilderness enthusiasts seeking escape from the hectic pace of urban life. William A. Barnhill Collection, Pack Memorial Public Library, Asheville.

hunters, and other transients cutting firewood. Even the view, which had overwhelmed tourists thirty years earlier, was often not enough to appease the new craving for raw nature, physical exertion, and untrammeled wilderness. The vista, one experienced climber noted after an easy walk to Mitchell's High Peak, "is a wonderland; but it is lacking in deep gorges and stupendous cliffs, and nowhere approaches the grandeur of [Mount] Adams or Bond [another notoriously rugged New England peak]."[18]

The Blacks fared better with bird enthusiasts. In 1888 William Brewster, a nationally known amateur ornithologist, spent two days near the summit (probably at the Sleeping Rock) cataloging animal life at the highest elevations. From those and other observations Brewster identified three distinct "faunal divisions" between the valleys and the highest peaks. At the base of the Blue Ridge, in the Carolinian zone, he found cardinals and mockingbirds singing among sweet gums and magnolias. From there he passed into an Alleghanian zone of oaks and hickories where he sighted thrushes and rose-breasted grosbeaks. Finally, in the upper reaches of the Blacks only an hour or two from the valley floor, he entered the spruce-fir Canadian zone, inhabited by winter wrens, golden-crowned kinglets, and red-bellied nuthatches. En route to the top he described two as yet unknown species: the Carolina junco and the mountain solitary vireo. Brewster's discoveries made headlines in ornithological circles, but his short expedition is probably more important as an early and, for its time, remarkable example of biogeography, the science that seeks to understand the worldwide distribution of plants and animals.[19]

Like most nineteenth-century ornithologists Brewster routinely shot birds and preserved them for study. Aided by various taxidermists, hunters, and friends, he eventually amassed more than 40,000 mounted specimens, some of which fell to his trusty pistol during his travels in western North Carolina. Bird collecting, as the practice was commonly called, was not without risks. In early June 1895 John Simpson Cairns, a twenty-five-year-old ornithologist from nearby Weaverville, set out with several friends and family members on a collecting trip into the Blacks. While his companions fished for trout—probably somewhere on the upper Cane—Cairns loaded his shotgun and went looking for birds. A few hours later his family found him dead on a well-marked trail, the right side of his face shredded by a load of bird shot. Investigators concluded that he had somehow tripped and accidentally discharged his weapon. An obituary listed his death as "a distinct loss to ornithology."[20]

As the Cairns incident suggests, bird collecting was a near cousin to sport (or game) hunting, another favorite pastime of those who took to the woods for recreation in the late nineteenth century. Unlike most meat hunters, sportsmen were well-read in natural history and much interested in dogs and guns. They hunted noble animals, primarily big game and birds, that offered a true test of their outdoor skills. Like most wilderness aficionados they also were "urban, wealthy, Eastern, and professional."[21]

In the 1880s two such men, K. M. Murchison and his brother, David, brought sport hunting to the Blacks. The Murchisons hailed from North Carolina but made their fortunes in New York City as merchants and factors for several businesses. For less than $2,500 the brothers purchased most of the vast 13,000-acre tract once owned by Amos Ray on the upper Cane River. That remote region, largely ignored by roving soldiers during the Civil War, already had a solid reputation among sportsmen. According to one description it was "a wilderness well stocked with bears and deer" and included some fourteen miles of "streams abounding in trout." For nearly thirty years the Murchison family kept the land as a semiprivate "hunting and fishing preserve" where they entertained sporting friends from across eastern America. To manage the estate in their absence the brothers opted for the best hunter available. They hired Big Tom Wilson.[22]

Big Tom worked on the preserve (locally known as the Murchison Boundary) as a guide and gamekeeper. He showed visiting sportsmen the best techniques for taking deer, bear, and trout and assisted local visitors. (Indeed, it appears that he helped the Cairns family recover the ornithologist's body.) The bear hunter also kept a sharp eye out for trespassers, poachers, and local miscreants who sometimes "explod[ed] powder in the streams to kill the fish." Shortly after he went to work on the preserve, Big Tom entertained one of his most famous clients, New England–born essayist Charles Dudley Warner.[23]

Perhaps best known for his collaboration with Mark Twain on *The Gilded Age*, Warner was one of a number of local color writers who began to trickle back into the region after the Civil War. Not to see Big Tom, Warner believed, "was to miss one of the most characteristic productions of the country, the typical backwoodsman, hunter, [and] guide." The writer was not disappointed. With his "splendid physique" and "iron endurance," Big Tom personified the new wilderness ethic. Warner dubbed him the incarnation of Leather-Stocking, the straight-shooting, woods-loving hero of James Fenimore Cooper's frontier novels. That description, published in Warner's 1888 essay *On Horseback*, contributed mightily to the developing perception of the southern mountains as an isolated region where the last pioneers lived free from the corrupting influences of modern society. As that image took hold in the minds of eastern Americans, Big Tom, like Mount Mitchell, became an Appalachian icon.[24]

The famous tracker regularly posed with his gun for visiting photographers and annually made pilgrimages on foot to Asheville, where according to one acquaintance, "his powerful figure with its massive head, snowy locks, and beard, and kindly eyes attracted great attention as he walked the streets." City folk reveled in his stories and "from the highest to the lowest vied with each other in doing him honor." When he died of natural causes in 1909 at age eighty-three, local eulogists remembered him as "one of nature's own princes."[25]

Appalachian stereotypes notwithstanding, Big Tom and his neighbors already had more than a nodding acquaintance with the larger world. In the half-century since Elisha Mitchell's death, they had witnessed the effects of modern warfare and the inquisitive spirit of Darwinian science. The increased presence of hikers and sportsmen also betokened the general angst of life in the nation's emerging cities. In the rural world of the Blacks those influences were sometimes hard to discern. Aside from a few small mica mines in Mitchell and Yancey Counties, heavy industry had not yet made its presence felt in the region. But that was about to change. Within five years lumber and paper companies would land in the Blacks

Big Tom Wilson, consummate mountain man and Appalachian icon. North Carolina Collection, University of North Carolina Library, Chapel Hill.

with a vengeance. Before long, those who remembered an old bear hunter and the allure of the nineteenth-century wilderness would wonder what went wrong.

SEPTEMBER

BUNCOMBE HORSE RANGE TRAIL

Morning haze is lifting as I park the truck in a roadside turnout and shoulder my pack. This is a day to savor, a last taste of summer before cold weather begins in earnest. A few blackberries still cling to rusty green stalks, and I hear clacking grasshoppers in the roadside ditches. Elsewhere, though, mountain ash leaves are already a deep ochre. The wet-weather springs that flow from this hillside in April have slowed to a trickle or disappeared completely, awaiting rejuvenation from a wayward tropical storm or, failing that, the first cold rains of winter. At woods' edge I detect the telltale autumn odor of rotting leaves and humus.

I half-walk, half-slide down a brushy bank, emerging on a well-defined path at the base of Potato Knob. For reasons still unclear to me, this route is known as the Buncombe Horse Range Trail. From here it winds lazily around the eastern face of the range, gaining only about 600 feet in elevation as it swings past some of the region's most famous sites. In most places it is wide, too, covered with gravel and loose stones, a veritable thoroughfare compared with the Big Butt or Crest Trails. I head north, moving easily around Clingman's Peak, Mount Gibbes, and Mount Mitchell. Even with an overnight pack and frequent stops for water and scenery, it takes just a few hours to trek the seven miles to Maple Camp Ridge, a prominent swell of gneiss and schist just below the peak now called Big Tom.

The easy grade between Clingman's Peak and Maple Camp Ridge is no accident. Workers hired by a Pennsylvania lumber company cut the path some ninety years ago when they blasted away the mountainside to make room for a narrow-gauge railroad. They chose the route carefully. As a quick glance at a topographic map will confirm, the roadbed hugged the mountains between the 5,600- and 5,800-foot contours, neatly bisecting the forest of red spruce that grew between 4,500 and 6,500 feet and affording access to some of the Blacks' most valuable timber. For more than a half-decade trains hauled logs from these slopes to sawmills in the Swannanoa Valley. Even now observant hikers occasionally find railroad spikes and rotting crossties along the route.

Like many of those who use the trail these days, I have mixed feelings about the lumbermen and railroads. It is difficult not to admire the ingenuity, perseverance, and sheer hard work of those who laid the track and cut the timber. Without that labor my hike to Maple Camp Ridge would be far more strenuous. Yet as I crawl into my tent and listen to the night wind whine across a landscape now devoid of large trees, I cannot help but wonder what it might have been like to camp among the tall evergreens and to sleep on a soft pallet of brown needles, where the air smelled like Christmas and chickadees and juncos twittered at first light. I take some comfort in knowing that I am not alone in my nostalgia for lost nature. During the early twentieth century some of the local people who witnessed the arrival of the railroad voiced similar feelings. But in a state still trying to attract outside capital to jump-start its economy, the needs of entrepreneurs and developers came first. In western North Carolina as in much of the rest of the nation, those who worried about the future of the mountain landscape did so with balance sheet in hand, carefully calculating long-term costs and constantly watching the bottom line.

Sometime late in the spring of 1909 (the year that Big Tom Wilson died), John Simcox Holmes, recently appointed state forester for the North Carolina Geological and Economic Survey, began preparations for a summer-long tour of the state's western counties. His task was to report to the state legislature on the current condition of mountain forests. It was an important assignment. Like many North Carolinians Holmes was concerned about the state's rapidly disappearing woodlands. Canadian by birth and an 1888 graduate of the University of North Carolina (where he earned a certificate in agriculture), Holmes spent the early part of his life farming family lands on the French Broad River, not far from Asheville. From there he watched with dismay as America's burgeoning timber industry rolled into the southern Appalachians.[26]

For most of the nineteenth century, American lumbermen had followed the railroads into the vast forests of Maine, New York, and Pennsylvania, working their way west toward the Great Lakes. They treated timber like miners treated coal, as an expendable resource. Hauling logs by rail and processing them at mills equipped with steam-powered circular saws, lumber companies typically took every usable tree from a site before moving on. By the 1880s, with America's great north woods fast becoming a memory, the nation's timber barons began to cast an acquisitive eye toward the still-verdant forests of western North Carolina. Because the region lacked an extensive railroad network, northern investors initially paid local people to cut high-grade timber and haul it by wagon to centrally located sawmills. However, in the 1890s, when the newly organized Southern Railway Company built lines from Asheville into South Carolina and Tennessee, lumber companies reverted to the same profitable but wasteful practices employed in the Northeast.[27]

Such irresponsible behavior did not go unchallenged even in the laissez-faire business climate of the late nineteenth century. From the 1860s on, a host of scientists, naturalists, writers, and even a few lumbermen condemned the reckless exploitation of forest resources. Indeed, in Buncombe and several surrounding counties, not far from John Simcox Holmes's farm on the French Broad, two of America's first professional foresters tried to provide loggers with more practical and profitable alternatives. In the 120,000-acre Pisgah Forest owned by railroad heir George Washington Vanderbilt, Gifford Pinchot and Carl Alwin Schenck instituted a program of selective cutting, fire prevention, and erosion control as part of an effort to show that conservation might ensure a continuous supply of marketable lumber. In 1898 Schenck established a forestry school on the Vanderbilt lands where he tried to teach would-be timber magnates to abandon the "cut out and get out" practices of the previous generation in favor of measures that would allow them to "cut and grow and cut again."[28]

Holmes was much impressed with the work at Pisgah and with the growing national interest in forestry. In 1903, with Theodore Roosevelt in the White House and "conservation" fast becoming a watchword of his administration, the UNC alumnus enrolled at the Yale Forest School, a fledgling, three-year-old institution founded through the philanthropy of the Pinchot family. Holmes finished near the top of his class in 1905, the same year that Roosevelt made Pinchot the first chief of the new U.S. Forest Service. Years later when Holmes looked back on his work in North Carolina, he noted that his admiration for Roosevelt and Pinchot played a key role in his decision to take up forestry. They were, he wrote, "two of the finest public servants this country has known."[29]

While Holmes studied at Yale, the North Carolina General Assembly decided to revitalize its Geological Survey to reflect the new interest in woodland conservation. The institution once headed by Elisha Mitchell received a new name and a $900 annual appropriation for "forestry investigations." Sometime between 1901 and 1905 it became the Geological and Economic Survey. The change was significant. At the national level the conservation movement had already begun to split into opposing factions. Utilitarian conservationists, typified by Pinchot and—maybe to a lesser extent—Roosevelt, argued that America's remaining forests should be managed to provide "the greatest good for the greatest number [of people] for the longest time." But others, the so-called preservationists, a group that included naturalist and writer John Muir, believed that certain wild lands should be left undisturbed as places of recreation and refuge from urban life. In sanctioning a survey that emphasized economics, North Carolina's leaders made their position clear. They sided with the utilitarians, and they sent Holmes to determine how the state's upland forests, including those of the Black Mountains, might best be used and managed. Over the next several decades, as he defined forest policy for western North Carolina, Holmes emerged as a key figure in Black Mountain history.[30]

Traveling with two representatives from the Forest Service, Holmes got his first close look at Yancey County's forests during the first days of summer in 1909. Though most lumber cut there still had to be hauled in wagons over rough roads, "the better grades of oak, poplar, and pine" had already disappeared from the valleys beneath the East's highest peaks. Farther up the slopes lumbermen had also culled ash, cucumber, and buckeye from the lower reaches of the northern hardwood forests, leaving less-valuable yellow birch and red maple.

Selective removal of high-grade trees was not the only ecological consequence of early lumbering. Because the South Toe and Cane proved too narrow and shallow for transporting timber, loggers usually pruned the limbs from trees where they fell and used horses or, more commonly, cattle to drag the logs to the nearest wagon road. This technique, called snaking, often destroyed seedlings growing nearby, making it difficult to replace the most sought-after species. The most serious threat to forest regeneration, however, came from fire. During dry seasons treetops and other brush left by local loggers provided enough potential fuel to kindle destructive blazes in lower-elevation forests, especially in spring when farmers routinely set fires to clear new land or refurbish old fields. Burning the woods remained "a most destructive practice," one that Holmes hoped to discourage.

Yet in 1909, despite years of cutting, snaking, and burning, some of the more remote Black Mountain forests remained untouched by axe and saw. Substantial stands of "good timber" flourished on the southern rim of the range near Clingman's Peak and Potato Knob. On the "headwaters of Caney Creek" the Murchison Boundary contained Yancey's "largest single tract of virgin timber." Sprawling east and west across the shank of the range, the great spruce-fir forest also stood uncut, though fires "said to be set by hunters" had recently destroyed about 10 percent of the trees on the South Toe side. Ever conscious of his economic mission, Holmes re-

Uncut spruce-fir forests in the Black Mountains, looking north from Mount Mitchell, across the upper Cane River Valley, ca. 1915. William A. Barnhill Collection, Pack Memorial Public Library, Asheville.

ported that such forests had "little commercial value at present." But, he surmised, "as transportation facilities improve, [the trees] will no doubt come into the market for pulp wood and lumber."[31]

It happened faster than even he could imagine. Before the forester could finish his lengthy report and submit it to the legislature, a bevy of northern corporations ("Carpetbaggers of the Woods," one historian has called them) began scrambling for the remaining timber. A Philadelphia syndicate calling itself the Carolina Spruce Company acquired 5,200 acres on the northwest flank of the range near the Cane River town of Pensacola. Dickey and Campbell, a company that had logged extensively in the northeastern states and Virginia, laid claim to the spruce-fir forest surrounding the East's highest peak. By 1912 Brown Brothers Lumber Company (also from Pennsylvania) had obtained logging rights to the Murchison Boundary, and the Mount Mitchell Company, headquartered in Chicago, had gained control of a narrow, nine-mile swath of spruce-fir in the upper reaches of the South Toe watershed. Champion Fibre, an Ohio firm that opened a mill at Canton, North Carolina, in 1908, coveted the Blacks' abundant supply of red spruce and American chestnut, both of which could be turned into pulp for making paper. In addition, chestnut provided an important liquid byproduct, tannin, used in processing leather.[32]

As elsewhere in eastern America, the great grab for Black Mountain forests began not with the thud of an axe or the whir of a saw but with overland surveys, dynamite, and the metallic ring of sledgehammers on steel rails. By 1911 new technology in the form of narrow-gauge track and low-geared Shay and Climax engines made it possible for Dickey and Campbell to begin construction on the Blacks' first logging railroad. Because narrow-gauge track could be laid over rough terrain without a wide cleared roadbed, it took only a year to complete the first section. From a double-band sawmill near the town of Black Mountain in the Swannanoa Valley the line ran north roughly fourteen miles, around Graybeard and Pinnacle, to the eastern face of Clingman's Peak. In 1913 Perley and Crockett, yet another Pennsylvania lumber company, bought the Dickey and Campbell operation and extended the railroad to the eastern slopes of Mount Mitchell. Branch lines eventually ran north to Maple Camp Ridge and across Stepp's Gap to the west side of the peaks. By then local people were calling it the Mount Mitchell Railroad.

Meanwhile at the other end of the range, the Black Mountain Railroad, a light-duty line that served the North Toe Valley, advanced south out of Burnsville to accommodate lumbering operations along the Cane River. Even before the new railroad arrived, the Carolina Spruce Company began laying track from its sawmill at Pensacola up Cattail Creek toward the spruce and fir on Celo Knob. Not to be outdone, Brown Brothers built a twelve-mile spur from Eskota, across the Murchison Boundary, to Stepp's Gap, where it connected with the Mount Mitchell line. Just five years after Holmes had pronounced Black Mountain forests inaccessible, a lumberman could board a train along the Swannanoa and ride across the East's highest mountains to Burnsville. The woods were open for business.[33]

In one sense the rapid assault on Black Mountain forests represented the fulfillment of Thomas Clingman's long-standing dream of attracting Yankee capital to western North Carolina. But it was also simply the local manifestation of a broader trend. Across the South logging operations reduced the region's forests by an estimated 90 percent between 1880 and 1924. Between 1914 and 1918 lumber companies—with considerable help from the railroads—removed more timber from Appalachia than in all the years before 1900. Yet the arrival of railroads and lumbermen in the Blacks seemed to hold special significance for North Carolinians. In a state

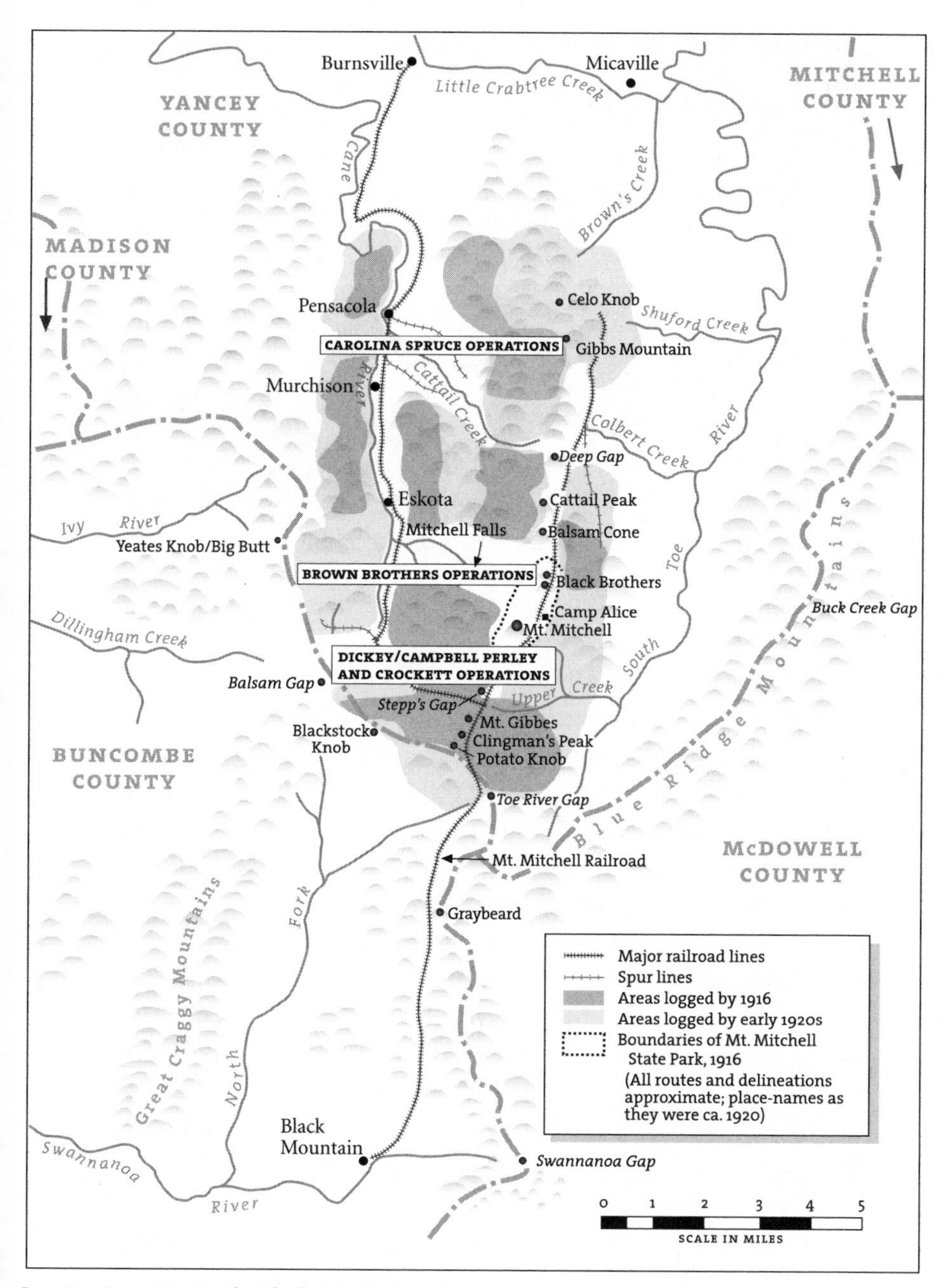

Logging Operations in the Black Mountains

The Mount Mitchell Railroad, engineering marvel, snaking across rough terrain near the Yancey-Buncombe County line, 1915. William A. Barnhill Collection, Pack Memorial Public Library, Asheville.

that had long struggled against a reputation for backwardness, the Dickey and Campbell line, quickly dubbed "the highest steam railroad east of the Rockies," became an important symbol of progress and prosperity. From Asheville to Charlotte newspapers heralded the "Mount Mitchell Conquest," calling the railroad "one of the most wonderful achievements in modern engineering."[34]

It was not just the timber trade that excited boosters and entrepreneurs. The railroad also had enormous potential for tourism. From the first, Dickey and Campbell had considered running passenger cars to the East's highest peak, a plan later taken up by Perley and Crockett. The advent of passenger service figured to be "a wonderful drawing card for [the town of] Black Mountain." The new line, the *Asheville Citizen* suggested, offered "more scenic beauty" than any other railroad in North Carolina. "It will be to Asheville and this section what the railway up Pike's Peak is to Denver and Colorado."[35]

Yet even as they hyped the railroad, a few publicists wondered about the long-term effects of logging in the Blacks. They fretted about increased erosion and feared that once the mountains had been cleared, the land might be incapable of producing another extensive forest of "spruce and balsam." They also worried about aesthetics. "From the standpoint of commercialism," one journalist noted, the logging operations afforded "a scene of intense interest" that "thrills the being with a sense of development and progress, turning nature's resources into money—the man-created standard of values." But with more than a hint of sarcasm the same writer added, "Think of it! Within another twelve months the magnificent forests of spruce and balsam on the slopes of Mt. Mitchell, the highest point east of the Rockies . . . and the pride of the entire eastern section of the United States, will fall before the axe of the lumberman to be turned into money."[36]

Such ambivalence was typical of the political climate in North Carolina in 1912. Since the turn of the century the state had increasingly come under the control of Democratic leaders who loosely defined themselves as Progressives. The Progressive movement was a complex phenomenon that scholars interested in turn-of-the-century America are still working to understand. At the national level it attracted a wide array of politicians and activists who, for various reasons, believed that government should seek remedies for the problems plaguing modern America. Though they had no quarrel with capitalism and free enterprise, many Progressives favored regulating business and reforming society in order to achieve orderly economic growth. They also believed that trained professionals should oversee such reforms. Indeed, the conservation policies of Theodore Roosevelt and Pinchot were prime examples of Progressive thinking designed to put experts (i.e., foresters) in charge of the nation's woodlands. Foresters could base their management practices on scientific study, not greed, and thereby ensure sufficient timber supplies for the future. As historian Samuel P. Hays has made clear, Progressive conservationists worshiped at the altar of science and preached "The Gospel of Efficiency."[37]

Such notions also appealed to southerners. Below the Mason-Dixon line Progressive reforms became entangled with the ongoing campaign for a New South, a movement that, among other things, sought to increase manufacturing by using the region's abundant natural resources. Most North Carolina politicians at least paid lip service to such ideals, but many of the state's prominent leaders were also fiscal conservatives, loath to raise taxes for forestry and other programs that served public interest. And, like most southern politicians, they believed that the power to regulate and reform rested solely with white men. Occasionally (as when they set out to eliminate discriminatory railroad rates) North Carolina Progressives passed legislation that placed tight restrictions on corporations operating within the state. But more commonly they preferred to encourage industrial growth, issue some mild warnings about its excesses, and then take moderate steps to see that it occurred in a neat and systematic fashion. Politicians who adhered to this philosophy, which modern historians call business progressivism, saw nothing incongruent in hiring a conservationist such as Holmes and then, at almost the same moment, throwing open North Carolina's most famous mountain to the railroads and the lumber companies.[38] Development was fine, even on Mount Mitchell, as long as it did not bring excessive disorder and as long as the state maintained a modicum of control.

At least it seemed that way in the summer of 1913 when recently elected North Carolina governor Locke Craig, a longtime resident of Asheville, came to speak at the official opening of the Dickey and Campbell/Mount Mitchell Railroad. By then the company had been cutting spruce, fir, and northern hardwoods for a year and had all but denuded the eastern face of Clingman's Peak. Craig was the consummate business Progressive. A Democrat who paid his dues in the state legislature, he rose to power in the 1890s during his party's "white supremacy campaign," a highly organized effort to rid North Carolina of "Republican-Negro rule." In typical Progressive fashion the governor had used his inaugural address both to praise the "spirit of progress" and to rail against the "rapacity of unlawful monopoly." On that July day in Black Mountain, standing before an audience made up largely of businessmen and local boosters, Craig began by lauding the railroad as "a great work" that required much "skill, patience, money, and time." Then, noting that Dickey and Campbell had asked him "to speak just what is in my heart," he continued, "When I look upon that mountain and see the havoc wrought by the woodsman's axe, I feel like a citizen of Rome who gazed upon his city burning and in ruins. I cannot but regret that it is not as it was."

If any members of the audience expected the governor to call off the lumbermen, however, those listeners left disappointed. Instead Craig placed his faith in the popular doctrine of utilitarian forestry. Though unsightly, the lumbering operations were only temporary, he said. Once the loggers left, he explained, "nature will re-create and re-forest this mountain and again man will cut trees from these slopes to build into his houses." Revealing a plan that both he and Holmes had apparently discussed for some time, the governor predicted that the railroad would one day transport visitors through a scenic park near the East's highest mountain. Finally, referring directly to Dickey and Campbell, Craig declared, "I am glad these folks have come among us," for what appeared to be a "mighty waste" was really the "beginning of magnificent things." "Nature," the governor concluded, "cannot be destroyed."[39]

True enough. But political rhetoric and popular promises notwithstanding, the logging operations represented far more than a mere momentary lapse in an otherwise sanguine relationship between people and land. Ecologically speaking, the lumber companies brought disturbance more sudden and widespread than anything in nature since the last days of the Pleistocene. Stated more practically, in the sort of language a Progressive politician like Craig might have understood, things were about to get messy on Mount Mitchell.

WINTER SOLSTICE

BOWLENS CREEK

Three days before Christmas, Bowlens Creek is a picture-perfect Appalachian community. Smoke from woodstoves and fireplaces hangs motionless in clear December air. Goats frolic in a winter pasture along state road 1109, the two-lane, tar-and-gravel byway that winds across nearby ridges to Pensacola and Burnsville. As I walk past houses decked with lights and seasonal greenery, a friendly dog wanders from a local residence, tail wagging in anticipation of a treat from my day pack. After I oblige, he follows for a short distance as I search out a secluded trailhead. Eventually I find it tucked away in the woods just off the paved road, near a large house plastered with signs warning hikers to steer clear of private property. A Forest Service marker notes that this is the Crest Trail, the rugged path that originates on Mount Mitchell and terminates here along the rocky stream from which this community takes its name.

I take a last look at the houses and begin the steep climb into the mountains. Two hours later, three and a half miles away, I stop for lunch on Grassy Knob Ridge, overlooking the Cane River Valley, just below Celo Knob, northernmost peak in the Blacks. I am surrounded by a thick stand of beech and mountain ash, but compared with the pastoral splendor of Bowlens Creek, this seems desolate terrain. Deer season ended a week ago, and the mountains are devoid of hunters. Aside from a light wind stirring dead leaves at my feet, I hear nothing. For the first time all day I notice the cold. Even at high noon, rime ice still hangs in the treetops. Small crystals of the frozen fog swirl past my face on every errant breeze. I am suddenly uneasy, keenly aware of the current contrast between valley and mountain, farm and forest. I eat quickly and start my descent, eager to be out of the Blacks before the second-shortest day of the year comes to an end.

Walking back, though, I force myself to remember that the forests on Celo Knob are to a large extent a creation of people who once lived at the base of the mountains. In the days when the railroads ran up the Cane from Burnsville, local workers helped cut red spruce from the forests on Grassy Knob Ridge. For nearly a decade

Bowlens Creek, Pensacola, and other communities that now seem so pleasantly rustic were, in fact, bustling boomtowns where residents enjoyed conveniences usually available only to urban folk. Thanks to the trees on Celo Knob, local people got to sample the good life in modern America.

On the trip home, shortly after 7:00 P.M. I stop to watch a full moon rise over the Catawba Valley. Because it coincides with the winter solstice, it is spectacularly luminous; according to astronomers it is the brightest moon in more than 130 years. As it climbs higher in the eastern sky, I see the familiar profile of the Blacks emerge on the far horizon, more than fifty miles away. Alone in the half-light, remembering my trek up from Bowlens Creek, I am struck by another irony. Due to Mount Mitchell's status as a North Carolina icon, the state kept a close eye on logging operations there. Eventually, when it appeared that a sacred landmark might be in danger, officials moved to protect the summit and surrounding peaks. Thus some of North Carolina's first efforts at conservation in the region resulted directly from the depredations of avaricious lumbermen. A building north wind sends me shivering back to the truck. I have a last look at the Catawba Valley and drive away, into the wind, again mindful of the tangled relationship between people and nature and the competing interests that helped shape the moonlit landscape in the distance.

In the first decades of the twentieth century (as in times past), political and economic developments in faraway places often determined what happened on the slopes of the Blacks. Escalating tensions in Europe and America's eventual entry into World War I created an unprecedented need for "industrial plants, shipyards, warehouses, ships, office buildings, and military training camps." Some of the hardwoods cut from the high peaks no doubt found their way into such structures. But during wartime the most prized wood came from red spruce. At Pensacola and at Black Mountain, sawmills sliced the largest evergreens into thirty-foot lengths of so-called clear spruce used for ship decking and for reconnaissance aircraft deployed in Europe. Smaller trees, those less than eight inches in diameter, went to Champion Fibre's Canton mill. There the fragrant wood was chipped, boiled in soda or sulfurous acid, and pressed into pulp to meet wartime demands for paper.[40]

As trees fell on the mountains, daily life in the valleys changed dramatically. In the town of Black Mountain the same "electric light plant" that drove the sawmill and illuminated the lumberyard furnished power to local residents. Pensacola got its own post office and movie theater. In addition Carolina Spruce provided "a full department store with ladies' and men's wear, a grocery store, hardware, seed and feed stores, a barber shop, and an up to date drug store complete with a modern soda fountain." The corporation also brought electricity and telephone service to Murchison, Eskota, and other communities along the Cane.[41]

Though the lumber companies used some immigrant labor (especially during the war years), many of those who spent their paychecks at company stores lived on nearby farms. In October 1913, when Perley and Crockett bought out Dickey and Campbell, the latter firm employed more than 200 men, primarily from Yancey County and the surrounding area. Housed in temporary camps (one near Clingman's Peak, the other on the east face of Mount Mitchell at a site now called Commissary Ridge), the men worked in shifts cutting trees and loading the trains. To get the logs off the mountains and into the mills, Perley and Crockett relied on six locomotives, seventy railroad cars, and thirty teams of horses. The company also owned at least two "skidders," gangly steam-powered log-moving machines equipped with more than 1,000 feet of solid steel cable. Lumbermen hauled the lines into the woods and attached them to felled trees. Then the skidders' noisy engines cranked to life, reeling in the lines and dragging the attached logs through the woods toward waiting railroad cars.[42]

Some of those trees were enormous. Newspapers reported that most of the spruces, firs, and hardwoods cut between 1912 and 1913 were "straight as an arrow and from one hundred to one hundred and twenty-five feet in height." The sawmill at Black Mountain chewed them up like matchsticks. Operating at peak capacity, the mill could consume fifty carloads of logs a day and turn out 110,000 board feet of freshly cut lumber. Yet even at that astonishing rate, the *Asheville Citizen* predicted, enough timber remained "in the neighborhood of Mt. Mitchell to require over twelve years cutting."[43]

The best and most accessible trees disappeared long before that. By 1914 Perley and Crockett had cleared 1,000 acres of spruce and fir from the eastern flanks of Clingman's Peak, Mount Gibbes, and Mount Mitchell. Following completion of its railroad across Stepp's Gap, the company acquired rights to the Murchison Boundary from Brown Brothers and began cutting a wide swath across the western face of the East's highest peak. Loggers spared only the most inaccessible tracts, namely some forests north of Maple Camp Ridge (beyond the railroad) and a narrow strip of spruce and fir at the very crest of Mount Mitchell and several neighboring peaks. Otherwise only a few isolated pockets of trees, primarily in deep chasms cut by the headwaters of the South Toe, escaped the skidders and logging trains. Meanwhile the Carolina Spruce Company hacked away at its 5,200-acre tract on the northern end of the Blacks. From 1912 on the company moved south from Grassy Knob Ridge and Celo Knob until its operations eventually met those of Perley and Crockett. In all likelihood Carolina Spruce also culled hardwoods from the forests on the other side of the Cane.

As trees fell along the railroads, the remaining high-elevation forests immediately became more susceptible to the influences of mountain weather, especially wind. Before the lumbermen arrived, wind had occasionally leveled small patches of spruce and fir, creating sunny clearings populated by yellow birch and mountain ash. But now the icy blasts of winter rolled unimpeded across bare ground and hit the high peaks with hurricane force. By 1916 "blow downs" and "bare patches up to five acres in extent" cropped up in the spruce-fir stands on the South Toe side, even north of Maple Camp Ridge where little logging had taken place. Indeed, the noticeable lack of "merchantable timber" in the disheveled, wind-battered forest atop Mount Mitchell might have been one reason why Perley and Crockett bypassed the summit as they moved across Stepp's Gap and into the Cane River Valley. Either they or Champion Fibre might easily return later to collect the smaller trees for pulpwood.[44]

Extensive clearing also left thin mountain soils exposed to the pounding rains of spring and summer. Erosion was especially severe in the southern reaches of the Blacks where Perley and Crockett deployed their skidders. As one observer noted in 1915, the "skidways on which the logs are dragged to the railroad . . . form great gutters and down these gutters the waters from the rains rush in torrents and cut deep into the soil through to the red clay and the rock." With the forest cover gone, stream levels fluctuated wildly, even in less-severe weather. In 1916 a surveyor who toured some of the logged areas on the eastern side of the range discovered that Upper, Lower, and Rock Creeks—South Toe tributaries originating on the deforested slopes of Clingman's Peak, Mount Gibbes, and Mount Mitchell—shrank "to nearly half their normal size during a drought." When the rains returned, the streams were instantly "swollen and filled with sediment." Yet just a few miles away, on the north side of Maple Camp Ridge where the railroad ended and logging ceased, Colbert's Creek and several smaller tributaries "var[ied] little from their normal size" and were "always clear," even during heavy downpours. Nowhere, the surveyor concluded, could one find a better example "of the opposite effects of cut over and virgin forest on the run off."[45]

Mount Mitchell, showing an area stripped by loggers. Winds blew unimpeded across such bare patches, battering the uncut spruce-fir forests that remained near the summit. Such clear-cutting also led to increased erosion and siltation in Black Mountain streams. North Carolina Collection, Pack Memorial Public Library, Asheville.

Though they now struck with unprecedented fury, wind and water had influenced the high-elevation forests for centuries. But the loggers also unleashed another, less familiar and more powerful force on the high peaks: wildfire. Hunters and tourists had occasionally set the hardwood forests ablaze during their travels, and burning to clear farmland had long been standard practice in the valleys. But along the highest ridges of the Blacks, some of the older spruce-fir forests had been untouched by fire for perhaps a thousand years. That changed when the first trees fell on Clingman's Peak. As workers pruned spruce logs for the sawmills, they cut away the branches and treetops, stacking them in great heaps six or eight feet high along the railroads and skidways. Baked by the sun, these huge mounds of slash (as the foresters called them) soon turned into blanched brushpiles of highly resinous, powder-dry timber. When the Shay and Climax locomotives chugged up and down the slopes spewing sparks and hot ashes from their wood- and sawdust-fired engines, the inevitable happened. The moment the weather turned dry—be it late spring, early fall, or the dog days of summer—the slash ignited and the Black Mountains began to burn.[46]

A few small fires flared on the eastern face of the range in 1913. The first serious conflagration, however, occurred the following year in late June. Kindled by a spark from a locomotive, the blaze apparently began early one evening along the railroad between Clingman's Peak and Mount Mitchell. By the time workers discovered the "wall of flame," it was roaring up the mountainside toward the Perley and Crockett camp at Commissary Ridge. There it destroyed "a boarding house and several shacks," forcing startled workers to don water-soaked garments and run for their lives "through the crackling sheet of fire." The flames then swept into an uncut stand of spruce and fir and reduced the trees to "gaunt and blackened stumps" in a matter of minutes. Though the total acreage burned is difficult to gauge, the inferno could be seen "for many miles," and Perley and Crockett estimated that the fire destroyed some "$10,000 dollars worth of cut and standing timber" on the slopes of the East's tallest mountain. A month later fire again ravaged the Perley and Crockett operations, and by autumn "an almost unbroken area of charred stumps and logs, blackened trees, and bare rocks" stretched "across Clingman's Peak, practically to the summit of Mount Mitchell."[47]

At the Division of Forestry of the Geological Survey, John Simcox Holmes kept close track of the Black Mountain fires, periodically making short visits from his Chapel Hill office to inspect the lumber camps and burn sites. As a disciple of Pinchot and utilitarian forestry, Holmes had little patience for the cut-out-and-get-out practices that encouraged wildfire. For five years since his initial appointment as state forester in 1909 he had lobbied the legislature to institute a state-run system of fire control and to provide funds for reforestation. Indeed, early in 1913, at the third annual meeting of the North Carolina Forestry Association (a group made up of primarily of businessmen interested in woodland management) Holmes had helped pass a resolution supporting the proposed state park on Mount Mitchell. Elsewhere in the United States the creation of state parks to save prominent landmarks had been gaining momentum. New York had moved to protect Niagara Falls by creating a state park there in 1885. Over the next twenty years Connecticut, Minnesota, Wisconsin, Massachusetts, Ohio, Idaho, and Illinois all established their first state parks. Holmes wanted no less for North Carolina, and he argued that Mount Mitchell, a landmark known to every citizen in the state, was the logical place to begin. True to his training in utilitarian forestry, Holmes also insisted that the proposed park should not be a wilderness preserve but, rather, a "demonstration forest" where visitors could observe the eco-

Black Mountain forests devastated by logging and fire, as seen from a passenger car on the Mount Mitchell Railroad, ca. 1915. Noting that tourists would not spend money to see such unsightly vistas, the Asheville Board of Trade lobbied the North Carolina legislature to create the state's first park on the summit of Mount Mitchell. William A. Barnhill Collection, Pack Memorial Public Library, Asheville.

nomic benefits of selective cutting, reforestation, and other conservation techniques.[48]

Though neither Holmes nor Craig ever said so, the plans for Mount Mitchell State Park probably stemmed, at least in part, from a heated argument over the uses of America's wild lands taking place at that moment in the U.S. Congress. That debate (the outlines of which are familiar to environmental historians but still unknown to most Americans) centered on a place thousands of miles from Mount Mitchell, a scenic valley in Yosemite National Park known by the odd-sounding Indian name of Hetch Hetchy. Since the 1890s residents of San Francisco had been pressing for a dam on the Tuolumne River, a wild, roiling stream that flowed through the heart of Hetch Hetchy. An energetic and vocal coalition of preservationists led by John Muir agitated against the dam, while the city and its allies, including Gifford Pinchot and other utilitarian conservationists, maintained that the tangible economic benefits of the dam to San Francisco's residents and businesses outweighed Hetch Hetchy's scenic value. The argument reached a crescendo in 1913 as Congress held hearings and debated the issue—the same year that Holmes began to press for a state park on Mount Mitchell. With at least tacit support from President Woodrow Wilson the proposal to dam Hetch Hetchy passed both House and Senate by substantial margins. In both chambers some of the dam's most ardent backers were southern Democrats, much like those who controlled the North Carolina legislature.[49]

Given the politics of the national debate over conservation, it comes as no surprise that Holmes and Craig tried to assure North Carolina lawmakers that the Mount Mitchell park would have a distinctly utilitarian bent. But it made no difference in 1913. Legislators simply turned down the request. The following year, as Hetch Hetchy faded from the headlines and fire again threatened the East's highest peak, Holmes renewed his efforts. In a press release prepared in October 1914, the state forester reminded North Carolinians that "the labors and tragic death of the beloved Elisha Mitchell" had made the name of the East's highest mountain "a household word" in the state. He then went on to compare Mount Mitchell to Niagara Falls, another natural wonder so recently threatened by unregulated tourism and development of water power. New Yorkers had rallied in defense of the falls and supported a state park. Surely North Carolinians could do the same for Mount Mitchell. Due to wartime inflation the lumber companies had paid a premium price for Black Mountain land. They would not, Holmes warned, be inclined to spend money on fire prevention. Nor could lumbermen be expected to abandon the shortsighted and highly lucrative practice of taking every tree from the mountain. Barring some action by the federal government (which Holmes thought unlikely due to the cost of purchasing land) the only solution was state control.[50]

As word of Mount Mitchell's plight spread, Holmes and Craig found allies across the Southeast. But park promoters needed something more than the loud voices of wilderness advocates. Hetch Hetchy clearly illustrated that when it came to protecting scenic landscapes, monetary concerns usually carried the day. In order to save Mount Mitchell, state legislators had to be convinced that a park could generate enough revenue to make up for the timber and other resources that might otherwise be turned into cash.

Civic leaders in western North Carolina had already laid the groundwork for such an argument. Beginning in the late 1890s businessmen in Asheville and other mountain communities had been lobbying the federal government for an Appalachian National Park. Indeed, as one of Asheville's most prominent citizens, Locke Craig had been active in the cause, insisting that a park in western North Carolina would not only preserve a scenic landscape but also help heal the lingering divisions between North and South. Craig and other supporters of the national park made little headway before the turn of the century. But by 1910 America's western parks—Yellowstone, Glacier, Yosemite, and Grand Canyon—had become major tourist attractions drawing thousands of well-to-do visitors who spent money on travel, lodging, food, and other amenities.[51] Perhaps, Craig reasoned, a similar argument might be made for Mount Mitchell. Though the East's highest peak did not have the cachet of Yellowstone or Glacier, it did have its own railroad and had always been popular with local sightseers. With a state park on its summit, the mountain might well attract eastern and southern tourists who lacked the time or resources to visit the West.

To help make that case to the legislature, Craig and Holmes called on the Asheville Board of Trade. A forerunner of the city's chamber of commerce, the board consisted primarily of merchants and businessmen, many of whom were the governor's political allies. Like Craig the board usually welcomed industry—including railroads and lumbering—to western North Carolina. When it came to tourism, however, the board insisted that "preservation of the forests on . . . Mt. Mitchell and nearby peaks" was a matter of vital concern. Most visitors, especially those from urban areas who came to western North Carolina to experience the outdoors, expected lush, tree-covered mountains. If they did not find them on Mount Mitchell, the board argued, sightseers would take their dollars elsewhere. Unless the lumber companies could be reined in, even the highest railroad east of the Rockies, so recently hailed as a technological marvel and a boon to tourism, would be worthless as a scenic attraction.[52]

In the summer of 1914, with the board trumpeting the economic benefits of a park, Governor Craig began negotiations with Perley and Crockett, hoping to convince them to curtail logging near the summit of Mount Mitchell until the legislature could act. Exactly what he said to the lumbermen and their representatives remains unclear. Craig promised that the state would pay a fair price for the land, but he also seems to have threatened Perley and Crockett with condemnation proceedings if they refused to cooperate. At any rate, the firm agreed to leave a tiny stand of wind-battered spruce-fir on the summit, though loggers continued cutting larger, more valuable timber on the mountain's western slope. A new park bill went to the legislature early in 1915. It called for $20,000 to buy land and establish a state park on Mount Mitchell. A five-man commission would oversee the purchases and report to the governor.[53]

Despite the tough negotiations, Craig continued to talk about conservation without defaming Perley and Crockett or any other timber company. Addressing the North Carolina Forestry Association in January 1915, the governor recalled his appearance at the opening of the Mount Mitchell Railroad almost two years earlier. "When I looked all around," he said, "where I had been bear hunting as a boy . . . I felt like a man that stood amidst the ruins of his home after the conflagration had destroyed it." Neglecting to mention that he had, in fact, welcomed Dickey and Campbell into the mountains, Craig instead challenged the state to repair the damage to Mount Mitchell. "The lumbermen—I am not criticizing them, but us—the lumbermen are destroying North Carolina," he intoned. "We cannot expect them to sacrifice their business for the public good. They have bought that timber. They are entitled to every stick of it." Concluding with a plea for the park, Craig explained that only "the people of North Carolina" could save the East's highest mountain, not from the lumber companies themselves (no business progressive dared suggest that), but from "the fire that follows the lumberman."[54]

Meanwhile Holmes kept the public relations machinery humming. In February 1915 *American Forestry*, the prestigious and popular journal of the American Forestry Association, published a lengthy article titled "Destroying Mount Mitchell." It blamed the deforestation, erosion, and fires solely on Perley and Crockett, Carolina Spruce, and the other northern corporations, noting that the "offices of the lumberman are generally located in another State and he has absolutely no interest in the mountain regions except in the profit which he will obtain from the timber and he naturally desires all the profit he can get." As for the proposed park, it represented a step in the right direction, but the limitations of the $20,000 appropriation would "not permit the accomplishment of as much as is needed at this time."[55]

The article, apparently written in consultation with Holmes and at least one member of the Board of Trade, appeared just as the legislature began deliberations on the park bill. By hinting that the state was reluctant to protect a precious landscape from the ravages of Yankee capitalism, the article struck directly at the legislators' southern pride. It also suggested that despite the progressive rhetoric of its politicians, North Carolina still had not shaken its backward ways. Such bad publicity was hard to stomach, especially for lawmakers from Asheville, Charlotte, and Greensboro, who wished to portray their cities as part of a progressive, vibrant New South.

Championed by state senator Zebulon Weaver of Buncombe County and by a full-time lobbyist sent by the Board of Trade, the park bill passed both houses of the legislature "by a large majority" on March 5, 1915. Recounting the incident a short time later, Craig did not mention the influence of the business lobby or the stinging commentary in *American Forestry*. Ever the politician, the governor insisted that the legislature's chief motivation was protection of what he and many others regarded as a sacred state landmark. "One of the considerations moving the General Assembly to the establishment of the park," he noted, "was to relieve this famous mountain from private control," so that "the people of North Carolina and tourists from all parts of the world might have the privilege of free access."[56]

A year later, after still more negotiations with the lumber companies, the state acquired an additional tract of land running across the very crest of the Blacks between Stepp's Gap and the peak today called Big Tom. The parcel included that last uncut stand of spruce and fir surrounding the summit of the East's highest peak. Four years after the logging began, North Carolina had its first state park, a fledgling tourist attraction, and a laboratory in which to practice utilitarian forestry. The same newspapers that had celebrated the arrival of the railroads immediately hailed the park purchase as a great victory for progress and conservation.[57]

The summer of 1916 brought more good news. In August, after nearly five years of political wrangling, the U.S. Congress passed the National Park Service Act. For nearly a half-century national parks had been managed haphazardly by various bureaus in the Department of the Interior. Now responsibility for park administration rested solely with the National Park Service, an agency designed to manage the nation's most popular public lands in "an efficient and business-like way." The new bureau also figured to provide guidance and an organizational model for state parks, including Mount Mitchell.

No doubt Holmes, Craig, and the Board of Trade liked what they read in the National Park Service Act's statement of purpose. The new legislation called on the agency not only "to conserve the scenery and the natural and historic objects and the wildlife" of the nation's parks but also "to provide for the enjoyment of the same in such manner and by such means as will leave them unimpaired for the enjoyment of future generations." It was in many ways a perfect expression of the ambiguities inherent in Progressive conservation, a utilitarian mandate that sought to protect America's natural wonders but also allowed for development, tourism, and presumably, profit. In years to come certain portions of the statement (especially its emphasis on making parks accessible while leaving their natural features "unimpaired") would be a source of endless debate about the true mission of the Park Service.[58] For the moment, however, the act seemed to settle the issue about what to do with America's national and, by implication, its state parks. North Carolina politicians could take heart that what they had done on Mount Mitchell was clearly in keeping with national trends.

But for Holmes and others at the Geological Survey's Division of Forestry, the excitement over the new park must have been short lived. Even as the state gained control of the East's highest summit, a new threat loomed on the horizon. Like the lumbermen, it came from the north and tore through the woods along the Atlantic seaboard with terrifying swiftness. But the new menace did not respond to the laws of the marketplace or to the carefully considered pleas of governors, foresters, and businessmen. It obeyed only the whims of nature, and before it ran its course, it would radically alter the forests on the lower slopes of the Black Mountains.

JUNE

HAULBACK RIDGE

Twisted branches of rhododendron claw at my ankles as I scramble through a sheltered ravine on the southeastern face of the Blacks. The sun has been up for three hours, but the vegetation still hangs heavy with last night's rain. Dark topsoil cakes on my boots and trouser cuffs. Behind and below I hear the main prong of the river roaring with spring runoff as it squeezes through a tight, boulder-lined chasm between the Blacks and the Blue Ridge. Some 2,000 feet above, shrouded in morning clouds, stands the pinnacle of Haulback Mountain (also spelled Hallback), an inconspicuous dome between Mount Mitchell and Mount Gibbes.

Several hundred feet off the valley floor I emerge onto an exposed hillock, and the forest changes abruptly. Here the woods are bathed in sunlight, and the soil is light and coarse, flecked with mica and other minerals. I hike north through an open stand of small red oaks and pignut hickories. Like much of the South Toe drainage, this ridge has been logged and burned repeatedly in recent decades. Nonetheless, the young hardwoods shade the forest floor, limiting understory growth. I step easily around red maples and mountain laurel as I make my way to the high point of the ridge, a prominent protrusion roughly 3,800 feet above sea level. Here I begin a search for another tree, one still fairly common in forests such as these. If place-names are any indication, this should be an ideal spot. According to my topographic map, I am standing squarely on Chestnut Knob.

For once, Black Mountain nomenclature runs true to form. In less than ten minutes I locate a shoot of American chestnut. It is waist high, with light gray bark still smooth with youth and spindly branches lush with newly sprouted, yellow-green leaves. Several other young chestnuts crop up nearby amid the shrubby dogwoods and laurels, overshadowed by taller oaks, hickories, and maples. To the untrained eye the chestnuts look healthy. But they will survive for only a short time. A few may reach a height of twenty feet and a diameter of four inches. Most, however, will be dead long before then and will be replaced by other saplings sprouting from their roots.

A century ago, perhaps about the time that Chestnut Knob got its name, these forests looked much different. On dry, exposed ridges such as this, high above the dark coves along mountain streams, roughly one of every four trees was an American chestnut. The trees thrived on disturbance and often sprouted profusely in the wake of a fire or windstorm. They grew tall and fast, standing 60 to 100 feet high and measuring 2 to 4 feet in diameter, forming a large segment of the forest canopy. On a late spring day like this, the chestnuts might well have been in full bloom, covered with myriad white flowers.

As devotees of natural history know, the reduction of American chestnut from a dominant canopy tree into a stunted understory shrub resulted from a deadly disease—popularly called a blight—accidentally imported into the United States from Asia. But this is not simply a tale of a native plant destroyed by an exotic pathogen. Across eastern America the disappearance of the giant hardwoods had enormous implications for people, including those who worked farms and cut timber in the shadow of the East's highest mountains.

Sometime in the summer of 1913 J. Van Lindley, owner of a tree farm and nursery near the Piedmont town of Pomona (not far from Greensboro in Guilford County) noticed curious yellow-orange blisters on some imported Japanese chestnut trees. Immediately concerned, the nurseryman conducted a hasty inspection of a few wild American chestnuts growing in the nearby woods. A quick look confirmed his worst fears. Some of those trees, too, exhibited the telltale spots and cankers. Panicked, Van Lindley contacted John Simcox Holmes at the Geological Survey. When the state forester heard the news, he promptly instructed the nurseryman to cut down, dig up, and incinerate every infected chestnut, Japanese and American alike, and clean and burn the ground on which the trees grew.

Van Lindley complied, and a few months later, after state inspectors found no spots on the remaining trees, Holmes breathed a cautious sigh of relief. Perhaps, he thought, North Carolina had survived its first encounter with chestnut blight.[59]

By then Holmes knew how devastating the contagion could be. Since the mid-1870s nurserymen from New England to California had imported thousands of chestnuts from Japan and China, selling them to American homeowners who craved fast-growing shade trees. During the last third of the nineteenth century Oriental chestnuts had been planted in many public gardens across the Northeast. Then, in 1904, Herman W. Merkel, a forester at the New York Zoological Park (better known as the Bronx Zoo) found odd-looking "cankers and necrotic depressions" on American chestnuts there. He immediately began pruning dead and diseased branches, but to no avail. Throughout the park, trees died in droves. Merkel suspected that imported trees might be responsible for the infection, an idea that gained credence in 1906 when William A. Murrill of the New York Botanical Garden found similar lesions on Japanese chestnuts. By 1908, five years before Van Lindley discovered spots on chestnuts in Guilford County, dying trees had been observed as far north as Massachusetts and as far south as Maryland.[60]

Holmes believed that fast action at the Van Lindley nursery might keep the dreaded pestilence from his home state. The odds, however, were against him. The blight was not the first threat to the American chestnut. Agricultural clearing and extensive logging in Piedmont uplands had already severely reduced the chestnut's original range in North Carolina. During the nineteenth century a root fungus infected many chestnuts growing at lower elevations, so that in the Piedmont at least, only a tiny fraction of the presettlement forest remained intact. But chestnut blight, also known as chestnut bark disease, was unlike anything Holmes or any other southern forester had ever seen. Infection stems from a fungus now known to scientists as *Cryphonectia parasitica*. It enters a living tree through breaks in the bark or fissures at the base of dead limbs and twigs. Once established there, it sends out myriad threadlike tendrils under the bark, disrupting the flow of water and nutrients. Eventually the blight effectively girdles the tree, killing all leaves and limbs above the point of infection.

As the parasitic fungus matures, it erupts outward through the bark, creating the telltale cankers and scattering hundreds of yellow-orange, pinhead-sized pustules across the tree's trunk and branches. Those pustules dispense billions of large dry spores (sometimes called winter spores because they often appear during cold weather) that travel on the wind and, through sexual reproduction, create new colonies on trees as far as fifteen miles away. But like most fungi, the deadly parasite also reproduces asexually. During wet weather the pustules disperse smaller "summer spores" that flow "in sticky masses from the bark cavities, like toothpaste from a tube, following warm-weather rains." Summer spores can be dispersed by precipitation or carried through the woods by birds, small mammals, and insects. (Indeed, some estimates suggest that a single small woodpecker can carry as many as 757,000 asexual spores, any of which might infect a healthy American chestnut.) In addition summer spores travel in shipments of chestnut lumber and adhere to the shoes and clothing of anyone who walks through an infected stand of trees.[61]

The rate at which the blight moves depends mostly on environmental factors, including rainfall, wind direction, migration patterns of birds and other animals, and most important, the availability of suitable host trees. In Japan and China, where Oriental chestnuts had evolved with the fungus, host trees gradually built up resistance to the blight, limiting its destruction there. It was also slow to infect Europe, where natural barriers such as the Alps kept it from spreading throughout the Continent. But in the chestnut-rich forests of eastern America, where host trees abounded and nothing in nature checked its progress, the fungus moved like wildfire. In southern upland forests the long growing season and warm wet climate provided near-ideal conditions for its propagation and spread.[62]

Holmes, of course, did not know all that in 1913. Until then the most determined effort to arrest the blight had occurred in Pennsylvania. Using state and federal money, officials there hatched an elaborate plan to halt the contagion by cutting and burning all chestnuts from a milewide strip across the state, essentially the same technique Holmes employed at Van Lindley's nursery. The Pennsylvania plan went awry, however, when foresters south of the isolation strip discovered that the blight had already leaped the proposed barrier. Within three years prevailing winds and the southerly migrations of birds helped push the fungus into the northern reaches of Virginia. Between 1913 and 1915, the Forest Service estimated, the contagion swept across that state at the rate of twenty-four miles per year, moving steadily south toward western North Carolina.[63]

Holmes's efforts in the Piedmont proved equally futile. In 1915 the blight reappeared in Guilford County not far from the Van Lindley place. But because American chestnut grew only sporadically in Piedmont forests, the disease seemed unlikely to spread west at a rapid rate. The real threat still came from the north. With trees dying across Virginia and no effective deterrent available, Holmes, like other southern foresters, began to make contingency plans. Starting in 1912, he used his share of federal funds and some meager state monies to keep two men in the mountains scouting for signs of early infection. While in the field, the agents also consulted with various lumber companies about current timber prices, railroad shipping rates, and the possibility of using dead or dying trees for lumber, pulp, and tannin. One of the men spent considerable time at Champion Fibre in Canton, the region's largest processor of chestnut. He discovered that Champion received some 300 cords of the wood per day. Much of it arrived by rail, having been harvested by farmers and lumbermen in forests up to 150 miles away, including the Black Mountains. Given such incredible demand, Holmes hoped that even if the blight could not be stopped, much chestnut might still be salvaged for paper and tannin. North Carolina had more than 3 million acres of chestnut forest, and the state ranked third nationally (just slightly behind West Virginia and Pennsylvania) in production of chestnut lumber. It would be a shame, the state forester often noted, to squander "one of North Carolina's greatest forest assets."[64]

He was not the only one worried about the future of the timber business. The best available estimates suggested that the blight would eventually take out 27 percent of the region's standing timber. According to the U.S. Department of Agriculture one of the most important "silvicultural problems" confronting the southern states in 1915 was determining "what other valuable species might be used to reforest the land once occupied by the chestnuts." In the utilitarian language of progressive forestry, "desirable species" included various oaks, hickory, white ash, yellow poplar, basswood, and white pine, all of which were well suited for lumber. "Less desirable" as potential replacements were scarlet oak, red maple, beech, and black gum. In a worst-case scenario, government scientists theorized, the lower ridges of the Blacks and other southern mountain ranges might be repopulated with "undesirable species" such as dogwood, gray birch, scrub oak, sassafras, witch hazel, and downy serviceberry.[65]

While various experts speculated as to how nature might restock the low-elevation forests, people living along the Cane and South Toe grappled with other concerns. For generations American chestnut had been vital to the region's livestock trade. In warm weather, when cattle browsed freely across mountain woodlots, the giant trees provided shelter from summer sun and sudden rainstorms. After the first frost, when the spiny chestnut burrs fell to the ground, farmers turned out their hogs to fatten on the oil-rich nuts. Chestnuts were also a source of ready cash for mountain people. Collected in autumn and delivered to local merchants, the tasty morsels were especially popular in eastern cities, where street vendors roasted and sold chestnuts as a winter delicacy. (Indeed, Holmes often worried that the vast numbers of "chestnut gatherers" abroad in the dry autumn woods might significantly increase the risk of forest fires in Yancey and other western counties.) Most local folk probably did not understand what caused the blight—especially since a prominent state newspaper erroneously reported that "chestnut beetles" were killing trees in Virginia—but they knew that it could put an end to an important seasonal business.[66]

The disease also threatened Black Mountain wildlife. Depending on elevation, American chestnuts flowered in June or July, usually long after the last frosts. Consequently the trees produced bumper crops of nuts even in years when cold spring weather restricted the supply of mast from oaks, hickories, and other early-flowering species. As a result chestnuts were regular staples in the diets of gray squirrels, wild turkeys, and white-tailed deer, especially in lean years when acorns and hickory nuts might be scarce. Bears were even more dependent on a regular supply of mast. When temperatures dropped and other vegetation became scarce, chestnuts scrounged from the forest floor were essential to helping the animals accumulate extra body fat needed for winter, when they began several months of semihibernation. No one knew for certain what the loss of such a dependable food source might mean for wildlife in southern Appalachia. Across the region sport hunters and other nature enthusiasts "gloomily predicted drastic reductions of deer herds, turkey flocks[,] . . . bear and smaller game."[67]

Despite their concern, western North Carolinians were, in the end, as defenseless as their northern neighbors against the alien fungus. Following the failure of eradication efforts in Pennsylvania, state monies for studying the blight dried up. After the U.S. entry into World War I, federal funds evaporated, too. With government research at a standstill, feckless entrepreneurs concocted a variety of expensive potions and powders that promised to check the contagion. A few well-to-do southerners invested thousands of dollars in such snake oil, only to discover that it did nothing to ward off infection. At the Division of Forestry, Holmes adopted a more rational and resigned approach. When he no longer had money to put agents in the field, he relied on local landowners to keep him apprised of the blight's progress while he considered the best methods for harvesting dead and dying timber.[68]

Due to sporadic reporting after 1915, it is difficult to know exactly how or when the alien fungus found its way into the Black Mountains. In all likelihood it arrived from the northeast, moving down and across the Blue Ridge from Virginia. Chestnuts growing in open areas near the valley towns or along the logging railroads were the first to become infected. Such trees were more exposed to windblown spores and, because of their proximity to human settlement, more likely to have cuts or other injuries that afforded the parasite easy entry. But remote stands of chestnut were vulnerable, too. Animals and birds carried asexual spores to even the most isolated regions, and given enough host trees, the disease might infect a considerable area before anyone noticed. Such spotty advance infections probably appeared on the lower slopes of the Blacks sometime between 1917 and 1920. From there the blight spread in a shotgun pattern up the ridges until it reached an elevation of 4,000 feet or so, the upper limits of the chestnut's range. By 1925, in parts of Yancey County nearly one of every ten wild chestnuts was showing signs of the disease. In Buncombe County, where more open country and the milder temperatures common to lower elevations probably helped propagate the fungus, the ratio was one in five. As Holmes noted, the disappearance of the majestic trees was "a foregone conclusion . . . not only a State but a National calamity." In the future, he predicted, "we shall have to take more care in cutting and subsequent management than has heretofore been given these forests."[69]

Though meant for all of western North Carolina, the caveat was especially appropriate for the Blacks. Everywhere the influences of the modern world seemed to weigh heavily on the East's highest mountains. In a mere half-decade the eastern slopes of Clingman's Peak, Mount Gibbes, Haulback, and Mount Mitchell had been cut out and burned over. Though the state now controlled the summit of the highest peak, lumbermen continued to mow down timber on the western flank of the range. With every rainstorm tons of loose topsoil washed into the South Toe and the Cane. Wildfire remained a constant threat to mountain vegetation and wildlife. And now, in the valleys and on the lower slopes, a lethal fungus that originated half a world away was slowly killing one of every four trees and changing the habits of people and animals alike.

Yet for John Simcox Holmes and others concerned with conservation, the news was not all bad. Changing ideas about wilderness, a new passion for raw nature, and a growing affinity for outdoor recreation were also fix-

Dying American chestnut trees in the North Carolina mountains. Chestnut blight probably made its first appearance in the Blacks sometime between 1917 and 1920. By 1925 one of every ten Yancey County chestnuts had been infected. N.C. Division of Archives and History.

tures of the modern American mindset. Though somewhat slow to take hold in North Carolina, such sentiments had significantly increased public interest in the high peaks and generated unprecedented concern about the destruction of Black Mountain forests. The new state park offered proof that enough public pressure could induce North Carolina's image-conscious politicians to support conservation measures, especially if they proved profitable. The State Reservation (as the park property was then known) was just a tiny oasis at the top of the mountain, designed to attract tourists and preserve the memory of Elisha Mitchell. But it afforded utilitarian foresters a tenuous toehold in a region that had previously been at the mercy of the railroads and timber barons.[70]

From 1915 on, as it became apparent that the new park would fall under the jurisdiction of the Geological Survey, Holmes decided to use it as an outpost from which to restore the Black Mountain landscape. In keeping with his earlier ideas for a demonstration forest, he immediately started to prepare a comprehensive plan for fire prevention and reforestation. As for the chestnut blight, he still hoped that the federal government's plant pathologists would come up with an effective fungicide or some other means of halting the contagion. If not, then he and other foresters would have to do their best to see that some other desirable species replaced the valuable trees. In the meantime prudence demanded that dying chestnuts be salvaged for lumber and tannin. It was an ambitious program and, due to the popularity of Mount Mitchell, one likely to be carried out in full view of the public. But Holmes never doubted that he and the Division of Forestry were up to the task. They would make the mountains green again and, in the process, show North Carolinians the benefits of practical conservation.[71]

Holmes had a forester's faith in science and a Progressive's obsession with management and efficiency. But at the core of those beliefs lay another assumption. Like others of his generation, he envisioned the Black Mountain landscape (and all of nature, for that matter) as a mostly passive entity, something to be controlled and manipulated by professionals for the public good. Apparently neither he, Locke Craig, nor the state legislature ever considered that the natural world might not comply with the carefully crafted designs of state government. In years to come, as Holmes and his associates worked to save the Blacks from the excesses of twentieth-century America, they would find out otherwise.

FIVE

Government

JANUARY

POTATO KNOB

After a week of mild weather, winter is back with a vengeance. Snowplows have cleared Route 128 so that a few year-round workers can drive up from the Blue Ridge Parkway to the state park offices. But as I step from the pavement, my boots sink deep into a ten-inch snowpack. A sharp northwest wind sends rime ice crystals swirling past my face. Walking briskly to ward off the cold, I begin a trek up a slippery, winding trail, toward the top of Potato Knob, the 6,375-foot rounded promontory at the bend of the Black Mountain fishhook. My plan is to hike up and across the eastern face of Clingman's Peak to Mount Gibbes, straight through the area once devastated by loggers. After twenty minutes in the wind and cold, however, I decide simply to wander around on Potato Knob, comfortably close to the asphalt road and my waiting vehicle. Eventually I find a snow-free seat on a granite outcrop in a thick stand of evergreens. Looking east, I can see the long outline of the Blue Ridge and the jumble of mountains beyond, a collage of gray-green timber splashed across a canvas of white.

Alone in the snow at 6,200 feet and surrounded by trees, I find it difficult to imagine how these slopes looked in 1915. But I have seen enough photographs of blackened stumps and ashen landscapes to know that the spruce and fir on Potato Knob are of recent vintage. From my seat in the woods I can also see another conifer. It has the same rust-colored bark and sharp-angled needles as red spruce. But this tree is narrower and more triangular in shape, with branches that droop toward the for-

est floor. Those traits suggest that this is probably a Norway spruce, a species that originated in the forests of northern and central Europe.

It is hardly surprising to find an Old World tree in the Blacks. Over the years a wide variety of alien vegetation has taken root here, just as in other parts of North America. But Norway spruce is a special case. It was one of several nonnative species brought in to replant slopes laid bare by the lumber companies. During the 1920s the exotic trees were part of a new plan for the East's highest peaks, a program of conservation, fire prevention, and forest planting that involved both the state and federal governments. Today the Norway spruce on Potato Knob stands as a reminder of those policies, living testament to government's faith in science and the Progressive zeal to rescue the Blacks from the perils of private enterprise.

By 1915 the North Carolina Geological and Economic Survey was not the only agency interested in Black Mountain forests. Authorized by the Weeks Act (passed by Congress in 1911) to establish national forests east of the Mississippi, the U.S. Forest Service began purchasing Black Mountain land in 1913. That acreage eventually became part of the mammoth Pisgah National Forest, so named because it included some of the lands once owned by George Washington Vanderbilt. The Forest Service acquired most of Pisgah at bargain prices by buying the land directly from lumber companies that had already stripped away the most valuable timber. The first tract stretched from Clingman's Peak and Potato Knob east across the South Toe and Blue Ridge, taking in Curtis and Mackey's creeks in the Catawba watershed. A second parcel included the western ridges of the Great Craggies that ran roughly parallel to the west side of the Blacks. Like all lands acquired under the Weeks Act, the purchases were designed to protect "the watersheds of navigable streams," in this case the Catawba, the Toe, the Cane, and the Nolichucky. The Weeks Act placed new restrictions on logging and mining, measures designed "not to injure, but rather to improve, the conditions of the forest." With fire prevention now a Forest Service obsession, the act also provided money to any state that developed a comprehensive plan to discourage wildfire.[1]

Meanwhile on the southern rim of the Blacks, care of the Swannanoa's headwaters fell to the city of Asheville. A sixteen-inch iron pipe carried water from an intake high on the Swannanoa's North Fork to the city reservoir, several hundred feet above the town. Fearful that logging, farming, and grazing in the valley might endanger its supply of "clean mountain water"—which, aside from the obvious benefits for residents, also proved a powerful lure for tourists—Asheville initiated its own program of watershed protection. Around 1900 the city began to buy land and erect fences along the North Fork to deter wandering cattle. A few years later the municipality began condemnation proceedings against many valley residents, forcing them to relocate. Local people raised loud protests against what they regarded as an unfair land grab, but they were powerless to stop it. By the late 1920s Asheville controlled the entire upper valley, including the trails that had once taken Elisha Mitchell and a host of nineteenth-century tourists up the North Fork to the East's highest peak.[2]

Though it raised the ire of Black Mountain people, the government takeover delighted John Simcox Holmes and others concerned with stopping the fires on Mount Mitchell. For five years the state forester had urged the legislature to establish a systematic program of fire prevention for North Carolina. In 1915, with the promise of federal money via the Weeks Act, lawmakers finally passed a measure that allowed for cooperation with the Forest Service. As usual, however, the new fire law carried no appropriation (other than the usual funds allotted to Holmes's salary and to maintaining the Division of Forestry), leaving conservationists to rely solely on a meager $2,000 offered by the federal government.[3]

Rather than spread those slim monies across the state, Holmes developed a plan (based on models from the American West) for small, voluntary "associations" that he hoped might protect forests in "a few large areas" of North Carolina. In July 1915 at a meeting at Montreat, a Presbyterian retreat in the Swannanoa Valley, Holmes and seven delegates from the Forest Service sat down with representatives from the Southern Railway Company, Champion Fibre, Perley and Crockett, and the Asheville City Water Department. By the end of the day the group had cobbled together a working outline for the Mount Mitchell Forest Protective Association. The plan called for five patrolmen who were to "put out such fires as they can, give immediate information as to others . . . [and] spend time building trails etc." to allow access to remote regions. A chief warden was to approach various landowners in the region (including those present at the meeting) and assess them for expenses based on acreage owned.[4]

The ambitious scheme immediately ran into problems. Because landowners were slow to pay their assessments, it was impossible for the association to hire the necessary personnel. Only two patrolmen, paid with federal money and assisted on occasion by employees from Perley and Crockett, worked the woods in the fall of 1915. Unusually wet weather made for a relatively calm fire season, though several small blazes erupted between Deep Gap and Celo Knob. The following spring, however, was

John Simcox Holmes, North Carolina's first state forester and a key figure in Black Mountain history. He hoped to use the state park on Mount Mitchell as a "demonstration forest" to show North Carolinians the benefits of fire prevention, resource conservation, and utilitarian forestry. N.C. Division of Archives and History.

dry, and during March and April one patrolman reported some eighty-four fires along the Mount Mitchell Railroad. Then, during the first week of May 1916, "the worst forest fire ever known in western North Carolina" broke out amid Perley and Crockett's operations near the Murchison Boundary. Pushed by gusty winds, the massive conflagration burned through the southwest portion of the state park and then tore across the clear-cut slopes of Clingman's Peak and Potato Knob into an untouched stand of spruce-fir along the Swannanoa's North Fork. For two days association patrolmen watched helplessly as forty-foot conifers, full of rising spring sap, exploded in flames. Fearful that the entire Asheville watershed might be destroyed, a thousand volunteers from the city cleared brush and set backfires along a seven-mile front to keep the blaze from advancing down the valley. By the time the winds diminished and the fire burned itself out five days later, it had blackened 1,000 acres of red spruce worth a half-million dollars.[5]

As if the fires were not bad enough, two months later, on July 15, a series of severe rainstorms swept into western North Carolina. The French Broad and the Swannanoa rose to record levels, washing out roads, destroying crops, and leaving thousands of mountain people homeless. Much of the water that swelled those streams came pouring off the denuded, burned-over slopes of the Blacks. As the storms intensified, the great troughs cut by Perley and Crockett's skidders collected rainwater and sent it thundering down the southern slopes of the range into the South Toe, the Cane, and their tributaries. Before the bad weather abated, torrential runoff from the logging operations took out several trestles along the Mount Mitchell Railroad and briefly halted operations at the Black Mountain sawmill. It was, as Governor Craig noted, "a great disaster" the likes of which most North Carolinians had never seen.[6]

This time, though, even nature's fury could not move the weak-willed legislature to action. Lawmakers approved a measure to acquire more land for Mount Mitchell State Park in 1918 (a tract that extended down the west side of the range from Stepp's Gap to Cattail Peak), but they provided no funds to acquire timber rights on the property. As a result, logging continued—even on park lands—for another four years. All the while fire remained a constant threat. Worried that the precarious state of the Mount Mitchell Protective Association might prompt the federal government to withdraw its funds, Holmes came up with yet another plan to stop the fires. Starting in 1917 he made arrangements for Champion Fibre and Perley and Crockett to buy usable pulpwood and saw-timber from wind- and fire-damaged forests in the park. Using money generated by the first such operations (which cleared fifty-three acres of dead or fallen trees, probably on the western slopes of Haulback and Mount Gibbes), Holmes secured Mount Mitchell State Park's first full-time employee, a forest warden named D. L. Moser.[7]

Considering that Holmes could afford only one person, Moser was a fine choice. A lithe, strapping man with a reserved but personable demeanor, Moser hailed from Asheville and had apparently worked on the Vanderbilt lands. He also had experience in fire prevention, having served as a federal patrolman under the cooperative provisions of the Weeks Act. For $75 a month Moser agreed to keep an eye on logging operations near the park and "do everything in his power to protect the property from fire." Except during January, he lived in a small, somewhat rundown cabin near the Sleeping Rock, the first in a long line of wardens and park superintendents to take up permanent residence on the East's highest peak.[8]

During his first years on the job Moser devoted much time to burning brush and slash on the western side of the range, hoping to create a firebreak between the park and the lumbering operations. When the weather turned dry, he patrolled the woods and the logging camps, checking for fires and clearing dead limbs and slash from the railroad beds. Aided by his son and other part-time employees, the new warden began collecting "cones from the spruce and balsam trees" for "an experiment in artificial regeneration of the spruce forests." It was at best a shoestring operation, a far cry from the great demonstration forest Holmes had envisioned. But by 1922 the state forester had found cause to be optimistic. Mount Mitchell, he noted, already stood as a shining example of "what can and should be obtained on the forests throughout the mountain region."[9]

Away from the orderly confines of the park, things were different. By the time the last logs found their way to the sawmills (probably sometime in the mid-1920s), much of the original high-elevation forest had been cut out, burned over, and severely eroded, leaving only a razor-thin layer of exposed mineral soil atop the ancient formations of gneiss and schist. Shortly after the loggers left—often within a year—"a dense, rank growth of blackberry" and other brambles flourished on the burn sites. The thick cover effectively curtailed regeneration of spruce and fir, and just as in times past when wind or some other phenomenon had leveled a stand of conifers, other trees moved in. The pin (or fire) cherry, that opportunistic tree whose dormant seeds sprang to life at the slightest disturbance, was first to arrive, followed closely by yellow birch and, a few years later, mountain ash. Fire also cleared the way for paper birch, a tree common on cutover lands in Canada and New England but apparently unknown in the Blacks until Holmes found a young specimen on the western face of Mount Mitchell in 1918. Within five years substantial stands of paper birch had cropped up on the east side of the range, mainly on Maple Camp Ridge.[10]

Visitors to the mountains were quick to note such changes. Paul M. Fink, a wilderness enthusiast who, with a friend, backpacked from the Cane River Valley south across the Blacks in the summer of 1920, wandered into a thicket of fire cherry somewhere between Grassy Knob Ridge and Deep Gap. The trees, he reported, ranged "from pencil size to as big as a man's waist," mingling with "blackberry briars in generous amounts" and other prickly stubble. The two hikers got a brief respite from "this hellish jungle" when they reached an uncut stand of spruce and fir on the crest of the range. But the next day, as they trekked toward Mount Mitchell, they again found fire cherry "so dense that even only twenty-five feet apart we would lose sight of one another." Worse still were the great holes created by fire and erosion "between the roots of the [burned] spruce stumps." Several times Fink and his companion accidentally stumbled into the pits (some of which were knee deep), twisting their ankles and bruising their shins as they made their way across the Blacks.[11]

On dry, exposed ridges where fires had not been particularly severe, rhododendron, laurel, and huckleberries took root, giving rise to shrubby "heath balds" not unlike the habitats maintained by grazing cattle a hundred years earlier. Where burns had been more intense, the demise of the spruce-fir forest cleared the way for grasslands. As one observer noted in 1916, "Where there is sufficient soil, grass comes in at first very scattered and sparse" among stumps and downed timber. Once the grass got a start, each additional fire enhanced its growth. Within five to eight years most competing vegetation simply disappeared. By 1920 several acres of lush grassland grew on the eastern face of Mount Mitchell (which had burned repeatedly after the spectacular fires of 1914) and on the bare slopes of Maple Camp Ridge near the railroad terminus. Some experts believed that, given time and shielded from fire, the grass and heath balds might again sprout spruce and fir. But such natural regeneration promised to be an excruciatingly slow process. According to contemporary estimates a natural shift "from grassland to commercial forest" might take 250 years.[12] To those concerned about watershed protection and the future of lumbering in the southern mountains, it seemed too long to wait. The Forest Service, which now controlled some 60,000 acres in the region, decided to see what might be done to move nature along.

Between 1923 and 1931, as the lumbermen retreated, researchers from the newly established Southeastern Forest Experiment Station in Asheville laid out seventy-seven small plots (each spanning about a tenth of an acre) at an elevation of 5,500 feet on the southeastern face of Clingman's Peak. Along with native red spruce and Fraser fir, foresters planted a variety of alien trees, hoping to find some fast-growing species that might be used to repopulate the fire-ravaged mountains. During the first years of the experiment they planted red pine and white spruce from the Great Lakes states. They imported white cedar from New England and Douglas fir, Sitka spruce, and lodgepole pine from the American West. From Scandinavia they brought silver fir, Scotch pine, and Norway spruce. From the Far East came Japanese larch, black pine, and red pine.[13] On the same slopes where André Michaux had once collected Black Mountain trees for European cultivation, foresters now reversed the process and planted exotic gardens of Old World specimens. It was the Progressive Gospel of Efficiency carried to the extreme. Not simply an attempt to recover what had been lost, it was an effort to improve the status quo, to make na-

State and federal foresters planting conifers among brambles and pin cherries on the fire-scarred slopes of the Black Mountains. In the 1920s and 1930s foresters set a variety of trees not native to the region, including Norway spruce, Douglas fir, and Japanese larch. They hoped to identify some fast-growing species that might help replace the forests lost to logging and fire. Courtesy of Coleman Doggett.

ture better, more efficient, and more profitable in *les hautes montaignes de Caroline.*

Several nonnative species—red pine, white cedar, and Japanese larch—showed promise. But Norway spruce was by far the most successful import. Having evolved in the near-Arctic climes of northern Europe, the trees adapted well to the milder Black Mountain winters, especially when planted on the wind-sheltered slopes of Clingman's Peak and Potato Knob. A decade later nearly 70 percent of the seedlings had survived and were growing at an average rate of nearly a half-foot per year. For some reason, however, the trees apparently did not sustain themselves through natural reseeding. Eventually foresters concluded that when it came to reforestation, neither Norway spruce nor any of the other exotics "demonstrated any superiority over the native red spruce and southern balsam [Fraser] fir." Of the two indigenous trees, the Forest Service preferred red spruce because of its commercial potential, though they acknowledged that extensive planting of Fraser fir might be of benefit for recreation and watershed protection.[14]

Meanwhile in the state park, Moser and his successors carried out their

In addition to planting exotic trees, foresters set native species, including red spruce and Fraser fir. This lone seedling, standing amid logging slash, pin cherries, briars, and other brush, suggests the monumental task that confronted the conservationists. Courtesy of Coleman Doggett.

own reforestation program, eventually replanting some 105 acres of state property. Less concerned about producing marketable timber, park wardens relied mostly on Fraser fir to refurbish Mount Mitchell. But they also set some red and Norway spruce, vestiges of which are still visible today within park boundaries. The Great Depression put an end to the Forest Service experiment and temporarily stalled park reforestation. But a decade later one could still find scattered specimens of Douglas fir, white spruce, lodgepole pine, and European silver fir growing on Clingman's Peak and Potato Knob.[15]

In 1929, at the height of the reforestation efforts, the experts at the Southeastern Forest Experiment Station prepared a detailed Planting Map for the Mount Mitchell project. It shows the eastern side of the Blacks from Potato Knob to a point just beyond Balsam Cone and lists the trees and other vegetation growing on various slopes from the summits down to about 5,500 feet. Standing over that map today is akin to looking at an aerial snapshot of the mountains in 1929. The work of the Progressive foresters shows up as small dark splotches with designations such as "RS" for red spruce, "F" for Fraser fir, and "NS" for Norway spruce, covering some 250 acres. Elsewhere, though, the cutover slopes were covered with a hodgepodge of vegetation: grass, pin cherry, yellow birch, assorted shrubby hardwoods such as locust and chestnut oak, and in the words of the frustrated foresters, "brush too dense for planting."[16] In the race to reclaim the mountains, nature clearly had the upper hand on science.

In the towns along the Cane, another sort of reclamation took place. As the lumber companies pulled out of Pensacola and Murchison, they tore up the railroad spurs that served the upper valley. Many of the workers left, too, headed for Asheville or more distant cities where they might still find the steady paychecks, regular hours, and easy credit to which they had grown accustomed. A forester who visited the valley in 1923 reported that "only a few houses are left where the mills once ran and the electric lights blazed and several hundred men worked in the town that was Escota [*sic*]. Corn is growing over most of it now." It was, he continued, "the old story of the lumber camps. The lumbermen come in, put up great mills, build railroads, cut off the timber, then go their way and the brief activity and prosperity of their day are things of the past."[17]

As foresters replanted the charred brushy ridges and the valley communities reverted to agriculture, others with a vested interest in the fate of the Blacks were already leaving their mark in the region. Since the establishment of the state park in 1915, vacationers from across the South had been pouring into the Blacks every summer. With considerable help from the Forest Service and other government agencies, the visitors began to reshape the Blacks in their image, turning the landscape into their vision of what nature should be.

MEMORIAL DAY WEEKEND

CAROLINA HEMLOCKS

Driving along the South Toe with the windows down, I can smell the campfires. Another half-mile down a twisting two-lane stretch of asphalt and I can see them, flickering like porch lights beside the tents, trailers, and motor homes. It is May; spring is in full flower, and the "campers," as they call themselves, have returned to the high mountains. I turn left into Carolina Hemlocks Campground, a gated, pay-to-stay facility, replete with running water and flush toilets, operated by the Forest Service. Hemlocks, as the locals call it, is now nearly seventy years old and showing its age. The parking spurs seem narrow and cramped. Trail signs are worn and crumbling. Many of the surrounding trees have been scarred by carelessly driven nails or gasoline lanterns hung too close to their trunks. The bathrooms sorely need fresh paint and new fixtures.

But at dusk on this holiday weekend the old campground recovers some of the magic that once made it the most popular recreational site in the Blacks. Brightly colored tarps and awnings sprout like mushrooms among the tall evergreens. Middle-aged white men in patterned aprons hover over glowing charcoal grills. At the choice campsites near the river, the unmistakable aromas of frying meat and potatoes mingle with the ever present smell of wood smoke. Flags proclaiming allegiances to area colleges and their athletic teams (Clemson and North Carolina State on this occasion) hang from clotheslines strung between giant hemlocks and poplars, fluttering alongside flannel shirts, blue jeans, and beach towels. Country music pours from portable stereos. Children on bicycles, hair still wet from an afternoon swim in the South Toe, tool along the campground's paved roads. A few self-described old-timers, retirees who know one another from past seasons, exchange gossip and show off new equipment. All the while in the background, hand-held axes ring against hardwood, converting downed timber into campfire fuel. The ambience is part county fair, part logging camp.

No doubt similar scenes featuring tourists of various means and social standing are unfolding elsewhere across the region tonight—at Black Mountain Campground, another Forest Service site farther upriver; at any number of privately owned operations; and at the small, nine-site tent campground in the state park. Even so, overnight campers represent only the tip of the tourist iceberg. Between now and the end of October tens of thousands of sightseers will pitch tents or park recreational vehicles along the South Toe. A hundred times that many will make day trips to the Blue Ridge Parkway and the East's highest peak.

Whether they stay for a few days or a few hours, the visitors share some important traits. Though a few come from out of state, most hail from cities in the North Carolina Piedmont—Charlotte, Hickory, Winston-Salem, Shelby, and Statesville. Most come to the Blacks for recreation, seeking escape from the urban workweek. They express a deep appreciation for nature but freely admit that they like to enjoy the outdoors without giving up the amenities of modern life. Like those who first camped at Carolina Hemlocks seventy years ago, they bring with them a machine that, when it comes to transforming the Black Mountain landscape, has proven as powerful as any log skidder or double-band saw: the motor vehicle.

In April 1910 Joseph Hyde Pratt, North Carolina's state geologist and Holmes's boss at the Geological Survey, bought a car. Built in Detroit by the Chalmers Motor Car Company, it was a powerful luxury vehicle boasting a thirty-horsepower engine and speeds of twenty-five miles an hour, the legal limit in North Carolina. But what Pratt really liked about his Chalmers-30 was its ability to negotiate rough terrain. "I have used it in climbing the Blue Ridge," he wrote the salesman in December, "and there it probably struct [*sic*] as bad roads as any machine has encountered."[18]

Pratt knew those roads well. For several years he had been the state's foremost spokesman for the Good Roads Movement, a well-organized campaign for better highways that had been building across the South since the 1880s. At first Good Roads advocates had focused on rural highway development, pushing for routes that might allow farmers access to railroad hubs. But during the early twentieth century, as cars like Pratt's rolled off Detroit assembly lines and into the hands of well-to-do Americans, the Good Roads Movement began to call for construction of long-distance "tourist highways" that might bring affluent visitors and a new wave of prosperity to the South.[19]

As always, such sentiments had tremendous appeal in North Carolina, especially in and around Asheville, where merchants coveted tourist dollars. In 1909 a group of Good Roads devotees from across the South gathered in the mountain city to plan a new tourist thoroughfare. Called the Crest of the Blue Ridge Highway, it was to span the southern Appalachians from Georgia to Virginia, passing through various resort towns along the way. As planners mapped the route across western North Carolina, they made sure the highway passed within a few miles of Mount Mitchell, where according to one Atlanta journalist, motorists might explore "the most picturesque section of the United States this side of the Rocky Mountains." Within three years Good Roads advocates raised over $10,000 for the project. Most of the money came from local railroad men who believed (correctly as it turned out) that better roads and increased tourist traffic would ultimately benefit both freight and passenger lines.[20]

Even with financial backing from the railroads, however, the exorbitant costs of blasting and grading a new route over rugged terrain eventually forced Pratt's partners to abandon the road with only a small segment completed. But for several years publicity generated by the project continued to bring car enthusiasts close to the Blacks. In 1912 Henry McNair, a New Yorker and onetime editor of the *Automobile Blue Book*, took part of his summer vacation in the upland South, driving from Greensboro to Asheville. Though he discovered that "on account of present road conditions" it was "not practicable" to venture farther into the mountains, McNair relished the prospect of someday driving to the pinnacle of the East. Within twenty-four miles of Asheville, he wrote, are "fifty-seven peaks . . . having an altitude of more than 6000 feet." Chief among them, he exulted, was "Mount Mitchell, the highest American Peak east of the Rockies, rising 418 feet higher than Mount Washington."[21]

Mountain motoring also got a boost from Governor Locke Craig, who in keeping with his Progressive politics immediately embraced the Good Roads Movement. As with other public works projects, he faced stiff opposition from the state legislature, which in 1913 rejected a proposed million-dollar appropriation for better highways. Still, just two years later the Asheville Board of Trade felt confident enough about local road improvements in Buncombe County to cite motor touring as one of the incentives for creating Mount Mitchell State Park. Noting that plans for a national park in the southern Appalachians had temporarily stalled, the board's chief lobbyist touted Mount Mitchell as the next best alternative. "Automobile tourists of the country," he wrote, "can motor to almost within a dozen miles of the summit of Mount Mitchell, to [the town of] Black Mountain." Comparing the East's highest peak to Yosemite and Yellowstone, the lobbyist explained that the proposed state park lay within easy driving distance of America's largest cities, "while the parks of the West are from three to five days journey from the population centers of the East." Yet even as they approved the state park, neither the governor nor the legislature had any immediate plans for a public highway to the summit of Mount Mitchell.[22]

Where southern politicians feared to tread, Yankee entrepreneurs rushed in, though at first they put their faith in trains as well as cars. From the time Dickey and Campbell drove the first spikes on the Mount Mitchell Railroad in 1911, they had considered using the line for hauling tourists as well as timber. When Perley and Crockett bought out the operation, they began limited passenger service during 1913 and 1914. Over the next two years, with the establishment of the state park, railroad tourism exploded. Between May and October 1916 some 10,000 sightseers made the trip from a station near Black Mountain to the eastern slopes of Mount Mitchell.[23]

The tourist boom owed much to advertising. To promote their scenic line Perley and Crockett hired Sandford H. Cohen, founder of South Carolina's Isle of Palms resort and a longtime champion of tourism in western North Carolina. Cohen peppered the southeastern states with pamphlets and advertisements designed to tap America's burgeoning interest in outdoor recreation. For $2.50 (the train fare in 1916) Cohen promised an excursion "Above the Clouds" with "Grandeur, Beauty, and Sublimity, Unequaled on the Globe," an "educational trip . . . pronounced by all who have taken it to be a scenic marvel of mountain magnificence." The Southern Railway Company, whose lines served the Black Mountain station, went even further, extolling the virtues of an overnight visit to the high

Passenger cars on the Black Mountain Railroad, with Perley and Crockett's sawmill in the background, ca. 1916–18. Even as the mountain became a prime attraction for railroad tourists, lumber companies continued logging the high-elevation forests into the early 1920s. North Carolina Collection, Pack Memorial Public Library, Asheville.

peaks. "Why not drop for a time your business burdens," one promotional booklet asked, "run away to this 'bit of Paradise,' pitch your tent there—and live!" In keeping with a popular theme of the time, the railroad touted wilderness as a panacea for the ills of modern society, a natural elixir that guaranteed "bodily vigor and mental alertness that the centers of civilization know not of." In the wilds of Mount Mitchell, the pamphlet boasted, "you will learn to rise with the lark and go to bed early—as you should. Insomnia . . . will vanish. . . . You will gain the appetite of your childhood days. . . . You will again become an optimist, and the world will look to you as it did in the days of your adolescence." It was perhaps the ultimate irony. The railroads, which brought modern business and industry to the Blacks, now offered respite from those very activities.[24]

The common denominator, of course, was money. To Perley and Crockett it made little difference whether their trains hauled timber or tourists, as long as the latter had money to spend on Mount Mitchell. During the first years of operation, most did. Because passengers usually had to drive to Black Mountain station, the line primarily attracted middle- and upper-

Camp Alice by rail, 1916. Visitors could eat in the dining hall, pictured here, and rent a nearby tent for an overnight stay in the Black Mountains. The surrounding forests show the effects of logging and fire. North Carolina Collection, Pack Memorial Public Library, Asheville.

class sightseers from Asheville and the surrounding resort towns, most of whom owned or had access to automobiles. All of the tourists were also white. No state statute officially barred African Americans from Mount Mitchell State Park, but in the Jim Crow South, where public facilities were segregated as much by custom as by law, all citizens understood (as North Carolina politicians liked to say) that the attractions of the East's highest peak were for whites only. Indeed, nearly three decades would pass before North Carolina set aside a single state park—Jones Lake near Elizabethtown in the southeastern part of the state—for use by African Americans.[25]

To provide their white, middle-class patrons with the sort of outdoor experience they sought, Perley and Crockett opened a rustic dining hall at a site known as Camp Alice (apparently named for the cook) on the eastern face of Mount Mitchell. From there tourists could hike a mile-long trail through the state park to the summit, climb a wooden observation tower (also constructed by Perley and Crockett), and visit Elisha Mitchell's grave. Those who craved the salubrious benefits of camping might rent tents and sleep in the woods while enjoying hot, home-cooked meals in the comfort of the dining hall.[26]

By the 1920s, however, Fred Perley, chief partner in the region's most prominent lumbering firm, realized that automobiles, not locomotives, would dictate the future of sightseeing on Mount Mitchell. As World War I wound down and the most accessible spruce along the railroad was logged out, Perley founded the Mount Mitchell Development Company, a corporation dedicated, among other goals, to turning the East's highest mountain into a prime vacation spot. Between August 1921 and June 1922 Perley's development company planned and built the first motor road to Mount Mitchell. Eighteen miles long, paved with cinders, and full of "hairpin turns and loops," it essentially followed the railroad bed from the Swannanoa Valley through the new Pisgah National Forest to Camp Alice. Gates at both ends allowed for traffic control. From 8:00 A.M. until 1:00 P.M. cars moved up the mountain; in the afternoon between 3:00 and 5:30 they came down. In the interim a "safety patrol" cruised the route to be sure no stranded or disabled cars from the morning interfered with the descent. Perley also closed the road to horse-drawn vehicles and "their attendant influence in slowing down the stream of traffic." Like the park, the new highway remained segregated. Perley's regulations stipulated that "no colored person . . . be admitted over the road except those going as drivers, nurses, or attendants, accompanied by employers." White drivers paid a dollar each for the privilege of taking their cars to Camp Alice. Passengers between the ages of five and twelve paid fifty cents.[27]

Among those allowed to use it, the motor road was an immediate hit, especially with wealthier seasonal residents of Asheville. No longer restricted by railroad schedules, excursionists could leave the city as late as 10:00 A.M., drive up the mountain, hike thirty minutes to the summit, and enjoy an early afternoon picnic on the East's highest peak before returning home that afternoon. By 1924 those who failed to pack a lunch could buy refreshments either at Camp Alice or at a "tea room" on the trail to the summit operated by the Mount Mitchell Development Company. (In exchange for the park concession Perley turned over 10 percent of its profits to the state.) Some 16,000 sightseers—more than half again as many as the railroad carried in 1916—made the trip during the turnpike's first summer of operation.[28]

Unable to ignore the white citizenry's insatiable appetite for "automobility," the North Carolina legislature finally embarked on a massive program of highway construction. Between 1921 and 1925 the state built some 7,500 miles of hard-surfaced highways that eventually connected all 100 county seats. Though many mountain roads remained rough and poorly

The automotive revolution: postcard showing the motor road to Mount Mitchell (*above*) and Camp Alice by car, 1920s (*below*). North Carolina Collection, University of North Carolina Library, Chapel Hill.

maintained, the state's western counties, including Buncombe and Yancey, became much more accessible.[29]

The growing affinity for cars also spawned a new craze in outdoor recreation: autocamping. Since the late nineteenth century, campers had ventured into the Appalachian woods carrying only such scant provisions as might fit into a knapsack. With a vehicle, though, one could enter the wilderness in style. Cars provided shelter from foul weather and offered plenty of space to store food and equipment. During the 1920s various companies began to produce a wide array of gear and gadgets designed especially for autocampers, including gas stoves, tents, chairs, clothing, and other amenities. By the early 1930s specialized "travel trailers," pioneered by the Airstream Company, offered affluent motorists a convenient way to take their houses with them into the wild. John Simcox Holmes was quick to note the potential impact of such developments on western North Carolina. "In these days of improved highways, rapid travel, and growing interest in the out-of-doors," he wrote in 1929, "the people are going to demand an adequate park system, not only for their own pleasure, but to help attract and interest the thousands of tourists who drop dollars wherever their cars stop."[30] Still, Holmes had little money to improve Mount Mitchell. While the state dawdled, a cagey local entrepreneur also set up shop on the East's highest peak.

From the time the first railroad arrived in Murchison, Big Tom Wilson's son, Adolphus Greenlee Wilson (Dolph to his friends), had operated a small hotel in the town. Like his father, Dolph catered to hikers and sportsmen, providing guests with country-style meals and regaling them with embellished tales of hunting in the Black Mountains. Seeing the success of Perley's toll road, Dolph and his son, Ewart, decided to build their own private highway from the hotel to Mount Mitchell. Completed in 1926 and built primarily with mule- and manpower, the new turnpike snaked up from the Cane River, following old railroad beds from Murchison to Stepp's Gap and from there to within a quarter-mile of the summit. Trading on his grandfather's legacy, Ewart named his thoroughfare the Big Tom Wilson Mount Mitchell Toll Road. A year later the Wilsons built "a road house" on their own land at Stepp's Gap, named the facility Camp Wilson, and began to serve meals and offer lodging to overnight guests.[31]

The Big Tom Road, though usually passable, attracted fewer cars than Perley's turnpike primarily because the grades were steeper on the western slopes of the Blacks. Even so, Perley worried incessantly about competition from the Wilsons, especially after Ewart proposed in 1927 that the two roads be joined at Stepp's Gap to allow tourists to drive all the way across the range. With competition for sightseer dollars keen, it was only a matter of time until trouble erupted. During the early 1930s Ewart Wilson charged that Perley was misrepresenting the Big Tom Road by telling tourists that it was dangerous and in poor repair. Perley, in turn, claimed that Wilson routinely came to Camp Alice soliciting business for his family's restaurant. Ewart was more than willing to take the argument to court. In 1932, as business dwindled during the early days of the Great Depression, he threatened Perley with a slander suit. For several months angry letters flew back and forth while Holmes and other state officials tried to mediate.[32]

By the mid-1930s, however, it became clear to all that the most important tourist highway into the Blacks would not be built by either Perley or Wilson but by the federal government. Franklin D. Roosevelt's New Deal planners wanted a road that would link the new Great Smoky Mountains National Park in Tennessee and North Carolina with Shenandoah National Park in Virginia. More than 400 miles long, the new highway was to be strictly a scenic route that catered to the needs of motor tourists and auto campers. The Department of the Interior called it the Blue Ridge Parkway.[33]

Eager as ever to attract visitors, the Asheville Chamber of Commerce (successor to the Board of Trade) enthusiastically endorsed the plan. Even North Carolina's conservative legislators, who were generally suspicious of the New Deal and still reluctant to spend money for public projects, were more than happy to have the federal government foot the bill for the parkway. The legislature immediately agreed to acquire the necessary rights of way. But as plans for the parkway took shape, the Asheville boosters began to worry that the government might locate most of the new road in Tennessee. To make sure that did not occur, Tar Heel businessmen and politicians invoked the appeal of the East's highest mountain. Appearing before the federal committee that ultimately determined the route, state spokesmen made it clear that the parkway should follow the highest contours of the Blue Ridge and provide access to Asheville. From there motorists might have easy access to other scenic attractions, including the phenomenally popular Mount Mitchell. With an assist from Josephus Daniels, a well-known North Carolina newspaper editor who had important ties to Roosevelt, the Asheville lobbyists got their way. Built with labor from the Civilian Conservation Corps (CCC), the road eventually followed much the same route as Pratt's original Crest of the Blue Ridge Highway. By 1939 motorists could drive the parkway (surfaced with crushed gravel) to the base of Potato Knob, just a few miles south of Mount Mitchell.[34]

With the new road a reality and federal largesse the order of the day, North Carolina also took advantage of government money to refurbish its first state park. The newly established Department of Conservation and Development (which subsumed the old Geological and Economic Survey) immediately requested a CCC work camp for Mount Mitchell. At first, though, the National Park Service (now in charge of the parkway) balked at the proposal because the roads to the summit were privately owned. Fred Perley, who seemed to sense that private enterprise was doomed on Mount Mitchell, immediately granted unrestricted access to Camp Alice. Ewart Wilson, however, had other ideas and stubbornly insisted on charging the government fifty cents per vehicle to travel the road from Stepp's Gap. Unwilling to let a cantankerous mountain man stand in the way of federal funding, the state Highway Commission announced plans to build its own connecting road from the parkway to the summit. Faced with that prospect, in 1936 Wilson signed an agreement that allowed free use of his route. Ever the tough negotiator, though, Wilson made the government promise to give him all "camp refuse or garbage suitable for hog or other animal feed."[35]

However, shortly after the CCC trucks began to roll, Wilson—apparently upset at the wear and tear on his road—reinstated the toll, this time charging a dollar per vehicle. Incensed at Wilson's defiance, Holmes got the legislature to pass a special bill prohibiting "the operation of privately owned toll roads within State parks." Condemnation of the Wilson road soon followed. True to form, Wilson hired lawyers and fought the proceedings at every turn until the Highway Commission finally agreed to pay him $8,000 in compensation. In 1941 the Big Tom Wilson Motor Road officially became part of Mount Mitchell State Park. But for the moment, as part of the settlement, the roadhouse and restaurant remained at Stepp's Gap. The state had not heard the last of Ewart Wilson.[36]

While the road controversy raged, the upper reaches of the East's highest mountain underwent a remarkable transformation. Fire prevention crews fanned out along the trails and parking areas, clearing away brush and standing dead timber. Other workers took charge of trail maintenance, refurbishing all the footpaths, including the well-worn and badly eroded trail to the summit. The CCC briefly revived reforestation in the park, planting additional Fraser fir and Norway spruce. Just below the summit workers cut and hewed red spruce logs for a new concession stand, a rain shelter, and a "building to house organized groups." By 1939 the park had flush toilets and a new water and sewer system. The use of red spruce in the buildings on Mount Mitchell was no accident. As in other state parks around the country, CCC workers used an architectural plan created for the national parks. In keeping with the National Park Service ideal of providing easy tourist access "without impairment of natural features," the structures on Mount Mitchell, with their spruce logs and shake shingles, were intended to blend into the surrounding forest and provide the state's new outdoor enthusiasts with modern conveniences in a rustic, wilderness setting.[37]

Even with all those improvements, however, the Blacks were not a place where motorists might vacation for several days. Until they settled with Ewart Wilson, state planners thought it unwise to invest in campgrounds or other improvements. For the moment that task fell to the Forest Service. From the time it first acquired land in the Blacks, the Forest Service had been interested in providing tourists with recreational facilities. That interest became a passion during the 1920s as demand for timber slackened and the Forest Service found itself in competition with the newly created National Park Service for government money. Before the New Deal, though, the Forest Service lacked the funds to do much more than establish a few primitive campgrounds in the Blacks. One of those was Carolina Hemlocks. Named for the native evergreen and situated on fifty acres along the South Toe, the site had long been a popular gathering place for campers from the surrounding region, most of whom simply pitched tents along a poorly maintained gravel road. By the early 1930s, however, new highways allowed for an easier drive between Asheville and Burnsville and made the valley much more accessible. The Depression itself also increased traffic along the South Toe. During the worst of the economic downturn, chronic unemployment led to the proliferation of free time among all classes of North Carolinians. By the end of the decade the forty-hour workweek (promoted by Roosevelt's short-lived National Recovery Administration) made it easier for urban folk to visit the mountains for a weekend.[38] To cope with the new flood of tourists, the Forest Service and the CCC set out to make Carolina Hemlocks a first-class recreational facility designed especially for autocampers.

Using a campground plan originally developed by the Forest Service (and later used extensively in the National Parks), workers constructed a long "loop road" through Carolina Hemlocks with parking spurs large enough to accommodate automobiles and trailers. The CCC built picnic tables from native hardwoods and cleared three miles of nature trails, including an old nine-mile path from the campground to the East's highest peak. An amphitheater and bonfire pit, where "a hundred or more people" could enjoy "a community singing or marshmallow roast," proved especially attractive to large groups. For those who relished a cold dip in the South Toe, the Forest Service installed a diving board and a "grand rock

Swimming hole at Carolina Hemlocks, 1944. Beginning in the 1930s, Forest Service improvements to the campground helped make it one of the most popular recreational sites in the area. N.C. Division of Archives and History.

beach" at a popular swimming hole. All in all the site provided the accessible, friendly outdoor experience that autocampers craved. As the *Asheville Citizen-Times* reported, the campground's "lacy canopy of leaves" let in just enough light to create "a glamorous, rather than gloomy, atmosphere." By 1940 some 30,000 motorists were bringing their cars, tents, and trailers to Carolina Hemlocks every year.[39]

Yet even as the press and civic boosters crowed about the virtues of the parkway and public campgrounds, local people complained that they, not the federal government, really paid the price for such facilities. Fred M. Burnett, whose family had been among the earliest settlers along the North Fork of the Swannanoa, paid more than his share. In 1927 the Burnetts lost their North Fork holdings when the city of Asheville took over the watershed. Hoping to remain nearby, Burnett purchased 200 acres on Bald Knob Ridge on the southeastern slopes of the Blacks. But he enjoyed only a few years of "peaceful and delightful possession" before the state took two-thirds of the new tract for parkway right-of-way, paying deflated Depression prices for his property. It was, Burnett sarcastically noted, only one of many such "crimes" committed "*pro bono publico* [*sic*]" in and around the Black Mountains.[40]

Meanwhile others protested that the government's real crime was robbing the Blacks of their rugged wilderness character. In late June 1937 a hiker from the Appalachian Mountain Club arrived from out of state to climb the East's highest peak. Hiking up from Murchison in the rain, he was much taken with the scenery but, upon arriving at Mount Mitchell, was disappointed to find a well-maintained road and "a party of ladies in high heels." Even motorists seemed to have mixed feelings about easier access. Commenting on the Highway Commission's plans for the connecting road between the parkway and the summit, a *Charlotte Observer* columnist complained that the new "scenic driveway" would take all of "the gusto and adventure" out of "the motor journey." Those who wished "to make the ascent by a picturesque, hard, and adventuresome route should do so now," he cautioned, "before it is too late."[41] What one thought of development at the state park and elsewhere in the Blacks depended, as ever, on how one defined those elusive ideals of nature and wilderness.

AUGUST

NEALS CREEK

Thunder rumbles in the distance, and raindrops slither from overhanging birch leaves as I watch my flyline unfold across a washtub-sized pool. The yellow fly, designed to mimic one of the native insects, teeters momentarily at the edge of a churning riffle then disappears in a gurgling splash. I raise the rod tip and watch a small fish dart back and forth until it finally leaps clear of the stream, nearly beaching itself on a tiny gravel bar. The aerial acrobatics instantly identify it as a rainbow trout. I bring it to hand quickly, admiring its cherry gill plates and the pinkish stripe down its flank before returning it to the water. Pulling on a jacket, I move upstream silently in the mist, casting carefully around the birches and rhododendrons. The next fish is larger, though at just ten inches hardly a trophy. It does not jump but instead bores headlong for the bottom, seeking refuge beneath a submerged ledge. As it comes closer, I can see that it is a brown trout, its red spots and golden flanks more subtle but no less striking than the bright markings of the rainbow. I twist the fly from its mouth and watch it swim away.

Neals Creek rises high on the western slope of the Blue Ridge, dropping a thousand feet or so through a steep, rock-strewn chasm before it widens and joins the South Toe at the base of Mount Mitchell. Like most of the region's headwaters, the creek is swift but easily waded, often narrow enough to step across. Its trout, too, are much like those to be found elsewhere in the Blacks: smallish fish, full of fight, nourished on the stream's abundant oxygen and ample insect life. The forests along Neals Creek are home to other animals, too. As I make my way back downstream in the mist at twilight, a doe and a fawn materialize suddenly in a lush meadow beside the stream. A few minutes later, sitting on the truck tailgate as I peel off my hip boots and store my gear, I hear the plaintive, haunting warble of a screech owl, a sound that, for me, always evokes powerful images of nature and the wild.

Along Neals Creek, though, such sights and sounds can be deceptive. These days the stream and the forests around it are, at least in part, man-made environments. The fish I caught this evening were born in the stream, but they are not truly native to the Black Mountains. The rainbows originated in the American West; the browns, in Europe. Their ancestors were stocked here in the 1930s during the heyday of wildlife management in the Blacks. Walking the stream in daylight, one does not have to look far for other reminders of those times. The cement troughs and stone walls of an old fish hatchery, originally built by the CCC, are still visible in weedy old fields beside the creek. Not long ago, while prowling the woods far upstream, I happened on an old metal sign that read simply, "Game Refuge—Dogs and Guns Prohibited." That sign was once attached to a single strand of wire that encompassed more than 20,000 acres in the surrounding national forest. Within that enclosure, known as the Mount Mitchell Game Refuge, wildlife managers raised white-tailed deer, wild turkeys, black bears, and quail for release into the adjacent woods. In all likelihood the deer and other animals here today descended from game stocked in the 1930s.

For those who remember it, The Refuge was a special place. It represented part of a grand scheme to undo the sins of the past, to restore fish and game lost during years of unrestricted hunting, logging, and road building. As the first state-run wildlife sanctuary in North Carolina, Mount Mitchell became the model for a host of similar facilities across the state. But like those who designed the parks and public campgrounds, state and federal wildlife managers adopted policies designed to appeal to a specific class of hunters and fishermen. In catering to their interests, government officials initiated policies that favored some species over others, and these decisions had profound implications for Black Mountain animal populations.

Early in the summer of 1927 four men walked into the Black Mountain Hardware Company and bought five sticks of dynamite and fifteen feet of fuse. Telling a suspicious clerk that they planned to blow up some stumps, the men climbed into a green sedan and drove up the Swannanoa's North Fork to the boundary of the Asheville watershed. A few minutes later nearby residents reported a thunderous blast near one of river's larger pools. That afternoon patrons at a local filling station again saw the green sedan. Two of the occupants, still "wet to their belt[s]" had a long stringer of trout that they proudly held up for passersby to inspect.[42]

A decade earlier such an incident might have gone unnoticed. Dynamiting streams had long been common practice in western North Carolina. Loggers working along the Cane and the South Toe probably lobbed explosives into likely-looking holes whenever they craved a meal of fresh trout. But by 1927 things were different. When Colonel Nelson Mease, a local man recently appointed game warden for the region, learned of the dynamiting on the North Fork, he launched a full-scale investigation, interviewing witnesses, collecting affidavits, and seeking advice from lawyers. When he could not find enough reliable eyewitnesses to identify "the ones who did the crime," he took his case to the newspapers. "The person who dynamites a stream," he told a local reporter, "is the enemy of all true sportsmen, lovers of outdoor life and the general public."[43]

Mease's comments reflected a recent sea change in attitudes toward fish and game in North Carolina. For much of its history the state had no comprehensive wildlife laws. Indeed, in the first decades of the twentieth century many rural folk annually participated in communal spring hunts for robins, blackbirds, bobolinks, meadow larks, flickers, and other songbirds, all of which made passable table fare. Likewise commercial hunters, primarily "from New York and New England," slaughtered Carolina shore birds, deer, and other game "in untold numbers, and packing their bodies in barrels of ice, shipped them to Northern markets." Though the general assembly had the power to curb such practices, state legislators had traditionally allowed county governments to set their own closed seasons and bag limits on various species. Most hunters simply ignored those regulations.[44]

In the early 1900s, as concern for diminishing wildlife mounted across the nation, North Carolina took its first halting steps toward game conservation. The General Assembly officially recognized the state Audubon Society (organized just months earlier at the North Carolina College for Women in Greensboro) and empowered it to regulate "the taking of game birds and animals." With meager funds provided by the legislature, the society issued hunting licenses and used volunteer wardens to enforce closed seasons and bag limits. But in a state where rural folk still relished a spring repast of robins, the society often found itself at odds with local authorities. State officials abandoned the program in 1909 due to a lack of funding and popular support.[45]

It took another decade and a half of lobbying by sportsmen, the Audubon Society, and other conservationists (as well as considerable political maneuvering by Governor Angus W. McClean) before North Carolina's irresolute lawmakers finally passed a state game law in 1927. The legislation distinguished between game animals (deer, bears, rabbits, squirrels, opossums, quail, doves, and wild turkeys), which could be legally hunted, and protected species (robins and other songbirds), which were not to be taken. It also set seasons and bag limits for every county. Similar regulations for fishing prohibited dynamiting, trapping, and other destructive methods and established size and creel limits for various species, including brook and rainbow trout. To facilitate enforcement the legislature created the Game Commission within the Board of Conservation and Development. The commission, in turn, hired regional wardens such as Mease to apply the law "without fear or favor."[46]

The state also made it clear that its chief concern was protection of game and fish for sportsmen. In a long pamphlet describing the new regulations, the Game Commission explained that hunting and "out-of-doors life" provided "the farmer, the professional man, the business man, and the worker" with "means for soothing their over-wrought nerves and diversions from their daily tasks." Noting that the nation's 7 million hunters spent more than $50 apiece annually on their sport, the Game Commission boldly predicted that the state might easily attract thousands of affluent sport hunters from across the South.[47]

However, when the new law went into effect, anyone visiting the Black Mountains might have been hard pressed to find any game worth taking. Years of intensive logging and habitat destruction had devastated local wildlife populations. In 1919 John Simcox Holmes noted that black bears, once so important to local hunters, had been driven out of the Blacks by the lumbermen. Deer, too, had vanished from the higher ridges. As whitetail populations waned, predators faced a critical food shortage. In 1917 an Eskota farmer killed the last mountain lion seen in the Blacks when it attacked one of his cows. One of the last confirmed sightings of a gray wolf occurred that same year.[48]

Things were no better on the lower slopes. There, logging of oaks, hickories, and chestnuts had led to a sharp decline in wild turkeys and other birds that depended on mast. Fish suffered, too. Removing the trees along rivers and creeks caused water temperatures to fluctuate wildly. Silt from denuded hillsides built up in the Cane, the South Toe, and their tributaries, killing aquatic insects that had once flourished on the rocky stream bottoms. By the early 1920s eastern brook trout had diminished noticeably in several Black Mountain streams, especially those that originated on Clingman's Peak, Haulback, and Mount Mitchell where fires and erosion had been especially severe.[49]

Well aware that without drastic measures North Carolina might soon have no animals to protect, the Board of Conservation and Development made sure that the 1927 bill carried a provision authorizing state-run refuges and preserves for wildlife. Before the law was two months old, the Game Commission announced plans to create the first such refuge near Mount Mitchell. To many it seemed an ideal locale. Most of the land in the region was now controlled by the Forest Service and the city of Asheville, meaning that animals could roam vast areas of contiguous woodland without straying onto private property where they might be killed. Moreover, the largest privately held tract in the Blacks—part of the old Murchison Boundary in the Cane River Valley—had recently passed into the hands of several well-to-do Piedmont businessmen who had converted it into a private hunting retreat. They were as concerned as the state about rehabilitating local animal populations.[50] Besides, what better place to begin wildlife restoration than in the Black Mountains, a region that had produced Big Tom Wilson, the state's most famous hunter and outdoorsman?

The idea of a safe haven for wild animals was hardly new. The federal government had outlawed hunting in Yellowstone Park in 1894, and by the time North Carolina got serious about wildlife management, game refuges had become prominent features in many of America's national forests. Indeed, in 1916 the Forest Service (with cooperation from the North Carolina legislature) had established its own animal sanctuary on the old Vanderbilt property not far from Asheville. Known as the Pisgah Preserve, it spanned some 90,000 acres and provided protection for deer and other big game.[51]

During the 1920s and 1930s, Forest Service interest in wildlife management owed much to the work of one man: an Iowa-born hunter and conservationist named Aldo Leopold. Today most Americans know Leopold as a founding member of the Wilderness Society and the author of *A Sand County Almanac*, a classic collection of conservation essays that championed a new "land ethic" based on a deep appreciation of nature and its complexities. Before he joined Henry David Thoreau and John Muir as one of the nation's best-loved nature writers, however, and before he emerged as the "Moses" of the modern environmental movement, Aldo Leopold worked for the Forest Service. After receiving a master's degree from the Yale Forestry School in 1909 (the same year Holmes undertook his first forest survey of western North Carolina), Leopold served briefly as a forester in Arizona and, in 1912, became supervisor in New Mexico's Carson National Forest. When a serious kidney ailment forced him to give up that rigorous duty, he turned his attention to managing wildlife on government lands. In the early 1930s, after leaving the Forest Service, he compiled his findings in *Game Management*, a text that for more than a generation served as a handbook for anyone interested in wildlife restoration.[52]

When he wrote *Game Management*, Leopold was already interested in the nascent science of ecology and its emphasis on balance and harmony, concepts crucially important to *Sand County Almanac* and later works. But to a great extent *Game Management* still reflected the utilitarian doctrines preached at Yale and practiced in the national forests. In essence Leopold argued that just as foresters culled trees and suppressed fires to produce usable timber, wildlife managers should manipulate food and habitat to increase populations of "shootable game." Refuges were vital to such efforts, especially in areas where game had been seriously depleted. "A game refuge," Leopold wrote, "is an area closed to hunting in order that its excess population [of animals] may flow out and restock surrounding areas. A refuge is at all times a sanctuary, and the two terms are synonymous." Properly placed, Leopold believed, a refuge could "prevent the extermination" of a declining species until it "recuperated sufficiently" to sustain itself over its normal range.[53]

For forty years *Game Management* was a bible for those who worked at Mount Mitchell. "Oh yes, we read it," recalled A. E. Ammons, retired supervising biologist for Western North Carolina. "It was the premiere textbook. Every wildlife biologist in the state had studied it. In fact, we still relied on it in the early [19]60s when I started to work." Following Leopold's advice, the state temporarily outlawed hunting throughout the Mount Mitchell Game Refuge and Wildlife Management Area, a 22,000-acre tract that stretched from the upper South Toe in Yancey County across the southern end of the Blacks and down Curtis Creek in McDowell County. Officials placed gates on all roads leading into the area and threatened trespassers with fines and jail sentences. To enforce the regulations, the Game Commission hired Colonel Nelson Mease, the well-known Black Mountain warden whose relentless pursuit of Swannanoa dynamiters (and other similar heroics) had earned him a solid reputation for upholding the new game laws. Mease lived on-site in a newly constructed warden's cabin along Neals Creek.[54]

At first, though, neither the state presence nor the new policies had much effect. The resident deer population had fallen so low that Mease and others believed the animals might be extinct in the wild. Fortunately, not far away on the federally run Pisgah Preserve, white-tails were thriving. Vanderbilt's foresters had stocked nine deer on the property sometime between 1890 and 1900, and by the late 1920s the Biltmore Herd, as it was commonly called, comprised some 3,000 animals. In 1930, starting with six fawns, Mease began transplanting animals from Pisgah to Mount Mitchell. It was slow and often frustrating work. Deer had to be trapped at the preserve, taken by truck to the refuge, placed in holding pens (along Neals Creek and atop Bald Knob Ridge), and fed until they adapted to their new surroundings. Over the next four years Mease (now serving as supervisor for all the state's western game refuges) and his fellow managers released about eighty-five deer at Mount Mitchell.[55]

About the same time, the wardens also liberated seven black bears imported from Pisgah and other locales. Game birds, too, were important to the stocking program. Using eggs acquired from eastern North Carolina and elsewhere in the Southeast, the state set loose 110 wild turkeys (16 gobblers and 94 hens) and 58 bobwhite quail between 1928 and 1933. Like the foresters who replanted the Blacks' upper slopes, wildlife managers also experimented with exotic species, releasing 59 ringneck and golden pheasants (birds native to Europe and Asia) into the hardwood forests.[56]

Some of the most publicized early game restoration work involved elk. Whether elk had ever lived in the region was (and is) a matter of considerable debate among wildlife experts. But restoring the stately animals had been a long-standing dream in North Carolina. A decade and a half before the Mount Mitchell Refuge opened, George Gordon, a sport hunter from Sinclair, Michigan, had stocked fourteen elk at Hooper's Bald, a mountaintop estate in the Great Smoky Mountains. Gordon fell on hard times during World War I and had to shut down his restoration efforts. But a few of the elk remained in captivity and somehow found their way to the Pisgah Preserve. In addition the managers at Pisgah imported a small herd of elk from Yellowstone National Park in 1917. Mease brought eleven of those Pisgah elk—four bulls and seven cows—to Mount Mitchell in 1932.[57]

To ensure that elk and other animals had more than a fighting chance, North Carolina not only restricted human hunters but also declared an all-out war on nature's predators. Strange as it seems today, most wardens once regarded elimination of "varmints and predatory's" (as the state called them) as an essential part of game restoration, something as crucial to effective wildlife policy as good habitat and an adequate food supply. Early in his career Aldo Leopold advocated such tactics. During the first two decades of the twentieth century, predator eradication was standard policy in America's national parks as well. There the list of "undesirable

Headquarters complex at the Mount Mitchell Game Refuge along Neals Creek, early 1940s (*above*), and a flock of wild turkeys released at the refuge in 1939 (*below*). N.C. Division of Archives and History.

animals" (that is, those that threatened visitors or preyed on "popular" species such as deer and elk) included cougars, wolves, lynx, bobcats, foxes, badgers, minks, weasels, fishers, otters, and martens.[58]

By the early 1930s some wildlife biologists had already begun to question such practices. Indeed, in *Game Management* Leopold readily acknowledged that nature's "great drama of tooth and claw" was highly complex and that in some cases killing off carnivores created more problems than it solved. Yet when it came to refuges, Leopold and most other experts still believed that controlling predators made sense. As he also explained in *Game Management*, "Breeding stock does not flow out from any area unless there is population pressure . . . [and] there can be no population pressure within a refuge without law enforcement and the control of predators, food, cover, or any factors which may be holding down productivity." As the wardens at the Mount Mitchell refuge understood it, the message from the experts was clear: If turkeys, deer, and trout were to flourish, their natural enemies had to die.[59]

Between 1928 and 1933 Mease and his colleagues killed 858 snakes, 127 skunks, 112 hawks and owls, 30 rats, 18 weasels, and 9 kingfishers (which threatened trout) at the refuge. So as not to make trouble with local residents, the wardens generally tried to capture stray dogs and return them to their owners. But free-roaming house cats got no such consideration. Officers simply shot them on sight. After the state established other refuges in western North Carolina, the wardens engaged in a friendly competition each year to see who could destroy the most vermin. Local newspapers took great delight in reporting the annual kills.[60]

In 1933 and 1934, after more than a half-decade of stocking game, feeding wildlife, and shooting predators, the Mount Mitchell wardens took a census of various animals living on the refuge. The wardens discovered that once delivered from their enemies, native birds had multiplied rapidly. An estimated 875 ruffed grouse lived on the refuge in 1934 and, aside from squirrels (which numbered 1,300), had increased faster than any other species. Exotic birds fared worse, though. Some 135 pheasants had survived at Mount Mitchell, but in keeping with their preference for open terrain, the birds clustered in fields and cutover areas instead of moving into the surrounding forests. The first wild turkeys, accustomed to warmer climes in the Piedmont and coastal plain, had difficulty coping with the harsh Black Mountain winters. Of the 110 birds set loose since 1928, only 16 remained. Elk, too, never reestablished themselves in the Blacks, partly because the animals brought from Pisgah were half-tame. By

1934 only 6 of the original 11 elk remained, and they stuck close to refuge headquarters where wardens often fed them by hand. Black bears did better. Without harassment from hunters, the bear population increased from a half-dozen or so in 1928 to 42 by 1934.[61]

When it came to judging the state's new wildlife policies, however, the animals that mattered most to Mease and the Game Commission were white-tailed deer. Deer hunting was by far the most popular pastime for sportsmen, especially those from Piedmont cities and out of state. Visiting deer hunters bought more expensive, nonresident licenses and spent large sums on lodging and equipment, all of which generated state tax revenue. Providing deer for sport hunters became even more important in 1937 when Congress passed the Pittman-Robertson Act. The legislation placed a federal excise tax on the sale of firearms and other sporting goods; the money was earmarked to aid various states, including North Carolina, in "management and restoration of wildlife." Joe Scarborough, an assistant warden who worked at the Mount Mitchell refuge in the mid-1960s, explained it this way: "You have to understand . . . back then [in the early days of wildlife management] deer paid for it all. That was where the money was at, that was where the interest was at. All the programs . . . federal money . . . everything that's here today, deer and the people that hunted them paid for."[62]

Fortunately for workers in the Black Mountains, deer came back strong. The 1934 census placed the refuge herd at 343 animals, a threefold increase in just six years. By the late 1930s, as Pittman-Robertson went into effect, deer had multiplied so rapidly that the state decided to allow limited hunting of white-tails on certain refuge lands. For a small fee ($2.00 in 1941), hunters with valid North Carolina licenses (and who had not already taken their limit of deer) could secure a permit that allowed them to hunt at Mount Mitchell for one to three days in early November. During that time they could legally take one buck. Hunters had to check in and out of the area every day and record their kills with the local wardens. Even with those restrictions the "managed deer hunts" at Mount Mitchell proved phenomenally popular. Before World War II, Mease had no trouble filling a quota of 200 permits per day during the short seasons. Indeed, the deer program worked so well that wildlife officials soon established similar short seasons for hunting black bear on the refuge. Strictly controlled hunts, Mease explained, kept refuge animals from "increasing beyond the carrying capacity of their range" and provided "a small return on the investments of wildlife management projects."[63]

When it came to fish, the managers had even greater success, at least in the short run. The Game Commission began restocking refuge waters in 1929 using trout raised in other mountain hatcheries. By 1933 wardens had planted 150,000 fingerlings in Black Mountain streams. Rainbow trout, better suited to survival in warmer water, were released into the South Toe, and hatchery-reared brook trout went into the tributaries. Mease might also have brought in a few brown trout during his first years at Mount Mitchell, though the exact date those fish became established in the area remains an open question. Stocking efforts were so effective that during the early 1930s, well before hunting resumed, the state opened the streams to fishing for a few days each summer. Like deer hunters, anglers had to check in and out so that wardens could measure and weigh the daily catch.[64]

Black Mountain fishing also received a huge boost in 1934 when Franklin Roosevelt's Public Works Administration provided funds for a new hatchery on Neals Creek. Capable of producing 300,000 to 400,000 fish per season, the facility was soon supplying brook and rainbow trout for all refuge waters. Sometime around 1940, in what he later called "a sort of publicity gag," Mease set aside Neals Creek as a stream "for women only." The special designation, apparently the first of its kind in the South, created an immediate sensation. On summer weekends when the stream was open, some fifty to sixty women lined its banks. A few years later Mease happily reported that "in spite of the large number of young men taken into the armed services," the refuge, with its women-only waters, had "more mountain fishing fans than ever before."[65]

Amid all the good press, however, problems persisted. Due to continued logging on the lower slopes and the advancing chestnut blight (which further reduced the supply of seasonal mast), deer, bear, and other game were still scarce outside the refuge. Lured by rumors of abundant white-tails, docile elk, and streams teeming with trout, residents of nearby communities frequently ventured onto government lands in search of meat for the table. During the first years of the Great Depression, when some mountain people began to rely more on wild game for sustenance, Mease's monthly reports to the Game Commission often read like a local police blotter. In August 1930 E. T. Robinson, from Busick, North Carolina, got a thirty-day jail sentence for trespassing on the refuge. Frank Hughes of Micaville, apparently caught fishing illegally in May 1931, paid a fine and court costs of $7.00. Horras [*sic*] Effler of Busick, apprehended in April 1932, spent two weeks in the Burnsville jail.[66] Long accustomed to hunting and fishing as

Wardens at the game-holding pens (*above*) and the fish hatchery (*below*) at Mount Mitchell Game Refuge, early 1940s. N.C. Division of Archives and History.

they pleased, many Black Mountain people saw no reason to comply with new rules written primarily for sportsmen and tourists. Illegal hunting and fishing remained serious problems at the refuge for years to come.

While local residents chafed under the new regulations, some conservationists raised other concerns. Starting in the mid-1930s, as ecologists began to unravel the mysteries of food webs and energy flow, the national parks "moved very slowly and erratically . . . toward a scientific understanding of predator and prey populations and the discontinuance of predator control." Aldo Leopold, too, began to advocate a more inclusive philosophy of wildlife management that stressed protection of predators and a balanced natural system. At Mount Mitchell, Leopold had a kindred spirit in fellow Yale alumnus John Simcox Holmes. Much interested in ecology and apparently well aware of changing policies in the national parks, Holmes openly criticized state laws that favored "the group known as game" and suggested that "instead of killing one species to favor another, nature should be allowed to do her own regulation." Though Mount Mitchell State Park bordered the refuge, Holmes steadfastly refused to allow any hunting, for game or predators, on park lands. "The controlling principle of wildlife management," he insisted, "should be to encourage the natural distribution of animal life."[67]

When it came to the state park, the Board of Conservation and Development had no problem with Holmes's ideas. But with Pittman-Robertson money at stake and sportsmen flocking to the South Toe for managed hunts, the board saw no reason to change policies at the refuge. Besides, as auto touring and car camping became the rage across North Carolina, the state drew up other plans for the land along Neals Creek. Still in competition with the Park Service for government dollars, the Forest Service, with support from the North Carolina Game Commission, launched an extensive program of scenic landscaping around the refuge headquarters. Works Progress Administration workers used native stone to construct the warden's house, checking stations, and other buildings. The CCC planted thousands of shrubs and wildflowers. To attract children and educate them about local wildlife, Mease put together an extensive taxidermy exhibit that he not only displayed for visitors but also took to the annual state fair in Raleigh. In addition the Forest Service built a small zoo for native wildlife that featured live fawns and black bear cubs. New Deal money also allowed for construction of a new camping facility, the forerunner of today's Black Mountain Campground, less than a mile from the refuge. More affluent visitors could rent private cabins along Neals Creek or reserve rooms at Gilkey Lodge, a hotel built partly with private money and named for longtime Board of Conservation member Joseph Quince Gilkey. By 1940 nearly 10,000 auto tourists came to Neals Creek every summer, turning the upper South Toe Valley into a nearly year-round recreational attraction.[68] Originally conceived of as a way to protect animals, the refuge, like Carolina Hemlocks and Mount Mitchell State Park, had become a sanctuary for people, too.

NOVEMBER

WINTER STAR

In late autumn, daylight comes slowly to the Blacks. At 7:00 A.M. under leaden skies the woods along the South Toe are gray and silent. At an elevation of 2,800 feet, where the trail begins, the air is still; the temperature is thirty degrees. Snow is in the forecast. I dress lightly, in a wool sweater and baseball cap. Judging from the closeness of the lines on my topographic map, the morning's walk will be plenty rigorous to keep me warm. Over the course of two miles or so I will gain some 3,400 feet in elevation as I hike the eastern face of Winter Star, a 6,200-foot mountain in the northern reaches of the range.

Despite its picturesque name Winter Star is one of the lesser-known peaks in the Black Mountains. Located well north of Maple Camp Ridge, it stood beyond the reach of the Mount Mitchell Railroad and the various lines that stretched up from the Cane. Because the mountain lacked railroad grades and major motor roads, any tourist who wished to see it had to hike up from Carolina Hemlocks or take the Crest Trail north from Mount Mitchell State Park. Today it remains relatively isolated, accessible primarily by the Crest Trail or this tortuous route up from the South Toe.

Ten minutes from the trailhead, still on relatively level terrain, I notice a number of old road cuts, now overgrown with yellow birch and red maple, that seem to dead-end into the mountainside. Judging by the four-foot mounds of dirt and rock still visible in the surrounding woods, this area probably once supported a number of small mines where local people dug mica and gemstones by hand for dealers in Burnsville and Spruce Pine. Not far beyond the summit of Winter Star, on the western flank of Celo Knob, one can still find old shafts and tunnels from the Ray Mine, a much larger operation worked by various corporations for more than 100 years. Indeed, as I walk out of the trees onto a rocky outcrop that overlooks the South Toe Valley, I can see the surrounding ridges of Yancey and Mitchell Counties, some of which are heavily scarred by the massive cuts, open pits, and deep quarries typical of today's mica, kaolin, and feldspar industries.

Above 3,000 feet the trail grows steeper. The mounds, old roadbeds, and smallish timber give way to a thick forest of red and white oaks interspersed with hick-

ory, yellow birch, and other hardwoods. These forests, like those at similar elevations throughout the Blacks, were once home to the American chestnut, and as I rummage around off the trail, I discover a few telltale stumps and sprouting saplings half-buried in a foot of recently fallen leaves. Here the forest seems to be repairing human damage, slowly erasing evidence of the blight that ravaged these slopes during the 1930s.

Climbing higher on Winter Star, I encounter fewer and fewer signs of people. At noon the skies darken and the promised snow swirls through a thick stand of red spruce near the summit. Walking back to the trailhead via a different route, I watch the snow turn to sleet and pause momentarily, listening to the tiny ice balls clattering on dead leaves and dry timber. As I sit quietly in the November woods with no roads, campgrounds, or old fish hatcheries to remind me of past practices, it might be easy to conclude that in the more isolated reaches of the Blacks the impact of government has been minimal. But here, as on Mount Mitchell or along Neals Creek, various state and federal agencies took a keen interest in the forests. And if one looks closely, Winter Star, with its old mines, chestnut stumps, and sprouting hardwoods, still bears the stamp of policies instituted more than a half-century ago.

Spanish explorers had seen mica and other minerals when they passed through the North Toe Valley, but mining in and around the Blacks did not begin in earnest until after the Civil War. Sometime in 1867 or 1868 Thomas Clingman, out of public office but still interested in attracting northern investors to the southern mountains, convinced a group of New York mica merchants to bankroll a prospecting expedition in western North Carolina. Clingman knew that the best mica deposits would likely be found in Mitchell and Yancey Counties. The counties straddled a giant band of pegmatites, those granitic, mica-bearing rocks that lay buried in ancient layers of gneiss and schist. Known as the Cowee–Black Mountain Belt, the formation stretched for 200 miles along the western flange of the Blue Ridge. In the Blacks it passed directly beneath Winter Star, Gibbs Mountain, and Celo Knob, the northernmost peaks in the range. Tapping into the Cowee–Black Mountain pegmatites, usually by means of a shallow shaft, Clingman and his prospectors unearthed large sheets of "rum-colored" muscovite or sheet mica, some of the finest specimens discovered anywhere in the world. Over the next several years, with help from various investors, Clingman helped open several mines in the region, including the Celo Knob operation that came to be known as the Ray Mine.[69]

Nineteenth-century prospectors found other valuable minerals in and around the Blacks, including feldspar, clay, kaolin, and various gemstones.

At first, however, mica proved the most profitable. Highly resistant to heat, sheet mica became an important component in stove glass, lamp globes, and various fireproof materials. By the early 1900s, as American industry became dependent on electricity to power machinery, mica served as an insulator. Electricians were soon using it in light sockets, spark plugs, rheostats, fuse boxes, and a bit later, telephones. In addition to high-quality muscovite, mines in and around the Blacks also supplied the nation with scrap mica, smaller—sometimes minuscule—flakes and shavings that might be culled from almost any granitic outcrop in the region. Also known as ground mica, it was used in wallpaper, fireproof paint, roofing materials, and tempered steel. By 1903 more than half the sheet and scrap mica mined in the United States came from western North Carolina. Burnsville, Spruce Pine, and Bakersville had no fewer than twelve mica processing plants.

Farmers, too, tried to cash in on the mica bonanza. Local folk living along the South Toe often ventured onto the slopes of Celo Knob and Winter Star, usually in late fall or early spring when crops and livestock required less attention. Equipped with picks, shovels, wagons, and mules, they dug mica and gemstones for sale to the processing plants. The exact number of independent miners in the Blacks is impossible to estimate, but by 1919 state geologist Joseph Hyde Pratt could write that "small mines or prospects worked intermittently by farmers" furnished "much of the output" for processing plants in western North Carolina, including those in Mitchell and Yancey Counties.[70]

Yet for all the early interest in the region's mineral resources, the mining industry never really gained a foothold in the high mountains. For one thing, much pegmatite lay deep in the extreme northern reaches of the Blacks, so it was difficult to locate and expensive to market. On the southern end of the mountains, where lumber companies owned both land and railroads, timber generally proved more profitable and easier to come by than mica and gemstones. In the years after World War I, changing markets and innovations in the mining industry also encouraged prospecting on more level terrain. Beginning in the 1920s cheap imported mica (brought to the United States from India) drove many independent miners and smaller processing plants out of business. To stay competitive, those who ran larger operations in and around Spruce Pine began to invest in modern machinery that allowed for greater exploitation of deposits in the Toe River Valley. Over the next several decades, as the market for domestic mica waxed and waned, power shovels ate into the surrounding hillsides, carving huge open cuts, some of which were hundreds of feet deep, into the Toe River landscape. By the 1940s "crude material" extracted from the cuts went to giant "washing plants" where a flotation process separated scrap mica, feldspar, and other valuable materials from the loose dirt, sending tons of sediment pouring into the Toe.[71] Even now, despite recent government regulation of the mining industry, restoration of open cuts and reduction of runoff remain important environmental issues in and around the Toe River towns.

The Blacks themselves, however, escaped relatively unscathed. The Ray Mine remained productive even as the market fluctuated, but because its principal products (sheet mica and gemstones) had to be extracted by hand, it seems to have remained mostly a pick-and-shovel operation. Descriptions of operations there are difficult to come by, but one account done in 1914 noted that the mine had a small 40-foot open cut and a single shaft running 110 feet into the mountain. Aside from the cut itself and the dumpings from the shafts (most of which were later hauled away and sifted for feldspar), the impact on the surrounding forests must have been slight. Two other mines, one on Cattail Creek and the other on the head of the Swannanoa's North Fork, apparently operated for a brief period sometime between 1910 and 1930 using shallow cuts and short tunnels to get at small deposits of muscovite.[72] Old roads and overgrown rock piles still visible at the base of Celo Knob (and much more distinct than those on Winter Star) indicate that miners continued to chip away at mica deposits there perhaps as recently as thirty or forty years ago. But on those slopes, too, maples and yellow birches spread quickly over the abandoned works, which suggests that mining generated only slight changes in existing forest patterns.

The federal government also played an important role in driving miners into the valleys. Passage of the Weeks Act in 1911 coincided almost exactly with the boom in the scrap mica market. Though nothing in the act specifically prohibited prospecting in national forests, would-be miners still had to get federal approval for their operations. With the market booming and accessible deposits available along the North Toe, they probably found it more profitable and expedient simply to avoid government land. By 1917, when the Forest Service began to purchase vast tracts on the eastern slopes of the Blacks, the market for scrap mica had already begun to diminish, and most local people eventually abandoned their seasonal operations on the lower slopes. The Ray Mine, located far out on Celo Knob, in one of the last areas acquired by the Forest Service, was the lone exception. It operated off and on until the early 1970s, an anachronistic monument to Thomas Clingman's dream of turning the Blacks into a treasure trove of minerals that might fuel southern industry.[73]

As the miners moved out of the Blacks, state and federal officials turned to another, far more pressing problem: the worsening chestnut blight. In the eyes of the Forest Service, trees had always been the most valuable resources of the high mountains, both for watershed protection and as a future lumber supply. Five years earlier John Simcox Holmes and others at the Department of Conservation and Development had been optimistic, hoping that the government's plant pathologists might develop a miracle cure. But now they faced a harsh reality. According to departmental estimates some 80 percent of Black Mountain chestnut stands would be infected by 1930. Yancey County alone figured to lose nearly 200,000 board feet of valuable hardwood timber to the lethal fungus.[74]

For Holmes and others schooled in the ways of utilitarian forestry, the possibility of so much wood rotting on mountain slopes was simply unacceptable. Spreading the word through circulars and press releases, the Forest Service urged lumbermen across the mountains to cut and sell as much chestnut as possible before the blight ran its course. Holmes knew that by encouraging such a policy he ran the risk of flooding the market and driving down prices. But since blight-killed chestnuts often remained standing for fifteen or twenty years (while the wood stayed relatively free of decay), he and other foresters believed the trees might be logged and sold gradually, thereby preventing a glut. In 1925, about three years after the various lumber companies left the spruce-fir forests, loggers began cutting chestnut in earnest on the lower slopes.[75]

Because it had the backing of the Forest Service and drew relatively little criticism from the press, we know far less about the logging of American chestnut than about the spruce-fir operations at the summits. Government foresters, local lumber companies, and even a few freelance loggers from nearby communities all cut chestnut on Forest Service land. During the 1930s the CCC probably got into the act, too. Elsewhere in the Appalachians the CCC cut live chestnut for sale and removed dead trees, or "snags," that might attract lightning and encourage forest fires.[76] Since there were several camps active in and around the Blacks, it seems almost certain that the CCC engaged in similar practices there.

In the mid-1920s, when the chestnut harvest began, many of the trees on the lower slopes had probably already been infected by the fungus. But at higher elevations, especially between 3,500 and 4,000 feet, some chestnuts probably remained relatively healthy. Perhaps due to cooler temperatures and generally drier conditions, the fungus seemed slower to invade higher ridges, especially those with a southern exposure. But to those who conducted the salvage operations (as the Forest Service would later call them), elevation and habitat made little difference. What mattered most initially was the size of the tree. Larger chestnuts were still in demand for construction lumber, furniture, railroad ties, and hundreds of other items. To no one's surprise, corporations that dealt in wood products quickly embraced logging of doomed trees. The American Forest Products Company, a leading producer of telephone poles (for which the taller chestnuts were particularly well suited) took a leading role in promoting the salvage operations.[77]

Over the next ten years, as big trees disappeared and the blight showed no signs of abating, chestnut harvesters in the Blacks also found a nearby market for smaller timber. At Champion Fibre's Canton plant, just fifteen miles west of Asheville, chemical engineers had perfected a procedure that allowed tannin to be taken from chestnut chips without significantly weakening the wood fiber. Once the tannin had been extracted, leftover chips could then be turned into pulp for paper, making for a doubly efficient operation. (Earlier methods had left the chips so brittle that they were of little use and had to be burned.) Thanks to this patented technique, the company emerged as a major producer of extract and pulp for the international market. By the early 1920s processing chestnut had become Champion's most lucrative enterprise.[78]

Running at peak capacity, the plant consumed an astonishing amount of chestnut wood: 275 cords during a single working day, or about 12 cords each hour. When it came to pulp and tannin, the size and condition of the trees made little difference. Healthy green wood worked best, but because a diseased chestnut retained acceptable levels of tannin and fiber for several years even as it succumbed to the fungus, Champion bought plenty of blighted wood, shipping it in by the freight-car load from suppliers up to 100 miles away. To protect against a slowdown in production, the plant kept at least 15,000 cords of chestnut (about a two-month supply) on-site at all times. During the 1930s, as salvage operations peaked, most of the chestnut cut in the Blacks eventually found its way to Canton, where according to the Forest Service, "it was chipped and cooked to extract tannic acid used to tan leather." Efforts to retrieve chestnut from the lower slopes slowed in the 1940s due to a shortage of manpower during World War II and an overstocked market. But Champion Paper and Fibre (the name change came in 1935) continued to buy blighted trees until the early 1950s, when the growing scarcity of chestnut and the invention of synthetic additives (which gradually replaced tannin) made extract and pulp operations unprofitable. As Reuben Robertson, an executive vice-president and later chairman of the board at Champion, explained, "The [good] timber was becoming more remote. We had to pay more to get it out; the tannin content had dropped . . . because there was a certain amount of decay [and] . . . we finally reached a point where it didn't pay anymore to use chestnut."[79]

Salvage operations in the Blacks mirrored those elsewhere in Appalachia. Indeed, during the 1930s some twenty-one extract plants operated in the southern mountains and furnished over half the tannin used in the United States, until they, like Champion, used up local supplies of blighted wood and shut down.[80] Given the rush to clear the lower slopes of usable timber, it might well be said that the lethal fungus introduced from Asia in 1904 only initiated the decline of American chestnut from majestic tree to understory shrub. Lumbermen, with the full cooperation of government foresters, finished the process.

Viewed from the present, with the precision of 20/20 hindsight and modern ecological sensibilities, the salvage operations might have been a mistake. For one thing, taking out the largest trees, especially those at higher elevations that would have been slower to succumb to the blight, removed some of the hardier specimens from the gene pool, thereby reducing the chances for American chestnuts to develop natural defenses against the fungus. No one can say for sure what might have happened. Beginning about 1950 in Italy, however, European chestnuts (also afflicted by blight) began to develop new "hypovirulent," or less lethal, strains of the fungus. Once established on an infected tree, the hypovirulent strains produced "healing cankers" that attacked the blight and, quite literally, allowed European chestnuts to cure themselves. Two decades later chestnut blight had ceased to be a problem in Italy, and scientists were successfully using hypovirulence to combat blight elsewhere in Europe.[81]

Attempts to inoculate trees in the United States with the healing European fungus met with mixed results, but in the late 1970s researchers discovered indigenous strains of hypovirulence in American chestnuts, first in Michigan—in trees planted outside their original range—and later in the blight-stricken eastern forests. American scientists still have much to learn about how hypovirulence spreads. (The actual "blight-debilitating agent" seems to be a ribonucleic acid contained in the cytoplasm or liquid cellular material of the parasite.) In the eastern United States, where the

Salvaged American chestnut stacked outside Champion Fibre in Canton, N.C., awaiting extraction of tannin, 1937. Some of the wood stored here probably came from the Black Mountains. Reprinted with permission from International Paper.

blight ran wild for seventy-odd years before the less toxic strains appeared, there are now more than 100 deadly varieties of the fungus, many of which do not yet respond to hypovirulence. A cure for American trees is probably a long way off. Even so, it seems clear that nature is slowly marshaling its defenses against the lethal fungus, and one cannot help but wonder if the story might have been different had lumbermen not taken so many trees for telephone poles, lumber, and tannin. As one careful student of the blight has written, "What natural blight resistance lurked in the vast gene pool of American chestnuts may have gone to carry wires for AT&T."[82] In the case of the Blacks, one might add, "to fill the extract vats at Champion Fibre."

As they supervised the removal of dead and dying timber, southern foresters continued to speculate about which trees might replace the American chestnut. By the late 1920s research conducted in Virginia and Maryland suggested that "blight-killed chestnut [had] been very largely replaced by oak," a species foresters found desirable because of its potential value for lumber. Over the next few decades, though, it became apparent that while certain oaks did occupy old chestnut habitat, the replacement process, as usual in the turbulent world of the high Appalachians, was far more complex and variable. At elevations between 3,500 and 4,000 feet, where chestnut grew more sporadically, removing the taller trees probably created only small openings in the forest canopy. In all likelihood certain oaks that flourished alongside chestnuts in such regions simply moved into the vacated space, just as the first studies suggested. Today red oak and chestnut oak are the most prominent trees in the upper reaches of the chestnut's former range. Farther downslope, though, on dry ridges where chestnuts were more abundant and salvage logging more extensive, changes were more pronounced. There, clear-cutting chestnut for tannin resulted in sizable clearings that invited invasion by a variety of competing trees and shrubs. White and scarlet oaks, red maple, black locust, yellow birch, hickories, rhododendron, and mountain laurel all cropped up among the chestnut stumps and sprouts, giving rise to the hodgepodge of mixed hardwoods now visible at the base of Winter Star and neighboring peaks.[83]

Progressive foresters had also worried that the chestnut's demise might spell doom for gray squirrels, black bears, wild turkeys, white-tailed deer, and other species that depended on the mast-bearing trees. In the high mountains, though, populations of those animals (except possibly squirrels) had already been decimated by logging and overhunting. It is difficult, therefore, to gauge the exact impact of the blight and the salvage operations on Black Mountain wildlife. Even at the Mount Mitchell Game Refuge, the one place where wardens carefully counted game animals, research seems to have focused on elimination of predators, artificial feeding programs, and the effects of various hunting regulations, not on natural food supplies. Still, the blight and, more important, the quick removal of healthy chestnuts (which might have produced valuable food for years to come) no doubt contributed to the general shortage of deer, bear, and turkey in the Blacks during the late 1920s.[84]

Loss of food and habitat from the salvage operations might be one reason why wardens had to import breeding stock from the Pisgah Preserve. The Forest Service conducted salvage operations there, too. But at Pisgah, where hunting had been restricted since 1916, deer and bear were not hounded by human predators and had begun to shift their diet more toward other available wild foods. That adaptation probably served the animals well when wardens moved them to the Blacks. It might, in fact, help explain why deer and bear flourished at the refuge in the 1930s but were slow to come back elsewhere in the range.

To foresters, though, the removal of dying chestnut and its apparent replacement by oak or other useful species were encouraging signs, the latest evidence that government had made considerable progress in the battle to save the Black Mountains. In 1939, as he contemplated retirement, John Simcox Holmes looked back with satisfaction on thirty years as North Carolina's chief conservation advocate. Up and down the Blacks, from the experimental plots on the brushy slopes of Clingman's Peak and Mount Mitchell to the campgrounds and fish hatcheries along the South Toe, state and federal agencies had left indelible marks on the mountain landscape. Many civic boosters reveled in the accomplishments. As the *Asheville Citizen-Times* declared regarding an interview with Holmes, the state's message to its citizens was now loud and clear: "You shall not desecrate this earth with your axes and your guns. Here the animals and trees shall live on equal terms with man."[85]

Caught up in the euphoria of the moment, the reporter apparently did not ask about problems inherent in the new conservation ethic. However, in the Blacks as elsewhere in the nation, government had done more than simply deliver a message. It had, in fact, taken over most of the region's resources. The U.S. Forest Service now controlled the bulk of Black Mountain land and timber; the state of North Carolina held title to Mount Mitchell. The city of Asheville had possession of the Swannanoa's headwaters, and an alliance of state and federal agencies managed the deer, bear, and other animals living on the game refuge.

Like any good Progressive, Holmes would have had trouble with that concept. Indeed he almost certainly would have denied that the Division of Forestry or any other government agency owned anything. Instead he would have insisted, as did most conservationists, that Black Mountain timber, water, and wildlife really belonged to all the people. The Forest Service, the Department of Conservation and Development, and the Game Commission only made sure that the people's resources—*their* property—remained protected for future use.

But that argument might have been tough to sell to local residents. From their perspective a mountain landscape that had once been open to common use was now closed. That the government could arrest and jail local hunters who ventured onto the game refuge offered proof of that. The closing of the forest commons also meant that livestock were now confined to pens and no longer so crucial to the local economy. As a result farmers increasingly devoted more acreage to cash crops such as burley tobacco. Better roads, access to motor vehicles, and the cash that came with factory paychecks meant that people who lived along the Cane and the Toe could now buy food and other items in Burnsville, Spruce Pine, or Asheville. Indeed, studies show that by the 1940s, throughout the southern mountains, consumption of canned goods and "store bought foods" had increased dramatically, while wild game and other traditional dietary staples — rye bread, buttermilk, and maple syrup — had declined in importance. As several historians have pointed out, the government takeover might have helped conserve Appalachia's resources, but it was also one of many factors that pulled mountain people away from their farmsteads and into America's industrial economy.[86]

The pervasive influence of government did not end there. As local folk moved out of the forests, tourists and sportsmen moved in. Due to the state and federal interest in promoting recreation, the Blacks increasingly became a place for leisure, not work. To the Progressive way of thinking, Mount Mitchell State Park, the wildlife refuge, and the various campgrounds, like the forests, belonged to all citizens (in the Jim Crow South that translated as all *white* citizens). But in reality the mountains had become largely the domain of middle- and upper-class city dwellers who had both the means and the inclination to spend their leisure time in western North Carolina. In years to come they, with considerable help from government, would determine much of what occurred in and around the highest peaks in the East.

Given the nature of Holmes's job and the tenor of the times, he could hardly have been expected to consider the ironies of Progressive conservation or to question a management philosophy that now seems more than a trifle elitist. Had he been asked, however, the savvy old forester could certainly have cited other, more practical problems that the conservationists had not solved. He might have mentioned the slow return of native trees to fire-scarred slopes, the ongoing feud with Ewart Wilson, a federal wildlife policy that sanctioned the killing of predators, or the dire predictions of hikers and other wilderness enthusiasts that automobiles and increased tourism could ruin the Blacks. Maybe, just maybe, as he looked back on three decades of work and considered the eccentricities of both people and nature, Holmes found reason to wonder whether the future of the East's highest mountains would be as bright as it seemed in 1939. If so, he was right.

SIX

Murphy's Law

SEPTEMBER

COMMISSARY RIDGE

Trudging slowly along the old railroad bed, head down against the wind, I hear the rain before I feel it. It sweeps in from the northeast and races up the South Toe Valley, pounding the leafy hardwood forest below. I drop my pack and struggle into a light jacket as the squall overtakes me. Like others hiking in the Blacks this Labor Day weekend, I did not anticipate bad weather. For nearly a month skies have been clear; the seasonal drought has all but extinguished many of the springs on the eastern slopes of the range. I could find no water to fill my bottles this morning and, strangely enough, standing in the middle of a downpour, I am thirsty.

I am also bone-tired. Last night I pitched my tent on Maple Camp Bald, a site not far from the old railroad terminus where shrubby vegetation has replaced the logged out spruce-fir forest. But for one of the few times while camping in the high mountains, I did not sleep. The wind began at midnight, roaring across the Blacks (I would later learn) at more than fifty miles per hour. Sometime in the wee hours a tent pole—supposedly made of high-strength "aircraft aluminum"—shattered, rendering my shelter useless and forcing me to spend the rest of the night outside, curled in my sleeping bag, with only a blackberry thicket for a windbreak. And now, the rain—the result, I will discover, of a hurricane that came ashore in eastern North Carolina late yesterday.

Depressed by the foul weather and lack of sleep, I opt for the shortest route out of the backcountry into the state park and the sheltered world of the automobile tourist. Taking an old trail from Camp Alice toward the summit, I climb directly up the highest mountain in the East, emerging finally in the paved parking lot near the summit where I left my vehicle. Inside the truck, with my wet gear stowed and the heater blasting, I drive to the Mount Mitchell Restaurant, a state-owned but privately managed facility advertising fine food and great views.

The restaurant is almost empty—most of the day tourists have been deterred by the bad weather—and I slump into a chair by a huge picture window overlooking the eastern slopes of the Blacks, including some of the route I walked this morning. Eating chili and a hamburger, I watch rain drumming the landscape below. A raven glides over a stand of mountain ash. A doe and a half-grown fawn stand sheltered in a grove of trees just below the building. Chickadees and juncos flutter about the windowsill. It is an odd sensation, sitting warm, dry, and well fed amid the remnants of a hurricane, gazing at nature under glass. But it is not unpleasant. I glance at my watch and am surprised to discover that an hour has passed. I order coffee to go and pause briefly in the restaurant's gift shop to look at T-shirts emblazoned with red-tailed hawks and black bears, boxes of "lifelike" rubber snakes, and plastic replicas of dragonflies and other insects.

Because I fancy myself a hiker, I am tempted to dismiss my lunchtime experience as artificial, to believe that real nature, the real Black Mountains, can be found only in the wind and rain, far from the world of fast food and tacky souvenirs. I remind myself that even the most remote parts of the Blacks (including the brush-covered slope on which I spent a restless night) have also been profoundly affected by humans and their machines. But in its most recent guise the Mount Mitchell landscape, where one can slip easily between auto and glass-enclosed restaurant, is a product of the recent past. It was born in the last half-century, when the state park began to offer middle-class suburban tourists a new kind of outdoor experience.

The state and federal money that flowed so freely into conservation projects in the 1930s disappeared when the first Japanese bombs fell on Pearl Harbor. As North Carolina and the nation geared up for war, tourist traffic on Mount Mitchell slowed to a trickle, especially after the federal government instituted mandatory rationing of tires and gasoline. The state park stayed open, and the warden, who remained in residence, "faithfully maintain[ed] his supervision." Aside from fire prevention and routine trail maintenance, however, he had little to do. Indeed, he probably had more contact with military personnel than with tourists. Throughout the war the U.S. Weather Bureau kept up its station on the summit, and several times during the early 1940s army units occupied the old Civilian Conservation Corps (CCC) camp and used the nearby observation tower "for secret training and experiments with radar equipment."[1]

At Mount Mitchell as in other state and national parks, military demand for resources put new pressures on previously sacrosanct supplies of timber and minerals. Apparently no one suggested cutting more red spruce—which was still useful for airplane construction—in the Blacks, but it does appear that lumbermen stepped up efforts to harvest chestnut and other hardwoods from the lower slopes. On the eastern side of the Blacks several Yancey County firms again investigated the possibility of mining mica on and around Mount Mitchell. A committee of state and federal officials that evaluated the plan quickly decided that opening new mines would be "unwise and unnecessary" since, as had been determined twenty years earlier, "more easily accessible prospects were available in the region." Taking their cue from administrators in the national parks, North Carolina conservation officials carefully affirmed their patriotism and support of the war, even as they tried to stave off the miners and loggers. Besides, before they could prepare for the return of tourists, park boosters faced a more immediate problem: They still needed a good road to the summit.[2]

Fortunately, in late 1945, as GIs returned from Europe and North Carolinians celebrated victory over Japan, the Highway Commission made good on its prewar promise to build a connecting road from the Blue Ridge Parkway to Mount Mitchell. With rights-of-way secured and much of the survey work already done, construction proceeded quickly. Dedicated in May 1948 at an elaborate ceremony attended by a host of dignitaries (including Governor Locke Craig's widow), the 4.7-mile strip of tar and gravel, with its easy grade and gently sloping curves, was a motorist's dream. Winding around the base of Potato Knob, Clingman's Peak, and Mount Gibbes, past Ewart Wilson's tourist camp at Stepp's Gap, and dead-ending into a parking area near the summit, N.C. Route 128 (as it came to be known) dramatically cut travel time between Asheville and Mount Mitchell. "Back in 1926," the *Asheville Citizen-Times* bragged, tourists were happy "to drive up the mountain and back in a single day. Now it takes only an hour to make the 33-mile trip . . . on an all-paved road." Governor R. Gregg Cherry offered a more practical assessment. "This highway," he predicted in a 1947 address to the Board of Conservation and Development, "will pay for itself many times over in gasoline taxes exacted from lowlanders who want to trod the highest spot."[3]

Cherry's remarks proved prophetic. More than any other single development, the road ushered in a new era in Mount Mitchell's history. Whether they came on foot, by rail, or via turnpike, tourists had always figured prominently in plans for the state park. But for conservationists such as John Simcox Holmes, replanting the Blacks' fire-scarred slopes and educating the public in the ways of utilitarian forestry had been equally important. Now Thomas W. Morse, first head of the recently created State Parks Division of Conservation and Development, oversaw a new program that emphasized tourism over conservation. Route 128, which sliced straight through the Forest Service plots on Potato Knob and Clingman's Peak, obliterating several old stands of Norway spruce and Douglas fir en route to the summit, was a perfect symbol of the coming transformation.[4]

During the early 1950s, general prosperity, the availability of new automobiles via easy credit, and the baby boom all helped make summer vacations and overnight car trips essential elements of the good life in postwar America. By the middle of the decade, interstate highways and white migration to the suburbs were also feeding the trend. Harried by the daily stresses of work and pressure to keep up with their materialistic neighbors, white middle-class families increasingly sought solace in nature. Indeed, the ability to take a vacation—to have the means to load the children into the car and set off to a national park or some other natural wonder—became in itself a mark of success and affluence. North Carolina moved quickly to cash in on the tourist bonanza. A new Travel and Promotion Division (initially under the auspices of the Department of Conservation and Development) touted the state as "Variety Vacationland" and invited visitors to enjoy sunny Atlantic beaches, gently rolling Piedmont hills, and cool, cloud-draped mountains, all from the comfort of the family car. By the end of the decade, tourism was a multibillion-dollar business and one of the state's most important industries.[5]

Mount Mitchell figured to be a leading North Carolina attraction, but at first officials seemed unsure about how to promote it. The Blacks had no natural lake or man-made reservoir, so swimming and boating—two of the most popular activities in eastern state parks—simply could not be developed in the high mountains (especially since summer temperatures rarely rose above seventy degrees). Moreover, because the state owned a mere 1,500 or so acres in the range, it could provide only limited opportunities for hiking and horseback riding, two sports long favored by tourists in the national parks. Camping had always been popular with Mount Mitchell's visitors, but experience showed that those who came by car preferred the milder nights and well-developed facilities at Carolina Hemlocks or Black Mountain, the two campgrounds operated by the Forest

Paving the wild: N.C. Route 128, late 1940s (*above*), and the modern entrance to Mount Mitchell State Park from the Blue Ridge Parkway (*below*). N.C. Route 128, which reduced travel time between Asheville and Mount Mitchell to one hour, engendered a new boom in Black Mountain tourism. N.C. Division of Archives and History and North Carolina Collection, Pack Memorial Public Library, Asheville.

Service. Traditionally, the higher peaks had appealed more to Boy Scouts and other organized groups who preferred backpacking to auto camping. With all that in mind, the Parks Division decided that Mount Mitchell should cater to "persons intending to make a short stay of one day to two weeks, rather than the all-summer visitor." By the late 1940s park planners had a wish list that included an inn with a restaurant, twenty log cabins, a campground and picnic area, "a museum with a naturalist in charge," and a system of trails and shelters "of the Youth Hostel type," as well as the "shops, garages, and storehouses" necessary for upkeep.[6]

Because the tragic tale of Elisha Mitchell remained a powerful lure for tourists, the state also began to rename various natural features in an effort to promote the story. In 1947 Governor Cherry petitioned the federal Board on Geographic Names to change the official designations for two peaks immediately north of Mount Mitchell. In the nineteenth century the northernmost pinnacle had been called Hairy Bear, but more recently local people had referred to the two peaks collectively as the Black Brothers. Now the mountain closest to Mount Mitchell (the East's second tallest peak) became Mount Craig, in honor of the former governor and founder of the state park. The northern peak, which the U.S. Geological Survey had recently measured at 6,593 feet, was christened Big Tom, in tribute to the region's foremost mountain man and finder of missing persons. Toe River Gap (where the headwaters of the South Toe coursed off the Blue Ridge and, not coincidentally, where drivers heading south on the Blue Ridge Parkway slowed for the turnoff to Route 128) was renamed Black Mountain Gap. Sugar Camp Fork, the stream into which Elisha Mitchell fell on that July evening in 1857, became Mitchell Creek, and the fateful waterfall itself was named Mitchell Falls.[7] Blue Ridge Parkway regulations prohibited gaudy signs and billboards advertising area attractions. But the new place-names effectively told Mount Mitchell's story to any motorist who could read a map.

Park promoters, however, found it far easier to change the names of mountains than to wring money from penurious state legislators. An initial appropriation allowed for a new refreshment stand in the summit parking lot, a nearby picnic area, a small museum, and a rustic tent campground, all of which were in place by 1953. (Updated versions of these facilities are visible in the park today; the campground sits on the site once occupied by the CCC.) But legislative wrangling and reapportionment of funds slowed construction after that.[8]

While politicians dragged their feet, sightseers flocked to the Blacks like never before. Thanks to Americans' new enthusiasm for outdoor vacations, annual visitation to Mount Mitchell soared, climbing to 196,000 by 1951, 362,000 in 1953, and a whopping 405,266 in 1954, a new record. Clyde "Hoppy" Hopson, who worked at the park in the mid-1950s, remembered one summer Sunday when 18,000 people showed up. It took "three or four of us," he recalled, "just to keep the traffic moving and to keep them from stopping and jamming up." Tourists battled one another for parking spaces and picnic tables and stood in line to use woefully inadequate restrooms. Upon seeing the crowds, many would-be visitors who drove up from the Blue Ridge Parkway simply turned around in frustration.[9]

Such troubles were hardly unique to the Black Mountains. Overcrowding became a chronic problem at many of America's national parks (including the Great Smoky Mountains), and congestion and traffic jams were common at recreational sites across North Carolina. But Mount Mitchell was the crown jewel of the state system, and as word of its troubles spread, politicians found themselves under pressure to take action. "It is a shame," a spokesman for the Parks Division lamented, "to advertise Mount Mitchell State Park and after people get there not give them a chance to see it." Hugh Morton, owner of a popular tourist attraction atop Grandfather Mountain and a member of the Board of Conservation and Development, echoed those sentiments, noting that "the Mount Mitchell situation . . . is about the worst we have." However, like so many times in the past, the loudest complaints came from Asheville merchants and business owners. While the legislature fought over money, the Chamber of Commerce again sent a delegation to Raleigh to argue that improvements on Mount Mitchell, including a new parking area and the much-anticipated inn, would inevitably boost the tourist trade and enrich the local economy.[10]

Just as in 1915, the business lobby carried the day. By 1954 the Highway Commission had carved out a new parking area that could accommodate an additional 350 cars. That same year the state completed construction of what it called "the restaurant wing" of the inn. Built north of Stepp's Gap, between the park entrance and the new campground, the restaurant featured a dining area and a room for group meetings but no facilities for overnight guests. Indeed, in the coming years, as spending for park facilities slowed, the Department of Conservation and Development eventually abandoned its original plans for an inn.[11] With the summit now so easily accessible by car, the mountain's chief attractions—the view, the weather, Mitchell's grave, and the privilege of standing on the highest ground in the East—could all be enjoyed in the space of a few hours. Most tourists simply did not need or want to stay longer.

Though it lacked overnight accommodations, the new restaurant (which still stands on its original site) reflected some of the latest trends in outdoor architecture. By the late 1950s America's national parks had embarked on a massive, multimillion-dollar expansion program known in park circles as Mission 66 because it was to be in place by 1966, the National Park Service's fiftieth anniversary. As defined by the Department of the Interior, the initiative "endeavored to enhance the quality of the visitor's experience through the development of modern facilities." Designers and landscape architects increasingly moved away from the rustic log structures so popular in the 1920s and 1930s and embraced "modernistic buildings" with "spacious lobbies and wide, floor-to-ceiling windows of plate glass from which the natural landscape could be viewed and interpreted."[12] In North Carolina as elsewhere, state park planners borrowed liberally from such designs. As a result the Mount Mitchell Restaurant, with its large glass panels, steeply sloping roof, open recreation room, and native stone accents, stands in sharp contrast to the cabinlike structures elsewhere in the park.

Modern architecture eventually found its way to the summit, too. Since the early 1850s, when Jesse Stepp and other local guides had cleared away trees to make room for a crude wooden platform, some sort of observation tower had graced the East's highest peak. In the 1920s private donations had allowed for construction of a rugged stone and masonry structure that served tourists for more than thirty years. However, as visitation boomed in the 1950s, sightseers often had to wait to climb the narrow stairway to the tiny observation deck. Finally in 1956, shortly after finishing the parking lot renovation, the state set aside money for construction of yet another tower. Built using native stone, with a sturdy steel stairway, a wide concrete terrace, and angled concrete support beams, it was the quintessential modern park structure, designed to provide quick, safe, and unobstructed views of nature.[13] The thirty-eight-foot tower stood on the summit, just north of Elisha Mitchell's grave. It became the mountain's most recognizable landmark.

Such renovations eased some of the congestion and helped make Mount Mitchell the most popular park in the state system. Yet as the decade drew to a close, most of the new facilities were still off-limits to a substantial segment of North Carolina's population. Like every other state-owned recreational facility, Mount Mitchell remained rigidly segregated. African

The Mount Mitchell Restaurant (*above*), originally conceived of as a lodge and completed in the mid-1950s, reflected the state park's new emphasis on modern architecture and visitor comfort. It stands in contrast to the concession stand at the summit (*below*), which is modeled on more rustic, cabinlike structures built in the 1930s. North Carolina Collection, University of North Carolina Library, Chapel Hill, and photograph by the author.

By the early 1950s Mount Mitchell's old stone tower (*left*), could no longer accommodate the hordes of tourists flocking to the state park. In 1956 the state authorized construction of a more modern structure (*right*), which still stands on the summit. North Carolina Collection, Pack Memorial Public Library, Asheville.

Americans could travel on Route 128 (just as they could on any other public road), but by custom and by law the restaurant and refreshment stand were open to whites only. Park officials also discouraged black visitors from mingling with whites in the picnic area and campground. Such policies were hardly unusual in North Carolina. Across the state integration of schools and other public facilities had proceeded at a snail's pace. In 1960, however, things began to change. Terry Sanford, a popular Democrat with moderate views on race, defeated the segregationist I. Beverly Lake in a hotly contested gubernatorial primary. Sanford went on to win the governorship in November, a victory that had important implications for Mount Mitchell.[14]

Sometime in the spring of 1961, shortly after Governor Sanford took office, he and Harlan "Skipper" Bowles, the new head of Conservation and Development, made a joint decision to integrate North Carolina's state parks. As luck would have it, Thomas Morse, whose job it would have been to oversee the change, had already decided to leave the Parks Division for a new position with the National Park Service. Responsibility for instituting Sanford's directive fell to his successor, Thomas Ellis. "When Skipper called me and told me of the plan," Ellis remembers, "I told him that it would be a delicate situation. But we agreed that it [integration] was the proper thing to do and Skipper promised to use force if he had to."[15]

Hoping to avoid such extreme action (and the bad press sure to follow), Ellis planned carefully, taking time to visit and talk with personnel in every state park and to encourage them to accept the change without fanfare. "A lot of people, including some park superintendents, thought I shouldn't be doing this because I was new," he recalls. "And Mount Mitchell presented unique problems. It was the only park with a restaurant and black people had never eaten there or at the refreshment stand." Indeed, one of the restaurant's waitresses, whom Ellis described "as a good and valuable employee," publicly vowed that she would never serve African Americans. But when it became clear that Bowles and Ellis intended to enforce Sanford's directive, she, like most park personnel, valued her paycheck more than Jim Crow tradition. By the end of the summer Mount Mitchell's facilities, including the restaurant and refreshment stand, were open to African Americans. "We had some close calls," Ellis notes, "but no real fights. We stayed out of court, and we went from totally segregated to totally integrated in one summer."[16]

Local people, too, seem to have accepted integration with few complaints. During the early 1960s mountain folk appeared far more concerned about the new park development program and the resulting flood of tourists. Fred Burnett, who had twice lost land to government projects, remained an incessant critic. "Mount Mitchell!" he fumed in a 1960 memoir. "From the best known and celebrated of our mountains, this rare primeval area is being turned into a something of a glorified Coney Island!" and all under the auspices of the Department of Conservation and Development. "Conservation? Development?" he ranted. "If the wanton destruction is not stopped, it will not be very long before the balsam on the highest mountain in the East will be just a dim memory."[17]

Burnett's attacks drew little attention, especially since he fired his verbal broadsides from the obscure pages of his memoirs. But between 1958 and 1961 Conservation and Development became embroiled in a controversy that eventually degenerated into one of the more unsavory episodes in Mount Mitchell's recent history. The dispute involved land, specifically fifty-one acres primarily on the Cane River side of Stepp's Gap, that the state had long coveted as an addition to the park. The tract had once been part of the 13,000-acre Murchison Boundary but was now owned by an old antagonist, Big Tom's grandson, Ewart Wilson.[18]

While the Parks Division had worked to make Mount Mitchell part

Ewart Wilson's inn, restaurant, and gas station at Stepp's Gap, late 1950s. After deciding that such privately owned facilities were out of keeping with its new emphasis on modern architecture and visitor comfort, North Carolina initiated condemnation proceedings against Wilson's property and eventually forced him to sell. The state later razed the buildings and erected a new park office near the site. Courtesy of Virginia Boone.

of Variety Vacationland, Ewart had quietly operated his own tourist attraction at Stepp's Gap. Comprised of an inn, a restaurant, a gas station, souvenir shops, and tourist cabins, the small complex was simply a refurbished, more elaborate version of the old Camp Wilson that had served visitors since the 1930s. But to park officials it was an eyesore. Sprawling along Route 128, Ewart's ramshackle, white frame buildings, with their brightly painted signs advertising cheap lodging and souvenirs, were out of keeping with the highly stylized concrete and stone structures recently erected in the park. Moreover, Wilson had the annoying habit of keeping "a great pack of bear hounds" penned on the property, dogs that he touted to tourists as a "special breed" differing "in size, weight, stamina, and courage from any other in the world." By all accounts Ewart himself was also one of the chief attractions at Stepp's Gap, since like his grandfather, he thoroughly enjoyed regaling visitors with tales of his hunting exploits.[19]

From the 1940s on, the state had repeatedly offered to buy the property. But Ewart steadfastly refused to sell, cannily biding his time and raking in what profits he could while mountain development proceeded at government expense. (Ewart apparently remained on friendly terms with park officials and often invited Ellis and others to the Wilson restaurant when they visited Mount Mitchell.) But in the late 1950s, with tourist traffic at an all-time high, the state decided to condemn the property and annex it under the law of eminent domain. With help from several local politicians Ewart appealed to the legislature, claiming that the park did not need the land and that the proposed purchase would constitute a needless expense for taxpayers.[20]

As usual those arguments had considerable appeal in Raleigh, and lawmakers responded with an act that prohibited the park from taking Wilson's land. But the Wilsons hardly had time to celebrate before a new legislature, seated after the elections of 1960, reversed the measure and forced him to negotiate. A year later, with his health failing and his family's enthusiasm for courtroom battles flagging, Ewart agreed to sell. True to form, he went down fighting, stringing out negotiations and telling the local papers that park officials intended to evict him. Worried about bad publicity, the state agreed to a price of $130,000, far more than Wilson might have gotten in a condemnation proceeding. A short time later the crusty mountain man and his bear dogs moved off Mount Mitchell and returned to family lands along the Cane. The state razed Camp Wilson, extended the park boundary to Stepp's Gap, and erected a new entrance and park headquarters on the site.[21]

As word of the settlement filtered into the newspapers, park officials resorted to the same arguments Locke Craig had employed in 1915, explaining that they had moved to save Mount Mitchell's "rugged beauty" from the excesses of commercial enterprise. But many North Carolinians, especially those who knew Wilson and remembered the Mount Mitchell of the 1930s, believed that something significant had been lost. Phillips Russell, a Chapel Hill writer, put it this way: Not so long ago "a visitor could imagine a bear and her cubs blundering into the open at [Stepp's] Gap, but not higher up on the peak where the State has smoothed things for tourists and their big cars." Now, with the takeover of Wilson's property, that "shaggy wildness [had] been all but barbered away." Thanks to the state, Russell concluded, "the tourist, with his popcorn and beer cans, has triumphed over the grand old mountain."[22]

So it must have seemed in the early 1960s. The old CCC camp, gravel roads, and ramshackle buildings at Camp Wilson had given way to carefully groomed picnic areas and campgrounds, paved highways, and sleek

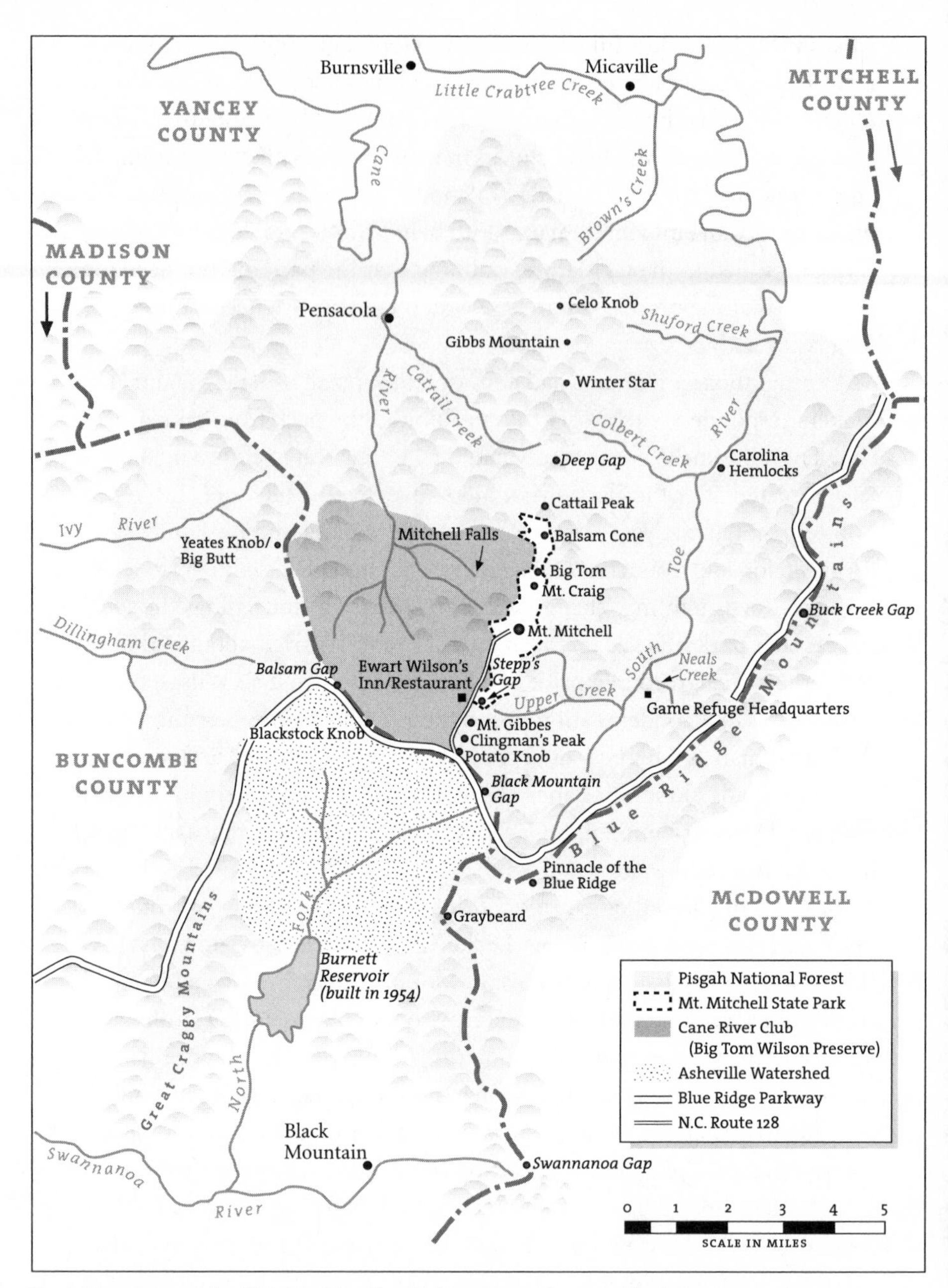

Black Mountains at the Height of the Tourist/Sportsman Boom in the Late 1950s

modern buildings. But even in the midst of the tourist boom, park officials served notice that they would restrain development if it suited their interests. In the early 1950s a group of winter sports enthusiasts from Asheville began lobbying the state to allow construction of a downhill ski slope on Mount Mitchell. The project had powerful supporters, including the Asheville Chamber of Commerce, and the state went so far as to permit clearing of an experimental run in 1956. But when studies suggested that putting in permanent facilities might increase erosion and bring myriad winter traffic problems, Ellis and his staff opposed the idea. "I was adamantly against it," Ellis remembers. "Our area would have been completely destroyed."

Fortunately for Ellis and the opposition, skiers got the chance to go elsewhere. As the southern Appalachian ski industry boomed in the 1960s and 1970s, new slopes opened in Gatlinburg (near Great Smoky Mountains National Park) and to the north of Mount Mitchell in Blowing Rock and Banner Elk. Ellis immediately seized the opportunity to make an argument he hoped the Asheville Chamber would understand. "I told them that we just couldn't open a slope at the park," Ellis notes. "It simply would not be fair for the state to compete with private enterprise [the other ski operations]." It was an odd and strikingly inconsistent stand for the park to take. After all, the park had few qualms about competing with private enterprise when it drove Ewart Wilson off the mountain. But Ellis believed it was the only way to make the case. He also got some help from budget-minded National Park Service officials who opposed the ski venture because the increased traffic to Mount Mitchell would necessitate keeping the Blue Ridge Parkway open in winter. It took considerable wrangling, but the chamber and the skiers finally relented and put aside their plans sometime in the 1970s.[23]

The controversy over the ski slope and the battle with Ewart Wilson were, in one sense, isolated incidents of little interest to anyone except the principals and a few concerned citizens. But in the larger scheme of things, events on Mount Mitchell reflected problems that had become all too familiar in American parks. Since 1916, when Congress created the National Park Service and directed it to accommodate visitors while preserving natural features unimpaired for future generations, the national and, by implication, state parks had faced an impossible task: how to provide access without destroying the very environment that attracted sightseers in the first place. The postwar travel boom simply made the paradox more apparent than ever.[24] On the East's highest peak, where sightseers could now enjoy the attractions from a car seat, a restaurant window, or a concrete terrace, it also set policy for the future. From the 1950s on, Mount Mitchell would be managed as a miniature national park that catered to day tourists and short-term visitors, nearly all of whom arrived by car.

NOVEMBER
FOREST SERVICE ROAD 472

The songbirds are gone, having fled the approaching winter. The summer campers, too, have moved on, leaving this popular South Toe thoroughfare deserted. Hiking the gravel road just after daylight, I hear only the raucous chatter of crows and the occasional rustle of chipmunks in leaf litter. I walk a mile to Black Mountain Campground, now closed for the season. There I rig a fly rod and fish upstream for an hour or so, hooking two small brown trout and spooking a half-dozen more by wading carelessly through the shallow water. White vapor rises from the stream banks as a warm November sun melts overnight frost from alders and rhododendrons. I shed my sweater, walk back to the vacant campground, and stretch out across a picnic table, dozing momentarily, caught in the spell of a perfect autumn morning.

The idyllic interlude does not last. At midday I am rousted by the roar of vehicles rumbling past the campground; cars, jeeps, and pickup trucks seem to appear from nowhere. Making my way back upriver, I discover that I am no longer alone. Dozens of men in camouflage clothing cluster in the forests along the road, stringing tents and plastic tarpaulins among the giant hemlocks and poplars. They are, from all appearances, a hard-case lot, experienced outdoorsmen who gather around fires kindled in fifty-gallon oil drums—trucked in expressly for that purpose—and prepare their meals on makeshift tables of flat river rocks. They are here for one reason: Deer season opens tomorrow.

I pack my gear and drive slowly past the roadside camps, stopping occasionally to engage a few hunters in casual conversation. Most of them are from cities and towns some distance away, but they have hunted here for years. They seem surprised that I have been fishing and look askance at the day pack, fly rods, hiking boots, and other paraphernalia stashed in my truck. The atmosphere is friendly enough, but still I feel out of place. Though no one says it, the inference is clear: My time in the mountains is ending; theirs has begun.

I smile, wave, and head home, musing about how regimented hunting and fishing have become in recent years. The man-made seasons are now as predictable as bird migrations and leaf fall. In the Blacks the regular routines and tight restrictions can be read as modern manifestations of policies developed in the 1920s and 1930s when the state made a concerted effort to attract visiting sportsmen. If today's crowd is any indication, the plan seems to have worked to perfection. Yet the story of people and wildlife in the intervening years is not quite so simple. Game managers, too, had to cope with a new flood of visitors with wide-ranging opinions about what made a good outdoor experience. And here, as on the mountaintop, modern visitors, be they fishermen or deer hunters, live with the consequences of decisions made—and, as it turns out, not made—in the decades after World War II.

Like the state park, Mount Mitchell Game Refuge lost its federal money when the war began. The documentary record for the late 1940s is murky, but the cutbacks seem to have all but ended efforts to educate the public about wildlife conservation. Much of the work done by the CCC, especially the scenic landscaping and other projects designed to appeal to summer sightseers, came to a standstill. Gilkey Lodge burned to the ground in 1949, and the state could find no money, public or private, to replace it. Officers of the Board of Conservation and Development (which had often used the lodge for meetings and generated much publicity for the refuge) now gathered elsewhere. The fish hatchery, originally a Public Works Administration project and a prime tourist attraction, operated only sporadically after the war and seems to have shut down for good sometime in the 1950s or 1960s.[25]

A shake-up in the state bureaucracy also affected activities at the refuge. In response to heavy lobbying by hunting clubs and the National Wildlife Federation, the North Carolina legislature disbanded the twenty-year-old Game Commission in 1947 and established a new agency called the Wildlife Resources Commission (also known as the Wildlife Commission, or in those post–New Deal days of alphabet abbreviations, the WRC). The new bureau made it clear that while it still believed in public education, its primary goal would be "restoration of game birds and mammals for the benefit . . . of sportsmen and other nature lovers." Four years later, in 1951, C. N. Mease, the feisty warden who had overseen much of the early work at Mount Mitchell, lost his job as supervisor of the state's western refuges. The circumstances surrounding his dismissal remain unclear, but it appears he somehow alienated a faction of the Wildlife Commission and was let go for political reasons. With his departure and the WRC's insistence that the needs of sportsmen came first, Mount Mitchell's managers spent less and less time on public relations. The Forest Service kept the zoo in place for several more years, but the area along Neals Creek now catered primarily to hunters and fishermen, not weekend tourists.[26]

As always, however, white-tailed deer got most of the attention from wildlife officials. In 1949 local estimates put the Yancey County deer herd at 2,300 animals, most of which could be found on lands adjacent to the East's highest peak. But by the mid-1950s hunters had begun complaining that deer had again become scarce. Moreover, the animals that could be killed seemed small and malnourished. Sportsmen blamed inbreeding (which appeared inevitable on a game refuge) and increased hunting pressure. But it seems more likely that the decline resulted from certain long-term changes in local forest composition.

Beginning early in the century and continuing into the 1930s, lumbermen had removed much of the hardwood timber, including American chestnut, from the lower reaches of the Blacks. For years after the loggers left, oak and maple seedlings, along with various weeds, grasses, and vines, flourished on the cleared slopes, providing tons of nutritious browse for deer. But by the late 1950s the woods had begun to mature. Larger trees again shaded the forest floor, and much of the brushy undergrowth had disappeared. (And because the various oaks that had replaced chestnut did not yet provide a reliable mast supply, deer could not resort to acorns.) Researchers at Asheville's Southeastern Forest Experiment Station, operated by the Forest Service, believed they had a solution. What the Blacks and other similar mountain habitats needed, the scientists maintained, were clear-cuts, or carefully selected patches of forest from which trees—virtually all trees—were periodically removed to allow for more browse and, presumably, the return of deer.[27]

The interest in clear-cuts also reflected some recent developments within the Forest Service. In the years immediately after World War II, managers at Pisgah and other national forests had primarily practiced "selective cutting," removing larger hardwoods from mature stands without damaging smaller timber. In the 1950s, however, faced with an unprecedented demand for wood and paper products, many government foresters came to favor clear-cutting. Removing all timber from a site not only allowed for a cheaper, more efficient harvest but also aided in the creation of even-aged stands of similar species that could be cut again and again for commercial purposes. But the new policy had its critics. As interest in outdoor recreation grew, national forests, like the national parks, had to accommodate hordes of hikers, campers, backpackers, and naturalists who came in search of old-growth woods and scenic forest vistas. Many of those visitors found the scraggly clear-cuts unsightly. Even within the Forest Service some officials argued that public woodlands should not be turned into tree farms. By the mid-1960s the intense debate over proper use of America's forests had captured headlines across the nation, especially in the West, where clear-cutting was more widespread and its long-term effects more severe.[28]

In 1964, partly as an effort to placate critics and generate support among hunters, the Forest Service began to study the effects of clear-cutting on deer at the Mount Mitchell refuge (or, as the WRC preferred to call it, the "management area"). In the Curtis Creek drainage foresters cleared five plots ranging in size from one to fifty-five acres, apparently the first clear-cuts ever in Pisgah National Forest. To find out how far deer might travel to browse the clearings and, ultimately, how far apart future openings might be spaced, two Mount Mitchell game wardens, Lee Boone and Joe Scarborough, began capturing and tagging live deer in the Curtis Creek area. At first the wardens used traps. But it soon proved more efficient to stalk the animals at night with tranquilizer guns and darts loaded with nicotine. After the deer collapsed from the effects of the sedative, the wardens placed identifying tags in the animals' ears so that their movements could be charted and researchers could note the location of any tagged deer shot by hunters.[29]

Scarborough remembers it as some of the most challenging and exciting work he did at the refuge. "We loved it," he says. "We were young and rough and tough and we would just dose them [deer] up just enough to make 'em wobble a little bit and we'd run 'em down and wrestle 'em and fight 'em . . . and put the tags in their ears—we even used color-coded streamers at one point—and let 'em go their merry way. We probably tagged 150 of them [over the course of the experiment]." He also believes that the effort aided the local white-tail population, though as he recalls it, hunters on the Curtis Creek side of the refuge eventually killed only seven of the tagged animals.[30]

Forest Service researchers, too, could never determine exactly how far deer ranged from the clear-cuts or whether the new practice significantly improved hunting in the Blacks. But one thing was certain. The plots produced browse in abundance—200 to 300 times as much as in the surrounding woods—and at the end of the study in 1967, the Forest Service concluded that with proper use of clear-cuts, "the timber grower can look forward to a stand of healthy vigorous trees and the wildlife manager to a prolific, healthy deer herd."[31]

Though it might have been as important to public relations as to the local deer population, the Curtis Creek experiment stands as a perfect example of the overall philosophy that guided wildlife management in the two decades after World War II. Like the conservationists of the 1920s and 1930s, most foresters and wildlife managers clung to the belief that with enough research, science could always make nature more productive and efficient. To a large extent such ideas reflected the overall optimism of the postwar era. American ingenuity had won a war, defeated polio and several other dreaded diseases, and created among its middle class a standard of living second to none. It was, as one careful student of Forest Service policy has written, an age of supreme confidence and a can-do mentality, the era of "the technological fix."[32] In the forests surrounding the East's highest mountain, clear-cutting, which promised to bolster local deer populations and improve the timber harvest, remained a standard practice for years to come.

Science also figured in the restoration of game birds. Finally recognizing that the region was not especially well suited to pheasants or quail, the WRC gave up on the earlier plans for stocking those birds and instead focused on wild turkeys. Studying the problems encountered by Mount Mitchell's wardens in the 1930s, wildlife biologists concluded that all turkeys restocked in the Blacks should come from similar climates. In the mid-1960s the WRC began a program of transplanting birds from several counties in southwestern North Carolina. While the turkeys adjusted to life in the Blacks, wardens continued to wage war on bobcats, skunks, hawks, great horned owls, or any other predator that might threaten young birds or eggs.[33]

Fisheries, too, got a dose of technology in the 1960s. When it seemed that nongame or "rough" fish had begun to outnumber trout in certain refuge waters, the Forest Service and the Wildlife Commission sanctioned a program of stream purging and restocking. Mimicking the National Park Service, which had instituted a similar policy in the Great Smoky Mountains, Mount Mitchell's wardens released rotenone, a highly toxic liquid poison, into the headwaters of several South Toe tributaries and effectively killed all fish, trout and rough species alike. Once the effects of the poison wore off, workers then restocked the streams with brooks, browns, and rainbows trucked in from nearby hatcheries. In an effort to ensure that rough fish did not migrate upstream, wardens erected a "rough fish barrier" in the South Toe River near the refuge boundary.[34]

To protect the fish and game it so carefully nurtured in its management areas, the WRC wrote regulations that clearly favored visiting sportsmen. Upon arrival, fishermen had to leave their licenses at a check station, and at the end of the day, wardens carefully inspected each angler's catch to be sure it complied with creel and size limits. Usually trout waters were open only on Wednesdays, Saturdays, Sundays, and certain holidays, a schedule intended to accommodate those who came from afar. Deer and bear seasons were short, usually two weeks or less. In the late 1950s or early 1960s the WRC also began to allow "wilderness hunts" at Mount Mitchell. After completing an elaborate application process and acquiring the proper licenses, a few lucky deer- and bear-hunters got to drive their vehicles along a Forest Service road to Camp Alice. There they could establish a base camp and then venture into the surrounding forests on foot in search of game. Bag limits were strict, usually one animal per person; but wilderness hunts became phenomenally popular, and many visiting hunters returned year after year.[35]

To enforce its regulations the WRC depended on Mount Mitchell's wardens. By all accounts the officers who patrolled the refuge in the 1950s and early 1960s gave no quarter when it came to protecting wildlife. J. J. "Jack" Kirkland, who preceded Joe Scarborough as assistant warden and worked the Curtis Creek area in the 1950s, had a reputation among local residents as a tough "mountain man who would just as soon shoot [violators] as not." Rumor had it that Lee Boone, who worked the South Toe side for most of the 1960s, was such an adept woodsman that he could sneak up on illegal fishermen and lift their lines from the water before they ever saw him. Boone once tried to stop two convicted murderers who had escaped from a South Carolina prison and driven to Mount Mitchell determined to kill a deer. The two men overpowered the warden, took his weapon, crashed his vehicle, and left him stranded in the woods while they went hunting. Undeterred and apparently unafraid, Boone simply made his way back to refuge headquarters, called the authorities, and later assisted in apprehending the felons.

But a collective reputation for toughness was not always enough to deter would-be poachers. The new network of roads, including the Blue Ridge Parkway, now afforded easy access to what had once been remote parts of the refuge, and it created a powerful temptation for local folk. "You had all these high deer populations just hanging around on the Parkway," Joe Scarborough remembers, "and it was just more than the people could stand." At certain seasons the parkway (especially where it met Route 128 near Mount Mitchell State Park) attracted scores of illegal hunters, some of whom shot night-feeding deer from their cars. In the early 1960s Scarborough, Boone, and other wildlife officers spent many weekend nights parked along the road with their headlights off, watching for slow-moving automobiles and listening for gunfire. Accosted by the wardens, some

Visiting sportsmen preparing for a bear hunt along the Cane River, 1945 (*above*), and sportfishing on the South Toe River, 1950 (*below*). N.C. Division of Archives and History and U.S. Forest Service.

violators inevitably tried to flee, and the Blue Ridge Parkway—by day a placid tourist thoroughfare—became a venue for high-speed car chases, impromptu fistfights, and in rare instances, potentially deadly shootouts. "It was wild west, cops and robbers," Scarborough said. "Seems like I caught everybody in Spruce Pine at least once and there was this one guy from Burnsville . . . I think I fought him just about every Saturday night."[36]

Though neither the WRC nor the officers realized it at the time, the confrontations along the parkway were some of the last serious skirmishes between refuge wardens and local people. By the end of the 1960s most Yancey and Buncombe County residents worked part time or full time in nearby mining operations, sawmills, and furniture factories. For them as for visiting sportsmen, hunting became mainly a diversion that offered welcome respite from the daily grind of industrial work. Indeed many local hunters organized themselves into clubs so that they, too, could take advantage of wilderness hunts and other privileges afforded by the special regulations. Other, more well-to-do groups, usually nonresidents, simply avoided the national forests altogether by leasing hunting rights from private landowners. The largest of these organizations, the Cane River Hunting and Fishing Club, held (and still holds) rights to land on that river's headwaters and controlled much of the original 13,000-acre tract that had once been the domain of Big Tom Wilson.[37]

Ideas about wildlife management were changing, too, albeit slowly. Ecology, a science still in its infancy when North Carolina established its first game refuges, now exerted a powerful influence on American thinking about game animals. Aldo Leopold's *Sand County Almanac*, published posthumously in 1949, plainly stated what both the author and John Simcox Holmes had hinted at in the 1930s, namely, that management geared to a few valuable species (i.e., white-tailed deer, trout, and commercially valuable timber) ultimately did more harm than good. Leopold, who remained something of a patron saint for many who worked at Mount Mitchell, openly criticized foresters who "gr[ew] trees like cabbages" and wildlife biologists whose "yardsticks of production [were] ciphers of take in pheasants and trout." He still advocated active management but urged a new approach, a morality-based "land ethic" that treated both predators and game animals (and for that matter, wildflowers, noncommercial timber, songbirds, and insects) with equal respect. The only ethical policy, he insisted, was one that preserved what remained of America's wilderness and promoted "the integrity, stability, and beauty of the biotic community."[38]

In 1962 Rachel Carson's landmark work, *Silent Spring*, a stinging indictment of America's pesticide and chemical industries, persuasively argued that the careless application of science and technology without regard for nature would ultimately lead to biological disaster. Carson's book, which popularized ideas about ecosystems, food webs, and energy flow developed by ecologists, marked the beginning of the environmental movement in the United States. Though no one said so directly, it called into question many techniques that had been applied in the Blacks. The new ideas were not accepted quickly, however, and in North Carolina very few foresters and wildlife managers completely embraced the new philosophy suggested by Leopold and Carson. But by the early 1970s the WRC could note that while its chief concern was still game for sportsmen, it would "actively oppose environmental degradation resulting from dangerous pesticides, pollution, wilderness area fragmentation, and damaging land use policies."[39]

In the end, though, economics, more than ecology, drove the commission to change policies at Mount Mitchell. With enforcement costs escalating statewide, the system of special regulations, wilderness hunts, and check-in/check-out procedures eventually became too cumbersome and expensive to maintain. In 1971, as part of a sweeping change that caught many sportsmen unawares, the commission decided to do away with all its management areas. Mount Mitchell and the other sites that had once been part of the refuge system became simply "game lands," the same designation they bear today. Many of the special regulations disappeared (though users did have to purchase a game lands permit), the check stations closed, and the boundary wire came down. The indiscriminate killing of predators ended; stocking of fish and game ceased. Wardens continued to patrol the parkway and Forest Service roads, but they no longer lived on-site. Hunters and fishermen could "come and go as they pleased in accordance with locally applicable seasons."[40] Inevitably and perhaps unwittingly, the policy shift, which kept the area protected but no longer so openly favored deer, game birds, and trout, put Mount Mitchell more in tune with some of the latest ideas in wildlife management.

However, sportsmen scarcely had time to take advantage of the new regulations before yet another proposal threatened to end all hunting in and around the Black Mountains. In 1976 Michael Frome, a prominent nature writer and environmental advocate, launched a public campaign to make the entire range part of a new Mount Mitchell National Park. By October Democratic representative Roy Taylor of Black Mountain (chairman of a House subcommittee on national parks and historic sites) succeeded in getting Congress to approve a "feasibility study" of the project. The study area spanned some 240,000 acres, including the Blacks, the Great Craggies, the Cane and the South Toe Rivers, and the North Fork of the Swannanoa.[41]

The push for a national park was "a last hurrah" for Taylor, who at age sixty-six was retiring from Congress. He had discussed the idea with various Asheville business owners for ten years without generating much momentum for the project. But now the time seemed right. As public concern for nature and ecology mounted, the federal government had responded with an array of legislation (notably the Wilderness Act of 1964, the National Environmental Policy Act of 1969, the Clean Air and Water Acts of 1970, and the Endangered Species Act of 1973) designed to preserve natural resources and protect scenic lands and wildlife. By 1976 both Congress and President Gerald Ford talked openly about doubling the number of national parks in the United States. But as Frome pointed out, much of the government's attention had been devoted to Alaska and the far West. Not since the 1940s, when the National Park Service established the Great Smokies and the Everglades, had there been any substantial initiative in the East. It was, Frome and Taylor insisted, time to even the score. Mount Mitchell would be a fine place to start.[42]

According to a columnist for the *Charlotte Observer*, the Taylor-Frome proposal landed in the public arena like "a two-ton boulder [dropped] into a quiet mountain pond." The heated debate that followed was in many ways Mount Mitchell's version of Hetch Hetchy, a classic battle between preservationists and conservationists. But the controversy (played out in dozens of local meetings and on the editorial pages of state newspapers) also generated some curious alliances. Environmentalists who favored protection of "spectacular vistas and unspoiled wilderness" made common cause with Asheville business owners who saw the park as an asset to tourism. On the other side, local people who had often been at odds with government agencies now joined the Forest Service and Mount Mitchell State Park in staunch opposition to the plan.[43]

As one of the park's leading opponents, the Forest Service insisted that it currently provided all the services visitors needed. It maintained two campgrounds, kept up nearly fifty miles of hiking trails, provided for backcountry camping, and with the state, accommodated more than 100,000 hunters and fishermen each year. The state park claimed that it already protected plants and wildlife on 1,500 acres and provided Mount Mitchell tourists with the same sort of experience they would get in a national park. Local citizens, especially in Yancey County, worried about increased tourist traffic and voiced traditional objections to government usurpation of private property. They also argued that if the area became a national park, many of the privileges they now enjoyed (access to firewood, the right to gather herbs and plants for sale, and especially hunting) would disappear.

As the controversy unfolded, politics and interagency rivalries proved more important than local opinion. When Taylor retired, his replacement, Lamar Gudger (also elected as a Democrat), sided with the state and the Forest Service. The North Carolina legislature passed a resolution stating its strong opposition to giving up the state park it had created in 1915. U.S. Senator Jesse Helms also weighed in, noting that a national park would bring unwarranted federal intrusion into the lives of local people, some of whom depended on income from forest resources (especially ginseng and other plants) to stay off welfare. The feasibility study went forward, but the outcome was a foregone conclusion. Without cooperation from state agencies and politicians, the proposal had no chance. In 1979 the Department of the Interior dropped all plans for a national park at Mount Mitchell.[44]

For the moment the resolution of the park controversy seemed to settle most of the questions raised in the postwar years. Park proponents had not gotten what they wanted, but they could at least be assured that the summit would remain protected and be managed much like a national park. The Forest Service and Wildlife Commission retained control of the lower slopes but, out of economic necessity, had to adopt a more ecologically oriented management strategy. Meanwhile, local folk and outsiders alike could fish and hunt without the severe restrictions of previous years. All in all, it seemed an agreeable compromise, one that might secure the region's future for years to come.

But security—at least as humans defined the term—had always been an elusive entity in the Black Mountains. As a host of earlier policy makers discovered, there seemed to be a principle akin to Murphy's Law at work in the area. Just when government believed it had a workable plan of management, something unexpectedly went wrong. Even as various officials pondered the future of tourism and wildlife policy, the forests atop the East's highest peaks were again under siege, this time by nearly invisible forces that posed the greatest threat to native vegetation since the chestnut blight. Within twenty years Mount Mitchell would be caught in the throes of an environmental crisis that would make most of its previous problems look tame.

JANUARY

POINT MISERY

Stepping from an ice- and snow-covered trail onto an open rock face, I finally relax. During the past forty-five minutes I made my way carefully through the winter woods, clambering over deadfalls, dodging low-hanging ice-encrusted branches, and struggling to maintain footing on the slick path. Though the day is not especially cold, the wind is gusty and erratic, periodically driving high clouds across the sun. As I struggle into a pair of insulated pants and light a small stove to make soup, it is easy to guess how this place might have gotten its name. But if one wants a bird's-eye panorama of the high-mountain forest in winter, this vantage point is perfect.

Point Misery is a 5,714-foot peak on the west side of the Cane River drainage, just south of Big Butt. Its elevation puts it squarely in the spruce-fir zone, though here as elsewhere in the range, logging, fire, and a host of other disturbances have allowed for an invasion of various hardwoods. However, as I look out across the Cane's headwaters, around the bend of the fishhook and down the long shank of the Blacks, I can still discern the general outline of what was once an almost unbroken band of evergreens. From old photographs and documentary footage I know that in the late 1950s and early 1960s this part of the spruce-fir forest had recovered nicely from the ravages of logging. Forty years ago it resembled a dark green cloth draped loosely across the high peaks, wrinkled here, folded there, clinging to the contours of the gneiss and schist below.

Now, however, all that remains are tattered remnants of what was once whole fabric. To the north I can see long gray-brown slashes across the summits of Big Tom and Balsam Cone. On the west-facing slopes of Haulback, Mount Mitchell, and Mount Craig, large blotches of pale gray and white appear randomly amid the green, as if someone had mistakenly sloshed bleach on the material. To the south the spots are more difficult to discern, but they are there, cropping up sporadically above 5,000 feet. What from here look like imperfections are, in fact, large patches of dead trees, mostly Fraser fir, killed during the last four decades. Now white with age and decay, they stand gaunt and lifeless on the peaks of the East's highest mountains, grim reminders of forests that flourished there after the lumbermen left.

Having had my fill of Point Misery (at least for this day), I drain the last soup from my cup, shoulder my pack, and hike south toward the bend in the Black Mountain fishhook. Along the way I pass through several stands of spruce-fir. I note brown needles and discolored bark on some of the larger Fraser firs and see the odd dead tree. But I also find seedlings that appear green and vigorous. The contrast is unmistakable, and it illustrates what, over the last two decades, has emerged as the most vexing issue in recent Black Mountain history: What killed Mount Mitchell's trees? Our search for answers begins among those ridges on the other side of the Cane in Mount Mitchell State Park.

Sometime in 1955 a Forest Service district ranger reported that a number of Fraser firs along Route 128 had turned reddish-brown and appeared to be dead. It took two years to organize and complete an investigation, but in October 1957 Charles F. Speers, an entomologist from the Southeastern Forest Experiment Station in Asheville, discovered "a white woolly waxy material" on the bark of twelve afflicted trees. He knew that the substance indicated the presence of a small insect, most likely an aphid. Speers immediately collected a bark sample and shipped it off to the Insect Identification and Parasite Introduction Section of the U.S. Department of Agriculture (USDA). Their finding was not encouraging. According to the USDA, the bug that had killed the trees along Route 128 was an exotic insect less than a sixteenth of an inch long. The agency identified it as the balsam woolly aphid.[45]

By 1957 government entomologists were already well acquainted with the pest. Native to central Europe, it had migrated across the Atlantic sometime around 1900, probably on evergreens imported as ornamentals or breeding stock. It made its first documented appearance in the United States in 1908 when it invaded a stand of balsam fir near Brunswick, Maine. Over the next forty years the tiny insect spread north and west across the continent, infecting various species of North American fir from eastern Canada to the Pacific Northwest. Speers's discovery was the first official sighting in the southern Appalachians, and it was cause for immediate alarm.[46]

Today when entomologists speak of the pest found on Mount Mitchell in 1957, they identify it not as an aphid but as a more primitive life form called an adelgid. The new classification results, in part, from a better understanding of its life cycle. Unlike true aphids, which move about continuously on various plants searching for food, the adelgid hatches into a "crawler," a small gelatinous larva, that is mobile for only two days. During that time the crawler must find a crack in the bark of a fir and insert its highly specialized feeding apparatus, called a "stylet," into the tree's inner tissues from which it extracts sap and other fluids. As the adelgid eats, substances in its saliva stimulate abnormal growth, causing fir stems to swell and twist at odd angles. Such changes slow the movement of water and minerals through the tree's heartwood so that the fir, in essence, starves to death, usually in three to nine years. Under certain conditions the salivary excretions themselves may also be toxic. Recent research suggests that an infestation of "100–200 adelgids per square inch on a fir stem" is sufficient to kill an otherwise healthy tree.[47]

Adelgid breeding habits are primitive, too (even by insect standards), but remarkably efficient. All balsam woolly adelgids are, in fact, females. They lay and fertilize their own eggs in small fiberlike nests that, as Speers discovered, look like tiny flecks of white wool. In North America, where the imported pest has no natural predators, entomologists estimate that under normal conditions a single adelgid laying an initial 100 eggs (only half the number of which she is capable) can produce a total population of 3,240,000 offspring in just three years. However, in the Blacks and in the rest of the southern Appalachians, conditions were far better than normal. In Canada and New England, as in northern Europe, cold temperatures kept adelgid populations in check both by killing off some of the insects each winter and by limiting reproduction to only two generations per year. But in the milder climes of the southern mountains more insects survived, and the warmer spring and fall temperatures allowed for three and sometimes four generations. At that rate it did not take long to build up the 100 to 200 per square inch needed to attack Mount Mitchell's native firs.[48]

The news got worse. Adelgids frequently migrate on the wind, and the East's highest mountain, a perpetually breezy summit located nearly at the center of the Fraser fir's natural range, was a perfect point of dispersal. Moreover, the unique "north-south orientation" of the Blacks, roughly perpendicular to the adjacent mountains, meant that windblown adelgids could easily move east and west to infest Fraser fir on the surrounding peaks. The balsam woolly adelgid also gets around by clinging to birds, animals, people, and vehicles. The Blue Ridge Parkway, which looped by the Blacks as it ran southwest to northeast along the spine of the Appalachians and straight through the spruce-fir forest, provided a convenient man-made corridor through which adelgids could migrate. As one scientist observed some years later, the balsam woolly adelgid could not have chosen a better site than Mount Mitchell from which to enter the southern Appalachian forests.[49]

As it became clear that Mount Mitchell might be a beachhead for a widespread adelgid invasion, government scientists began casting about for the initial source of the infestation. When further study revealed that the insects had probably been present on the peak at least twenty years prior to Speers's discovery, foresters confronted a difficult question: Was it possible that the pest had arrived on some of the exotic trees used to reforest the mountains' fire-scarred slopes in the 1920s and 1930s? Suspicion immediately focused on a stand of European silver fir, "a most likely host" planted on Clingman's Peak between 1923 and 1931. But in tracing the origins of the nursery stock, foresters found that the imported trees had been grown from seed in Pennsylvania. Because that area had no natural fir population, it seemed unlikely that the seedlings could have been infested there. Instead the Forest Service argued that the insect probably came from "similar plantings . . . on other lands or in some of the nurseries in the Mt. Mitchell area, in which aphid infested stock was used."[50] Officially the early conservationists were off the hook, but the exact source of the infestation remains a mystery.

By the early 1960s the Forest Service estimated that the adelgid had already occupied 7,000 acres of Black Mountain forest. Mortality in the afflicted areas ranged from 20 to 90 percent of all trees infected. As predicted, the pest had migrated to nearby peaks. Foresters found the adelgid on Roan Mountain in 1962. A year later it appeared on Grandfather Mountain and on Mount Sterling in the Great Smokies. In those areas—indeed throughout most of its range across the Southeast—Fraser fir had not been logged for a half-century. Since the adelgid did not attack red spruce or hardwoods, its impact on the timber industry would be minimal. But government agencies no longer measured the value of their forests simply in terms of board feet. Roan and Grandfather, like Mount Mitchell, attracted several hundred thousand visitors each year; the Smokies drew twenty times that number. Though no one could put an exact monetary value on "public attraction to the beauty" of spruce-fir forests, state and federal officials agreed that the prospect of whole mountaintops turning red-brown with death would not help the tourist trade.[51]

Moreover, since the 1950s the North Carolina Agricultural Extension Service had been actively encouraging farmers in mountain counties to cultivate Fraser fir for the commercial Christmas tree market. Because the trees took eight to ten years to reach marketable size, Christmas tree farmers usually planted a new plot every year, so that after the initial harvest they had trees available for each succeeding season. As a result, thousands of acres had been planted by the early 1960s, but many farmers, including some in Yancey and Mitchell Counties, had not yet realized any

Dead Fraser firs near the summit of Mount Mitchell appear as rents and rips in the forest fabric. Initially foresters believed such mortality resulted solely from an invading exotic insect, the balsam woolly adelgid. By the early 1960s state and federal officials had begun to worry about the possible effects of dying trees on mountain tourism. Photograph by the author.

significant return on their investment. If the balsam woolly adelgid could not be checked, the insect might well put an end to Christmas tree production before it really got started, effectively destroying what had already been heralded as a new multimillion-dollar industry for western North Carolina.[52]

Given prevailing attitudes about insect control, the state's reaction was swift and predictable. For decades the USDA, the Forest Service, and a host of other state and local agencies had been engaged in a full-blown war against injurious pests. In the late 1950s their weapons of choice were chemical insecticides. According to historian Edmund Russell such poisons were, in fact, products of war, spawned by the same research that provided chemical weapons—especially nerve gases and incendiaries—for use against human enemies in World Wars I and II. With the onset of the Cold War, the U.S. government equated its campaign against harmful insects with the crusade against communism. Just as the military had to build new and better weapons to stay ahead of an aggressive Soviet Union, so government scientists had to develop new and better pesticides to stay ahead of the advancing hordes that threatened American forests and jeopardized prosperity on the home front. To apply the poisons, foresters relied on military tactics, primarily aerial spraying and dusting done by World War II surplus aircraft.[53]

In 1958, with help from the Forest Service, North Carolina launched its own chemical offensive against the balsam woolly adelgid. Speers first experimented with a mixture of soluble oils and water, which he sprayed on afflicted Fraser firs. When that proved ineffective, he tried a variety of other toxic paints and powders, some of which had to be applied directly to tree bark or sprinkled on the soil. By 1960 he had treated roughly 104 acres of Fraser fir at a cost of $1,500. "The degree of aphid control," the Parks Division noted dryly, "was unsatisfactory." One year later, though, the state began testing a substance that looked more promising. It was a new compound, born of war-related research and billed as particularly effective against mites, lice, and aphids. Scientists knew it as a chlorinated hydrocarbon called hexachlorocyclohexane, a form of benzene hexachloride, abbreviated as BHC. Today we know it as lindane, a powerful insecticide with chemical properties similar to those of DDT.[54]

State park workers began spraying lindane in earnest in 1962, primarily in a designated 300-acre "protection zone" accessible from Route 128 or other access roads. (Hoses from the park's sprayer reached only 500 feet.) The protection zone included Fraser fir stands near the campground, restaurant, and picnic areas as well as on the summit—in the words of one official, "the most 'seen' area from the standpoint of visitors to the park." The state treated the zone annually for about five years until operations temporarily ceased, apparently due to funding cutbacks. During that time foresters discovered that lindane, applied directly to infected trees, killed aphids by the millions. But for the poison to be effective, each Fraser fir had to be thoroughly sprayed, coated top to bottom, trunk and branch. And the trees had to be treated every year. Firs missed or partially sprayed soon attracted adelgids from the surrounding forest. Therefore extensive aerial spraying was useless. Lindane could (and eventually did) prove highly beneficial to Christmas tree growers who tended small plots that could be sprayed by hand. But maintaining scenic stands of the evergreens at Mount Mitchell or elsewhere in the high mountains would be an expensive and time-consuming endeavor.[55]

Cost was not the only consideration. Within the USDA and the Forest Service, scientists and entomologists had long been aware that chemical pesticides could endanger people and animals. That was especially true on the East's highest mountain, where spraying an area frequented by tourists meant that visitors might easily come into contact with the poison. From the beginning the state proceeded with notable caution, constantly taking water samples to check for contamination and conducting various tests to determine whether lindane might be toxic to park wildlife. Many of those early experiments were crude. Park naturalists simply drenched or injected small mammals with lindane to determine what constituted a lethal dose. Another tactic involved trapping and marking rodents, releasing them in the protection zone, and then capturing them again after spraying to look for signs of distress. When it came to birds, researchers simply checked to see if juncos returned to their nests and resumed "general bird activity" after trees had been sprayed. Though park naturalists acknowledged that none of their studies measured the "residual effects" of pesticides, they concluded that "at field levels," lindane "produced no measurable undesirable effects to either plants or mammals."[56]

Even while they were testing insecticides, state and federal entomologists also experimented with nonchemical methods of control by releasing several predatory beetles (imported from Europe) into Black Mountain forests. Though the beetles eventually became well established in the region and actively fed on the adelgids, the exotic predators could not slow the destruction of Fraser fir. In the southern mountains the adelgid simply reproduced too quickly for natural enemies to have much impact. Chemicals appeared to be the only solution, and the state decided that as soon as money became available, it would again spray lindane.[57]

Nationwide, however, attitudes about chemical pesticides were changing. By the mid-1960s Rachel Carson's *Silent Spring* had dramatically demonstrated that chlorinated hydrocarbons—especially DDT—not only lingered in soil and water but also accumulated in the fatty tissues of plants, animals, fish, and humans, greatly increasing the risk of cancer and genetic defects. The threat of poisoning might not be immediate, but it was real. Largely because of Carson's work, all hydrocarbons, including lindane, came under new scrutiny. At Mount Mitchell a group of scientists from North Carolina State University began a systematic five-year study to determine the long-term effects of the insecticide on park soils, water, and wildlife. Meanwhile, spraying continued. Though records are incomplete, state foresters apparently applied lindane sporadically in the late 1960s and then resumed annual treatment of the protection zone between 1972 and 1974.[58]

In 1974, however, the five-year study showed that while significantly less persistent than DDT, lindane still lingered in the mountain environment for years following an initial application. The chemical was especially slow to break down in the soils of spruce-fir forests where cold and darkness slowed decomposition of all organic material. Samples taken from the fatty tissue of small mammals trapped in the park also revealed traces of the lethal chemical. In the ensuing years additional research conducted by the Environmental Protection Agency (EPA) (established by the federal government in 1970) found that under certain conditions, even a single application of lindane could prove highly toxic to birds and fish. By the mid-1970s the EPA had determined lindane to be a carcinogen and placed severe restrictions on its use. Because of the potential health risks to workers, U.S. companies stopped manufacturing the chemical in 1977, though it could (and can) be purchased from foreign suppliers.[59]

Officials at Mount Mitchell took careful note of the new findings, but at the state park, as at the game refuge, economics, not ecology, drove management decisions. The new restrictions on lindane and the difficulty of acquiring it after 1977 meant that treating scenic areas would be increasingly more expensive. Moreover, a substantial portion of Black Mountain fir forest remained in the hands of private individuals or other government agencies, including the vast tract held by the Cane River Club and the North Fork of the Swannanoa, controlled by the city of Asheville. Unless those and other landowners could be convinced to treat their trees, adelgids would continue to invade government forests. Despite all the problems, park officials never really gave up on chemical pesticides. Though the state and the Forest Service did not spray the protection zone after 1974, both agencies seriously considered using lindane in the early 1980s when the trees along Route 128 again became heavily infested. Apparently efforts to secure funding for that enterprise came to naught. To this day lindane remains a legally registered pesticide and, if available, can be applied in accordance with government guidelines. The North Carolina Cooperative Extension Service currently lists the chemical as an effective means of controlling adelgid infestations on Fraser firs grown for Christmas trees. No law prohibits its future use on the East's highest peak.[60]

The high costs, economic and otherwise, of lindane did incline some foresters to look to insecticides from the pre–World War II era, including fatty acids and biodegradable soaps, both of which showed promise when applied directly to infested trees. In 1975 Charles F. Speers, the entomologist who originally discovered the adelgid on Mount Mitchell, offered a more comprehensive remedy. Visiting what remained of the old Forest Service plots on Clingman's Peak, Speers found that Norway spruce,

Japanese larch, and European silver fir had all "shown promise as timber-producing species" when transplanted in the Black Mountains. Perhaps, he suggested, one or more of these might serve as a replacement for the dying native trees. He was particularly interested in European silver fir, which had "good survival, growth, form, and vigor" and had long since demonstrated its resistance to adelgid attack.[61] Nothing came of Speers's proposal, perhaps due to lingering suspicions that European silver fir might have been the original source of the adelgid infestation. Who could say what new plague well-meaning foresters might unleash this time?

After more than twenty years of research, state and federal agencies came up short in their battle against the balsam woolly adelgid. Like wildlife managers, park officials had learned the hard way that a technological fix did not always work. Indeed, by 1980 most researchers believed that the Fraser fir's best hope was its own ability to develop natural defenses against the pest. Young firs, seeded from cones of dying trees, continued to spring up, even in heavily infested areas. During their first years, for reasons not completely understood then or now (perhaps because younger trees have fewer cracks in their bark that allow adelgid crawlers access to the inner tissues), the seedlings appeared more resistant to the adelgid than did older trees. In time, some scientists reasoned, a new strain of native fir naturally resistant to the adelgid might be replanted on Mount Mitchell.[62]

But even as foresters and park naturalists grudgingly moved toward a hands-off approach to dealing with the adelgid, research into Mount Mitchell's dying trees was about to take a startling new turn. Within a decade the discussion of lindane would seem largely irrelevant. Despite all the efforts to understand and control the adelgid, some researchers would contend that the imported pest was simply a symptom of a much larger problem. Murphy's Law was alive and well.

JUNE

MOUNT MITCHELL STATE PARK

Despite seasonal droughts that dry up springs and stem the flow of mountain streams, there are times in the Blacks when it seems as if it might rain forever. For three days now, my site at the park campground has been shrouded in impenetrable fog. Thunderstorms have rolled in and out, sometimes dropping more than an inch of precipitation per hour. Twice, at suppertime, I had to leave food on the fire and dive for my tent to avoid a soaking. Somehow I have managed to keep my sleeping quarters dry, but dampness now permeates every piece of gear I own. Potato chips and pretzels go soggy if left outside their sealed plastic bags for more than a few minutes. Even when the storms cease, water droplets from the low-hanging clouds collect on my jacket, on the brim of my cap, and on my glasses. A quick walk to the restroom or the water pump is enough to saturate my clothes.

I have my reasons for enduring the implacable wetness. As numerous writers and visitors to Mount Mitchell have pointed out, in weather like this the East's highest peak takes on an ethereal, almost macabre character not evident at other seasons. I begin to sense it as I drive up Route 128, past the dead trees on the west-facing slopes. It grows stronger as I walk alone on the trail past Elisha Mitchell's grave. I hear it in the sharp echoes of my footsteps on the stairs of a deserted observation tower. I see it as I stand shivering on the concrete balcony, gazing at what is left of the mountaintop forest. For as far as I can see (given the fog), all the larger trees are dead. Some stand upright, bare limbs outstretched, like pallid skeletons pointing east in the direction of the prevailing winds. Others, torn up by their rotting roots during wind and ice storms, lie randomly on the ground like so many cadavers. I am reminded of the cemetery scenes prominent in old black-and-white horror films, where slow-moving fog advancing through blanched headstones always portends something terrible and unexpected.

But what exactly? The balsam woolly adelgid is still present, still largely unchecked, and still killing Fraser firs. Yet over the last two decades some researchers have begun to question whether such a tiny insect can account for such widespread desolation. The real cause of death, they say, may lie in the fog and rain that wreathe the high peaks on days like this. Some recent studies show that the moisture-laden clouds, so vital to life on this mountaintop, are now laced with toxins, airborne poisons generated by modern industry and automobiles, substances as detrimental to trees as to human lungs. But as the history of the high peaks so aptly demonstrates, conventional scientific wisdom—whether it involves the height of mountains, the origins of life, or in this case, the balsam woolly adelgid—is not easily overturned. In the last twenty years the new theories linking air pollution and forest decline have generated a scientific and political furor unlike anything seen in the Blacks since the days of Elisha Mitchell and Thomas Clingman. Just as in the 1850s, the controversy revolved around a professor from one of the state's leading universities.

In 1982 Robert Ian Bruck, a young plant pathologist at North Carolina State University, launched his own investigation of the dying Fraser firs on Mount Mitchell. Bruck had recently worked in Germany's Black Forest, where he and a team of American scientists had observed vast forests of deformed and dead trees, both conifers and hardwoods. The culprit, some scientists believed, was air pollution generated by largely unregulated factories and coal-fired power plants in Eastern Europe. The die-off had begun at high elevations, in spruce and fir regions not unlike those in the Black Mountains. Bruck knew, of course, that southern Appalachia had been invaded by the balsam woolly adelgid. Still, he wondered, might pollution somehow be contributing to the problems on Mount Mitchell?[63]

What he discovered over the next few years forever changed the debate about the dying trees. Working with several atmospheric chemists and a host of technicians, he found that between 1960 and 1980 the mountain had become a "garbage dump" for toxic pollutants. For twenty years nitrogen oxides from car exhausts and sulfur dioxides from coal-fired power plants (mostly in the Tennessee and Ohio Valleys) had been accumulating in the earth's upper atmosphere, miles above the Black Mountains. There, in the presence of moisture and sunlight, they formed nitric and sulfuric acid, which mixed with precipitation and fell on the Blacks at all seasons. Science calls the process acid deposition. Most Americans know it as acid rain.[64]

On Mount Mitchell those pollutants became especially concentrated in fog and rime ice. Bruck's early research showed that at times the fog on the high peak had a pH of 2.1, indicating an acidity level somewhere between battery acid and lemon juice. In June 1987, after clouds with a pH of 2.8 (roughly a thousand times more acidic than unpolluted fog) had enveloped the mountains for sixteen hours, Bruck and his fellow researchers observed that needles on some of Mount Mitchell's conifers had been burned all the way through. Later laboratory analysis showed the needles had been impregnated with sulfur (or more properly, sulfates), presumably absorbed directly from the toxic fog. For an average of 268 days per year, Bruck theorized, polluted clouds and precipitation saturated the high-elevation forest, bathing the trees in what amounted to "a toxic chemical soup."[65]

Poisonous metals—lead, cadmium, and mercury, among others—which were also byproducts of air pollution, had accumulated in Mount Mitchell's soils; indeed, the concentrations of toxic metals on Mount Mitchell proved frighteningly similar to those in the devastated forests of Eastern Europe. Then there was ozone, another pollutant created by chemical reactions between nitrogen oxides and sunlight. Some ozone occurs naturally, high in the upper atmosphere where it helps absorb potentially damaging ultraviolet rays. But closer to the earth and especially on mountaintops, increased ozone levels slow photosynthesis and inhibit the processes by which trees make and store food. In the Blacks, peculiar summer wind patterns around the fishhook-shaped range tended to keep ozone levels especially high during the short growing season, a trend that magnified the pollutant's debilitating effects on trees.

As a result of those factors, Bruck concluded, Mount Mitchell's Fraser fir and red spruce had been under constant stress for more than twenty years. Core samples from trees near the summit showed that their growth had slowed markedly in the late 1950s and declined steadily during the ensuing decades. No doubt Fraser fir had suffered severely from the adelgid, but even that, Bruck and his fellow researchers thought, might be partly owing to pollution. The worst infestations had occurred on the west-facing slopes of the Blacks, mountainsides openly exposed to winds that carried airborne toxins from the industrial Midwest. Perhaps, Bruck speculated in 1987, the trees had been so weakened by man-made poisons that they had become especially vulnerable to the invading insect. But the most compelling evidence, he believed, came from Mount Mitchell's red spruces. Although completely immune to the adelgid, those conifers appeared to be dying, too, and in unprecedented numbers up and down the slopes, even to their point of contact with the hardwood forest below.[66]

With a young scholar's enthusiasm Bruck began publishing his research in peer-reviewed scientific journals. Gregarious by nature, he also talked to anyone who seemed interested, including the popular press. The press listened. During the late 1970s, research in New England indicated that air pollution might be threatening sugar maples and that acidification of streams and lakes in the Northeast and West might be a major cause of fish mortality. In 1980 Congress had created the National Acid Precipitation Assessment Program (NAPAP), an independent organization charged with coordinating acid rain research and keeping government informed about the issue. As acid rain became big news, Bruck emerged as something of a celebrity scientist. He testified before Congress on several occasions. His work on Mount Mitchell showed up in *U.S. News and World Report*, *Audubon*, and the *New York Times* as well as in local newspapers and the phenomenally popular state magazine *Wildlife in North Carolina*. In countless interviews he carefully cast his research as preliminary, noting again and again that high-elevation forest decline was a complex phenomenon that required more study. But he pulled no punches in his overall assessment of air pollution. In 1988, in a follow-up to an earlier interview, he pointedly told *Audubon* that he was "ninety percent certain" that airborne pollutants played an important role in the death of red spruce and Fraser fir on Mount Mitchell. *Audubon* published the statement in an article titled "Acid Murder No Longer a Mystery."[67]

Skeletonlike dead firs on Mount Mitchell. By the early 1980s some scientists suspected air pollution and acid deposition as underlying causes of spruce-fir decline in the Black Mountains, and Mount Mitchell soon became a symbol of environmental degradation. As the larger trees died and the evergreen canopy thinned, blackberries and various shrubs invaded the open ground, creating brushy habitats that favor rabbits and white-tailed deer. Photograph by the author.

Bruck's research and his willingness to talk about it helped trigger a spate of new studies on southern spruce-fir forests. Perhaps as many as 100 scientists and 300 technicians from more than twenty universities conducted experiments on Mount Mitchell in the 1980s. Others worked on nearby peaks, including Roan Mountain, Virginia's Mount Rogers, and various sites in the Great Smokies. As one newspaper put it, researchers studying forest decline "seemed to be tripping all over each other" in the southern Appalachians.[68]

It soon became apparent, however, that the scientific community would not be so quick to accept Bruck's ideas. For one thing, exactly how air pollution affected mountain vegetation remained unclear. One theory, tested and found viable on Mount Mitchell, held that acids leached various nutrients from the soil and, at the same time, made it much easier for trees to absorb aluminum, an element naturally present in the Appalachian environment. Excess aluminum, in turn, adversely affected the roots and limited the ability of the trees to take in food and water. Other scientists focused on ozone that entered trees directly through stomata, or tiny pores, in their leaves and needles. Another group pointed to the effects of excess nitrogen. In large quantities that vital element initially triggered rapid growth, so rapid in fact that it sometimes weakened cell structure, which eventually led to a sickly, malnourished tree. Too much nitrogen also allowed red spruce to grow well past its seasonal limits, leaving it open to injury from early frosts and freezing temperatures. As if that were not enough, no one knew for sure just how those toxic metals accumulating in mountain soils affected roots or nutrient absorption.[69]

Nature's complexity only compounded the problem. At other sites, notably Mount Rogers, spruce-fir mortality was not nearly so high, indicating that the East's highest peak might be an atypical case. Additional work by other chemists suggested that some of the original figures concerning cloud acidity and deposition on Mount Mitchell did not reflect conditions elsewhere in the Blacks. Shepard M. Zedaker and Niki S. Nicholas, two scientists from Virginia Polytechnic Institute who conducted systematic studies of the forest canopy in the Blacks, concluded that severe droughts and ferocious ice storms in the mid-1980s had contributed significantly to "high recent mortality" in the region. Nicholas also refocused attention on adelgid destruction of Fraser fir, speculating that as the firs died and the canopy opened, red spruce might suffer from increased sunlight and lack of moisture, a phenomenon known as "thinning shock."[70]

Other researchers argued that past land use should be taken into account. Red spruce, they insisted, had a natural tendency to flourish for fifty or sixty years after logging ceased. In the ensuing years, though, growth slowed dramatically. Perhaps the trends Bruck had noted after 1960 simply reflected a delayed reaction to the departure of Black Mountain lumbermen. Core samples from other locales showed that spruce growth had also dropped off sharply between 1850 and 1870, making "it much more difficult to claim that the growth reduction [since the 1960s] . . . is truly abnormal."[71]

Bruck never denied the validity of the new research. On more than one occasion he noted that it would be impossible to put a plastic bubble over Mount Mitchell, to eliminate nature (or the history of land use) from the equation and demonstrate an absolute cause-and-effect relationship between air pollution and forest decline. His disagreement with his peers, he insisted, was largely a matter of emphasis—whether one chose to look for underlying factors such as pollution that might be weakening the trees or for more immediate causes of death such as severe weather or the adelgid. He also argued that regardless of what could or could not be proven, enough data had accumulated to warrant new standards of air quality.[72]

The dispute over dying trees was never just an arcane discussion among scientists, however. From the first the controversy had distinct political and economic overtones that helped put Mount Mitchell and its dying trees squarely at the center of a national debate over air quality. For the most part environmentalists and others who favored stricter controls on automobile and power plant emissions lined up behind Bruck (who openly described himself as "an environmentalist"). For them the devastation on Mount Mitchell became yet another example of human arrogance and callous disregard for nature. In contrast, public utilities, automobile manufacturers, and other industries burning fossil fuels liked the work of those who emphasized natural forces as the reason for forest decline. Meanwhile the Forest Service blamed familiar enemies such as the balsam woolly adelgid, apparently in the hope that it might one day be controlled with biodegradable insecticides or some other technological fix. Just as with the chestnut blight, foresters seemed most concerned with what sort of trees might replace red spruce and Fraser fir and whether future woodlands might be of commercial or scenic value. At Mount Mitchell State Park, where stories of foul air, toxic chemicals, and dead trees threatened to drive off tourists, officials also seemed to favor explanations that focused on natural phenomena. Indeed, in the late 1980s the park still had signs that identified insect pests and the harsh climate as primary reasons for the spruce-fir die-off.[73]

The scramble for money to conduct acid rain research only fueled the debate. Much of Bruck's early funding came from the EPA, and his admitted sympathies for the environmental movement led some critics to brand him an "alarmist" more interested in regulating industry than in conducting rigorous scientific experiments. On the other hand, some of those who questioned Bruck's work had ties to the Forest Service and the Tennessee Valley Authority (TVA), an agency whose coal-fired power plants had been identified as major sources of pollutants on Mount Mitchell. Moreover, one of the studies that disputed some of Bruck's original conclusions about cloud acidity was conducted by a scientist working for General Motors, hardly a disinterested party when it came to the effects of automobile emissions.[74] All the research appeared in professional peer-reviewed publications, but cynics on both sides remained skeptical. Small wonder, they said, that the various experts had reached such strikingly different conclusions.[75]

The political clamor reached its climax between 1987 and 1990 as NAPAP completed its ten-year investigation of acid rain. "Interim reports" issued in the late 1980s acknowledged the importance of Bruck's studies and the marked decline of Black Mountain spruce-fir forests. But in the summaries to NAPAP's extensive documents, writers focused on the work of those who downplayed the role of air pollution. As NAPAP saw it, ice damage, drought, the woolly adelgid, and other natural factors could easily account for tree mortality in the Blacks and elsewhere. Convinced that much of the science (including his own federally funded contributions to NAPAP's study) had been ignored, Bruck penned a lengthy and pointed rebuttal. When the final government report appeared in 1990, it was more balanced and acknowledged that acid deposition did, in fact, have detrimental effects on high-elevation forests in eastern America. But those forests, NAPAP's writers reminded policy makers, made up only a tiny fraction of the country's forested landscape (indeed, only about 0.1 percent). The great majority of the nation's woodlands had still not been demonstrably affected by pollutants.[76]

All of that was true, of course, and was supported by rigorous, peer-reviewed research. But what bothered environmentalists was the government's apparent willingness to use the disagreement among scientists as an excuse to do nothing. When NAPAP officials began to talk about the report to the press and at various scientific conferences, they insisted that as long as scientific opinion remained divided, they could identify no crisis, no need for immediate action. Without further research, scientists could not conclusively link acid deposition to forest decline, either in the Blacks or anywhere else. Over the course of his work on Mount Mitchell, Bruck had often compared Fraser fir and red spruce to the canaries once carried by coal miners. If the air in the mines went bad, the canaries died first, giving the miners time to get out. The high peaks' declining high-elevation forests offered government the same opportunity, the chance to do something about air pollution before damage became widespread. But as Bruck noted in one of his many interviews about NAPAP's work, government had looked at the evidence and concluded it could ignore the problem. After all, only the canaries were dying.[77]

In retrospect NAPAP's conclusions probably had less to do with the air on Mount Mitchell than with the political atmosphere in the nation's capital. Under Ronald Reagan the Republican Party had distanced itself from the regulatory environmental legislation of the 1970s, much of which had been (at least tacitly) supported by Republican congressmen and signed into law by Republican presidents. From the mid-1970s on, the party became a haven for what one historian calls "the environmental opposition," a loosely organized coalition of groups from various industries (timber, mining, ranching, and petroleum among them) that actively "sought to reverse many environmental gains of the preceding years," including the growing power of the EPA. In the early 1980s Reagan slashed the EPA's budget and turned over much of the responsibility for enforcing the Clean Air and Clean Water Acts to the states, where "industrial development often took priority over environmental protection." By initially arguing that acid deposition could not be positively identified as the primary cause of forest decline and that no emergency warranted federal action, NAPAP simply moved in lockstep with Reagan's antiregulatory agenda.[78]

By the end of the decade, though, Bruck and the environmentalists had heard at least a smidgen of good news. George Bush took office in 1989, and a year later, after much political wrangling, Congress passed and the new president signed several amendments to the Clean Air Act. The new legislation provided tougher restrictions on vehicle tailpipe emissions and called for sulfur dioxide emissions to be reduced nationwide by 10 million tons, something Bruck publicly called "a damned good beginning." In 1991 research by the U.S. Geological Survey suggested that despite difficulties involved in enforcement and the lapses under Reagan, federal regulation, enforced by the EPA, had worked. Over the preceding eleven years, the survey found, sulfate concentrations in rainwater had dropped significantly at many monitoring sites, including five locations in North Carolina.[79]

But the new regulations made little difference on Mount Mitchell, primarily because the problem there was not rain but fog. As Viney P. Aneja, an atmospheric chemist from North Carolina State, concluded after several years of studying cloud vapor on Clingman's Peak, "a natural cycling process" allows "material from a large volume of air" to become concentrated "into a small volume of cloud water." Put another way, clouds on the high peaks functioned as atmospheric scavengers, "vacuum cleaners of the sky," sucking up and storing pollutants. As a result the fog hanging on the Black Mountains might remain highly acidic (and potentially toxic) even as sulfate levels declined elsewhere. Aneja's research uncovered another disturbing trend. Although he identified "industrialized areas of the Ohio Valley and the Great Lakes region" as major sources of sulfuric and nitric acid, vehicular exhausts from cars in Asheville, Burnsville, and on Interstate 40 also appeared to accumulate around Clingman's Peak, making the air there strikingly similar to that of a large metropolitan area. (More recent studies, too, suggest that pollution from local cities and highways affects the Appalachians far more than scientists once believed.)[80] In a nation addicted to fossil fuels—so much so that it went to war in 1991 to protect oil supplies in the Middle East—the prospect for new and cleaner sources of energy seemed bleak. Given the rapidly declining forests on Mount Mitchell, the implication, though unspoken, seemed clear: A state park created to accommodate the automobile might well die from it.

Park officials, however, had a more pressing problem: how to answer tourists' queries about the dying trees. Bruck's work had gotten so much attention that many visitors to Mount Mitchell now had at least a rudimentary understanding of the acid rain issue. In 1991, as the park prepared to celebrate its seventy-fifth anniversary, the state finally authorized new signs and brochures designed to answer the question that had nagged scientists for forty years: What's happening to the trees? With help from Bruck, Aneja, and others, the park's information specialists tried to synthesize the existing work on Fraser fir and red spruce into something the average visitor might understand. The new placards and pamphlets cited the boreal climate, the adelgid, and natural cycles of life and death; but they also noted that air pollution—in the form of ozone and acidic fog—might weaken the trees and make them more susceptible to the destructive effects of insects and weather.[81]

It was an eclectic interpretation, obviously intended to appease a number of agencies and interests. Even so, the new signs represented what still passes for truth in the scientific community. By the mid-1990s many of those involved in spruce-fir research began to speak of a general syndrome of forest decline, a kind of death spiral, in which a variety of factors, natural and man-made, combine to kill mature trees. Airborne pollutants were likely part of the lethal mix, though no one could (or can) say for sure exactly what role they played. As Bruck explained in a summary article, isolating air pollution as a cause of forest decline "is made extremely difficult" due to "the complexity of [species] competition, climate variations, and the potential interactions . . . among anthropogenic pollutants such as acid deposition, gaseous pollutants, and trace metals." Likewise, the Forest Service conceded that pollution may contribute to the decline of red spruce, though it still blamed the majority of tree deaths in the Blacks on the balsam woolly adelgid. In a 1994 précis written for a TVA workshop, Niki Nicholas accepted that premise, acknowledging that "atmospheric deposition may affect red spruce." To date, however, she sees "no evidence that air pollution has any impact on Fraser fir mortality."[82]

As cautious scientists grope their way toward uneasy consensus, Black

Mountain trees continue to die. The most conservative estimates put mortality among Mount Mitchell's larger Fraser firs and red spruces at 50 percent. On the west-facing slopes the death toll perhaps exceeds 80 percent. In some areas, especially on certain western ridges, virtually all spruces and firs with an average diameter of four or more inches are now dead. A growing body of research also indicates that the general pattern of forest decline is spreading to lower elevations. Some environmentalists point to air pollution as a major cause, though just as with spruce and fir, a clear link has not been demonstrated. Several species of deciduous trees appear to be suffering damage from ozone. The lower slopes have also been invaded by another insect pest, the hemlock woolly adelgid. Apparently an Asian import that arrived in the 1920s, the hemlock adelgid at first seemed no threat to eastern forests. But its numbers have exploded in recent years, perhaps because it thrives on the excess nitrogen generated in trees under stress from pollution. Like its cousin the balsam woolly adelgid, the new pest injects deadly saliva that interferes with natural growth processes. An infestation can kill a healthy eastern hemlock in just one to four years. In addition an exotic and deadly species of anthracnose, a fungus that attacks dogwoods, seems poised to wreak havoc among those trees.[83]

It is too soon to know how all this will affect Black Mountain forests. Among the stumps and dead trees on Mount Mitchell, Fraser fir seedlings continue to sprout and grow. Some foresters still hold out hope that the trees will develop natural resistance to the balsam woolly adelgid. But for now, when the larger conifers die and fall to the forest floor, other opportunistic trees and plants also move in. During the last five years yellow birch, mountain ash, and fire cherry have all become more prominent above 5,000 feet. Higher up, where colder temperatures limit growth of hardwoods, thickets of blackberry, raspberry, and other sun-tolerant shrubs flourish on what were once heavily forested mountaintops. On Mount Craig, Big Tom, and Balsam Cone various grasses have sprung up, creating "heath balds" similar to those once maintained by grazing. On Mount Mitchell the brushy thickets and young Fraser firs give the summit—which just thirty years ago supported one of the last stands of uncut timber in the Blacks—the strange appearance of an infant forest. Invoking the language of Frederic Clements, Bruck labeled it a "dysclimax," forest succession in reverse.[84]

Preliminary studies also suggest that the opening of the spruce-fir canopy might dramatically affect wildlife. Blackburnian warblers, ruby-crowned kinglets, black-capped chickadees, and saw-whet owls that for-

In some areas near the summit, where larger trees have died, young Fraser firs continue to sprout and grow, and some foresters hope that the spruce-fir forest may one day reclaim the high peaks. For that to happen, scientists will have to solve a wide range of human-related problems. Photograph by the author.

age and nest in the branches and cavities of taller trees will likely seek other habitat. The proliferation of sun-tolerant shrubs on the forest floor may discourage juncos, wrens, and other birds that dwell close to the ground. Recent surveys in the Blacks suggest that catbirds, towhees, and other species that favor deciduous forests have increased significantly in the last five years. Populations of small mammals, including deer mice and red-backed voles, seem to be on the rise, too, perhaps because of a decline in owls and other birds of prey. The proliferation of blackberry and raspberry thickets (and the scarcity of predators) seems to favor various species of rabbits. White-tailed deer find the grassy balds and brushy thickets attractive, and by 1994 they "appear[ed] to be the most numerous large animal" in the state park.[85]

Nothing, however, illustrates the ecological effects of forest decline like the current plight of the spruce-fir moss spider, that tiny, rare arachnid discovered on Mount Mitchell in 1923. As the evergreen canopy disintegrated, allowing more sunlight to reach the forest floor, the damp, mossy mats beneath the conifers gradually disappeared. Springtails, the diminutive moss-loving insects on which the spider feeds, grew scarce. As a result, spider numbers, never large to begin with, went into sharp decline. By

1996 the spruce-fir moss spider had been listed as an endangered species. Some experts believe it is now extinct in the Blacks.[86]

The impact of the dying trees extends well beyond what ecologists can measure. In the very recent past Mount Mitchell has become a pulpit from which environmentalists routinely deliver jeremiads about the perils of air pollution. Walter Cronkite stood among its dead trees as he narrated a 1994 Public Television documentary titled *The Search for Clean Air*, which featured Bruck and other scientists who advocate more federal regulation of air quality. In 1995 Defenders of Wildlife, a national environmental organization that claims 120,000 members, designated southern spruce-fir forests, including those on Mount Mitchell, as the second most endangered ecosystem in America. (Only the embattled Everglades seemed in worse shape.) That same year Charles E. Little, a noted environmental writer, included a chapter on the mountain in his highly regarded book, *The Dying of the Trees*, which chronicled the effects of bad air on America's forests. Borrowing from Shelley, Little concluded that "hell must be a place very much like Mount Mitchell." *An Appalachian Tragedy*, a 1998 Sierra Club book edited by Little and Appalachian State University anthropologist Harvard Ayers, put images with those words, juxtaposing glossy pictures of Mount Mitchell's "sepulchral" landscape with dramatic photographs of smokestacks, chemical plants, and coal-fired generators.[87]

When it comes to tourism, Mount Mitchell remains the most popular state park in western North Carolina, though other sites in the Piedmont and coastal plain now draw more sightseers. By the mid-1990s, estimates put annual visitation at more than 300,000, a slight increase since the early 1960s and (if earlier figures can be taken at face value) a sharp downturn from the 1950s. Local people especially seem less interested. In 1991 state park officials estimated that some 75 percent of visitors came from out of state. According to John Sharpe, park superintendent, many have never heard of Elisha Mitchell, Locke Craig, or anyone else associated with the mountain's history. They just drive up from the Blue Ridge Parkway out of curiosity. Clyde "Hoppy" Hopson, who recalled the days when the park had "more people than parking places," thought that the disappearance of the large trees had driven away visitors. In the 1950s, he remembered, "it was gorgeous. . . . On the forest floor, we had some awful pretty moss growing out from the picnic area toward Celo Mountain. It was shady and stayed damp all the time under the balsams. . . . It makes you sick at your stomach to see what has happened."[88]

Over the last five years interest in the state park appears to have picked up somewhat. Even so, Hopson's simple statement seems to capture what many North Carolinians feel when they consider the current state of Mount Mitchell. It is the maddening frustration of Murphy's Law, of something gone strangely and drastically wrong. There is also a sense of finality, of resignation, that, barring a miracle, the once-majestic spruce-fir forests are doomed. The East's highest mountain, so long a source of state pride and regional bragging rights, has become an icon of degradation, a place that evokes images of pollution, disease, death, and hell.

Given the present state of the summit forests, it is easy (and tempting) to end here, to offer up Mount Mitchell and its dead trees as yet another example—a "tragic symbol," the *Winston-Salem Journal* once called it—of our flawed relationship with nature and our failure to come to grips with the current environmental crisis.[89] But to take that tack would, in large measure, abrogate the historian's responsibility. For as serious and dramatic as it is, the decline of the high-elevation forests is only one very recent piece of a larger and more complicated story. To gain some perspective on what has happened on Mount Mitchell and the surrounding peaks, we must look back well beyond the last fifty years. And we must ask some tough questions: What, if anything, have people done right during their long sojourn in the Black Mountains? Why have the best-laid plans of state and federal agencies so often gone awry? Most important, can we learn anything from the past that might be relevant for the future? When we began this chronicle of human experience in the Blacks, we noted that it would be difficult to separate ourselves from our history—that almost inevitably, our current perceptions of nature would intrude on the narrative. But we also acknowledged that we would eventually have to decide whether the story is one of progress or decline, ascension or declension, or maybe something in between. That time has come.

CONCLUSION

Stories from Four Thousand Feet

SUMMER SOLSTICE

MOUNT MITCHELL, EAST FACE

Ten minutes into the woods, sweat soaks through my shirt. Gnats swarm around my face, unfazed by the insect repellent smeared on my cheeks, forehead, and hat brim. It is not especially hot, but the morning humidity is high and the trail, as usual, is steep. Fortunately my pack is light, holding only fishing gear, water, and lunch—ample provisions, I think, for what has become a ritualistic journey. This narrow path extends more than two miles up the east flank of Mount Mitchell and ends on the headwaters of a small stream that spills off the high peak in a chain of long cascades and plunge pools.

Not far from the trail's end I stash my pack in a rhododendron thicket, slip into wading shoes, and ease into the chilly water. From here I tread the creek bed, wading against the current and clambering over rocks until I recognize a familiar yellow birch tree. It grows on a tiny island created by a fork in the stream, but it looks for all the world as if it somehow sprouted straight from the water. When I spot the "water tree," I know that I have climbed above 4,000 feet, into the northern hardwood forest. It is my favorite spot in the entire Black Mountain range. I make it a habit to be here on or near the summer solstice, for I know of no better place to pass the longest days of the year.

Ostensibly I come to catch the eastern brook trout that hold in knee-deep riffles along the granite ledges bordering the creek. But usually I spend as much time wandering the woods as casting a fly. As at most other sites throughout the range, evidence of the past abounds. The trail I walked this morning was once a logging road.

Most of the hardwoods along the creek look to be fifty to eighty years old, which suggests that nearly all the timber was once taken from these slopes. Chestnut saplings sprout from stumps; yellow birch and other sun-tolerant species dominate the canopy.

Deer are abundant, and rainbow trout flourish in the lower reaches of the stream, living reminders that this forest was once part of the Mount Mitchell Game Refuge. Most telling, perhaps, are the rules that govern fishing here. This tributary of the South Toe is now classified as "Catch and Release" water. To quote the signs posted streamside, "Single-hook, artificial lures only; No fish may be harvested or possessed." It is the ultimate sportsman's regulation, instituted in the last twenty years and designed solely for recreation, not sustenance.

I know, too, that were I to climb 2,000 feet farther up, I might hear the whine of automobiles on asphalt roads and find forests under stress from pollution and the balsam woolly adelgid. Yet even with all these distasteful signs of human influence, Mount Mitchell always manages to surprise me. While standing at the water tree, I watch in silent amazement as a large bobcat crosses the creek not twenty yards ahead. The animal strolls casually across a fallen hemlock and seems unconcerned with my presence. An hour later a routine cast into a glassy pool produces a brook trout nearly eleven inches long, the largest I have ever seen, much less caught, in this or any other Black Mountain stream. Though it can hardly be described as pristine, this place retains an essential element of wildness, a sense of nature unfettered and unpredictable, that draws me back year after year. It is, perhaps, an ideal site from which to ask our final questions, to evaluate the long story of people in the Black Mountains, and to decide what we might learn from it.

Let us begin with the good—and the obvious. That this isolated mountain cove exists, that it remains a place where one can find woodsy solitude on the first day of summer is, in itself, remarkable. Buncombe County, which takes in the southern rim of the Blacks, is among North Carolina's ten most populous counties, and Asheville, just thirty-five road miles away and with a population exceeding 66,000, is the state's eleventh largest city. Over the last thirty years Yancey County, which encompasses the bulk of the range, has become a premiere mountain recreational and retirement community, a haven for seasonal visitors from other states, many of whom build summer homes in the area. That trend is not immediately apparent in the county's official population count, which hovers around 18,000 and usually does not include second-home owners. A more telling statistic is the enormous increase in what economists call "wage and salary employment," primarily low-paying jobs in construction and service industries that cater to seasonal residents. Since 1970 the number of such positions in Yancey has more than doubled, a growth rate comparable, North Carolina statisticians say, to some of the state's most populous urban areas, including the famed Raleigh–Durham–Chapel Hill Research Triangle.[1]

What all this has meant for the region surrounding the Blacks is, in a word, development, and lots of it. Outside Asheville, fast-food restaurants, convenience stores, and other trappings of urban (and suburban) sprawl stretch right to the boundaries of the Blue Ridge Parkway and Pisgah National Forest. Forest Service Road 472, the gravel thoroughfare that serves hikers, fishermen, and hunters, now joins a paved highway that passes through Mount Mitchell Lands and Golf Club, a large resort built in the early 1970s. According to its promotional literature, the club has been "carved out of the Pisgah National Forest and around the flowing South Toe River." It provides upscale patrons with "beautiful mountain scenery, pine trees, laurel and rhododendron, rivers and rocks, sand and bent grass as plush as a new Cadillac."[2]

On the other side of the range the upper Cane as yet has no golf courses, primarily because members of the Cane River Club retain an interest in hunting and fishing. But it does have its share of summer homes, and both the upper Cane and the Asheville watershed, which the city has kept relatively free of development, are off-limits to the public. Yet on the east face of the Blacks the woods and streams of the national forest are open to anyone able and willing to walk, an accessible public oasis amid the shopping centers, bent-grass fairways, and riverfront homes. For that we must give credit where it is due: to the Forest Service and to the state conservationists who, in the early twentieth century, insisted that North Carolina do something to protect its diminishing woodlands and wildlife.

In general terms the same holds true for Mount Mitchell State Park. On a clear day the view from the observation tower includes man-made reservoirs, resort communities, and the scarred ridges of open-pit quarries and mines. At night lights from Asheville, Marion, and the town of Black Mountain flood the eastern valleys and illuminate the skies across the southern flank of the range. That the high peak has not yet been swallowed up by all this—that it still provides an opportunity to camp and picnic above 6,000 feet, to see ravens, red-tailed hawks, deer, and the occasional black bear—is also testament to the zeal of Progressive foresters and politicians, notably John Simcox Holmes and Governor Locke Craig. Had they not badgered state legislators to take Mount Mitchell away from the lumbermen, the summit might have no park.

This tale of salvation—the rescue of a precious landscape from the perils of urban-industrial development—is one that the state and the Forest Service love to tell. It has become an important part of their institutional histories, one of the narratives by which they define themselves. As one drives along Route 128 up from the parkway, the story is impossible to avoid. Forest Service signs along the road recount the fires that ravaged Clingman's Peak and note that conservationists planted most of the trees now visible on the mountain. At the state park visitors learn that Indians passed through the region on hunting expeditions and that the first white settlers worked hard to eke out a living in the valleys along the Toe and the Cane. But the real action began in the 1850s when Elisha Mitchell's explorations and the heroics of Big Tom Wilson helped turn the Blacks into a prime tourist attraction. Then, in the years before World War I, greedy Yankee lumbermen launched an all-out assault on the range, threatening to ruin a precious landmark. Governor Craig, however, would have none of it, and since then Mount Mitchell and the surrounding peaks have been preserved for the people of North Carolina. The new signs put up to explain the recent decline of summit forests suggest that the Blacks still face problems, but overall the message remains positive and hopeful, a progressive, ascending narrative.

There is truth to that story. We can appreciate the good work of state and federal agencies every time we walk the woods, fish the streams, or visit the park. But this upbeat account—what we might call the rescue narrative—is incomplete. For one thing, it does not emphasize the peculiar qualities that made the Black Mountains worth saving. From the mid-nineteenth century on, Americans have been curiously drawn to high mountains, in part because such rugged terrain evokes a sense of the supernatural. As several historians have noted, one need only consider the places chosen for our first national parks—Yellowstone, Yosemite, Grand Canyon, Rainier, and Zion—to recognize America's enduring fascination with vast, powerful landscapes.[3] In terms of sheer elevation the Black Mountains did not measure up to the more dramatic peaks in the Northern Rockies or the high Sierras. Indeed, North Carolinians have even had to concede that Mount Mitchell is not, as they once believed, the highest mountain east of the Rockies. Technically, that title belongs to Harney Peak in the Black Hills of South Dakota. (Surveys using satellite technology have also begun to establish new elevations for a number of Black Mountain summits. Elevation at the top of Mount Mitchell's observation tower was listed at 6,719.9 feet.)[4] But simply because the Blacks contained six of the ten highest mountains in eastern America (now defined as "east of the Mississippi River"), the landscape had immediate standing in the public imagination.

By the 1850s, however, the Blacks had something that other Appalachian mountains and, for that matter, the Rockies lacked: the dynamic saga of Elisha Mitchell. That story—exaggerated, embellished, and promoted by Mitchell's friends—became ingrained in the consciousness of North Carolinians, inseparable from the terrain on which it occurred. Were it not for this sense of cultural significance—the new meaning imposed on the land after the professor's death—would conservationists have worked so hard to save the East's highest mountains? Or would the Blacks, like so many less-majestic and less-meaningful peaks nearby, have been left to the lumbermen, miners, and real estate developers?

We do not have to look far to identify other problems with the rescue narrative. Indeed most historians of southern Appalachia tell a quite different story about the region's past. Generally speaking, they argue that the mountains were at their best when human influence was minimal, when local people, be they Indian or white, maintained control of their most valuable commodity: land. The big trouble began in the late nineteenth and early twentieth centuries with the arrival of northern capitalists who plundered mountain resources like pirates ravaging a treasure ship. Local people became enmeshed in the nation's emerging industrial economy and fell victim to wage labor, easy credit, and a host of other modern evils. Government conservationists only made matters worse by evicting mountain residents to make room for national forests, parks, and various recreation sites. Gradually the southern Appalachians became a seasonal playground for America's upper and middle classes, a trend that continues to this day with the construction of golf courses and summer homes.[5]

There is truth to that story, too. In the Black Mountains one might point to thousands of native inhabitants killed by war and infectious disease, to land-hungry white settlers using local forests to raise livestock for distant markets, to rapacious lumber companies, to a stingy state legislature that ignored conservation except when it figured to aid the economy, to aggressive government agencies enforcing wildlife law, to business interests promoting the paving of mountaintops, to the exclusion of African Americans from park facilities, to the recent ravages of pests and pollution and Mount Mitchell's emergence as a symbol of environmental degradation. This is no story of salvation, of mountains and woodlands rehabilitated. Instead it is a sordid account of a landscape ruined, of paradise lost to the sins of capitalism and industry, a tale of disintegration and decay, or to use the historian's vernacular, a narrative of declension.

These two conflicting stories, both of which can be culled from the historical record, suggest just how difficult it is to come to grips with our first question: What, if anything, have people done right during their long sojourn in the Blacks?[6] The very landscape that we admire and enjoy—where we hear tales of Elisha Mitchell's exploits, walk in forests recovering from logging, and catch trout in mountain streams—is, in fact, far from perfect. Moreover, its salvation involved certain assumptions about nature and its uses that many of us (if we take time to consider them) can no longer accept. As Yellowstone Park historian Paul Schullery reminds us, in conservation as in most other endeavors, human nature never goes "on holiday." Those who created America's parks and national forests "were not exempt from greed [or, we might add, the class and race prejudices of the twentieth-century South] any more than they were immune to wonder."[7] It is appropriate to applaud North Carolina conservationists for what they did and count it a good thing that the Blacks have not yet been steamrolled by development. But we must also take the bitter with the sweet and remember that much has been lost over the years, and that those who moved to protect Mount Mitchell preserved it primarily for people like themselves.

The search for answers gets no easier as we move to our second question: Why did the best-laid plans of state and federal agencies so often go awry? Here we must again remember to give nature equal time and status in the Black Mountain story. It is a step that historians of Appalachia (like historians in general, perhaps) have been reluctant to take. For the most part the mountain past has been written using the natural world as little more than a scenic backdrop for human activity. In those rare instances when nature does come to the forefront of the historian's narrative, it often takes one of two forms. In Appalachia's early history, nature is primarily an obstacle, the proverbial harsh environment that settlers must overcome. Later it usually appears in the form of various disasters—fires, floods, or famines—most of which result from nefarious human activities such as railroading, logging, or war. Apart from people, the natural world has had little life of its own.[8]

The history of the East's highest mountains suggests otherwise, especially when one evaluates government policy. When state and federal agencies set out to remake Mount Mitchell in their image, they constantly tried to tilt nature to their advantage, to make the Blacks conform to their vision of an ideal landscape. But no matter what sort of management technique they employed, nature stubbornly moved to its own unpredictable rhythms, writing, if you will, its own chaotic story. As a result foresters, game managers, and other experts found that their schemes for improving the landscape had unintended consequences, many of which we still live with today.[9]

Despite well-thought-out and generally successful plans of fire control, the native conifer forest did not quickly recover from logging, and large sections of Mount Mitchell (especially the area around Camp Alice and Commissary Ridge) are now covered with grass, pin cherries, mountain ash, and shrubby vegetation. Salvage logging of chestnut on the lower slopes probably slowed the development of natural resistance to the blight and helped turn once majestic trees into little more than understory shrubs. At the game refuge an obsession with providing sport for visiting hunters helped justify clear-cutting and the systematic elimination of predators, practices that continued (even in the face of criticism from ecologists) as long as federal and state money held out. According to estimates made by the Wildlife Commission in the 1990s, the area around Mount Mitchell now has one of the largest deer herds in western North Carolina. Indeed, the number of white-tails in that part of Yancey County appears dangerously close to exceeding the available food supply.[10] The sighting of an occasional bobcat suggests that as deer become more numerous, predators might again move in. But for the moment, white-tail numbers seem destined to grow, perhaps to unhealthy proportions.

Fisheries, too, still show the effects of policies aimed at sportsmen. In the Blacks as in other parts of the southern Appalachians, brown and rainbow trout have shown a marked tendency to outcompete native brook trout for food and habitat. As a result, in the lower reaches of the South Toe, the Cane, and many of their tributaries, brook trout have all but disappeared. No one is sure exactly why the introduced fish have done so well. Their success may stem from their more aggressive and territorial natures, their ability to tolerate warmer water temperatures, or simply the general tendency of exotic species to dominate a new environment. Whatever the reason, over the last thirty years, even after the Wildlife Commission stopped stocking rainbows and browns, brook trout have constantly retreated upstream.[11] They still survive, but only in remote headwaters on the highest Black Mountain slopes. Indeed the catch-and-release regulations recently put into effect on such streams are designed to ensure that anglers who fish remote parts of the range do not add to the brook trout's troubles. In water as on land, nature continues to confound the experts.

Defining nature as an active agent in the past means that the story of the Black Mountains—and, by implication, the rest of southern Appalachia—becomes even more complex. The natural world demands that we put aside the simplistic, human-centered tales of unmitigated progress or steady decline that have dominated our histories. Instead we must accept a far more nebulous story of people in the Black Mountains, a tangled and equivocal record of achievements mixed with setbacks, a narrative that periodically rises, falls, and occasionally flattens out, and one in which nature, as much as or more than human nature, dictates the ups and downs.

Such ambiguity may not sit well with those who worry about the future of the southern mountains. We would perhaps prefer a simpler account that clearly separates success from failure, good from evil. That sort of story would make it easier for us to judge the Black Mountain past, to decide what has (and has not) worked and apply that knowledge in years to come. But the messy history of the East's highest peaks affords no such option. As a result we approach our final question—Can we learn anything from the past that might be relevant for the future?—knowing that easy and satisfying answers may be difficult to come by.

As a number of environmental historians have suggested, future managers may have to let go of an idea that has influenced thinking about the Blacks and other scenic American landscapes for centuries: the notion of wilderness. In the early 1800s the belief that western North Carolina was a savage place in need of civilizing influences inclined the state to promote exploration and development of the high peaks. Conversely, by the first decades of the twentieth century the mountains had become a haven for sportsmen, hikers, birders, and auto campers, a wilderness refuge for persons seeking respite from the workaday world of urban America.[12]

This "wilderness ethic" is closely tied to the myth of Eden, that powerful and pervasive notion that holds that at some time in the past, nature reached an idyllic state of bountiful equilibrium. In the Blacks as in similar places across the United States, some environmentalists have maintained that the only way to return to Eden is to institute a hands-off policy for the future, a scheme that would turn back the clock and drastically reduce the widespread influence of people. In contrast other interested parties, including the Forest Service, have tried to return to Eden via management, essentially arguing that with enough technology, nature might again be made perfect. But whether would-be policy makers attempt to tame the mountain wilderness or to rehabilitate it, whether they seek to "recover" Eden or "reinvent" it with technology, they share a common conviction: People and nature are separate entities, often at odds and usually working at cross purposes.[13]

Recently, however, a number of scholars have argued that such notions have outlived their usefulness, especially in regions such as the Blacks that have a long and uneven history of development mixed with conservation. Increasingly, those who ponder the current state of the American environment seek some middle ground, some way of accepting and accommodating the human presence without unduly compromising nature. Environmental writer Michael Pollan has proposed one such alternative model. He suggests that we replace the timeworn metaphor of wilderness with a new "gardener's ethic." Simply stated, we should think of the natural world as a working garden, cultivated ground, a place to be tended, managed, and—yes—changed by people. But this does not mean that "anything goes," that humans can do what they please when they please. Instead, like the best gardeners, we should learn to distinguish between "kinds and degrees of human intervention," to accept the natural world's quirks and whims, and, finally, to ask, "How can we get what we want here while nature goes about getting what she wants?"[14]

One immediate advantage of the gardener's ethic is that it enables us to make peace with the Black Mountain past. Thinking of the region as a working garden instead of a despoiled or rehabilitated wilderness allows us to see the landscape for what it is: a place in which natural and human history are so entangled as to be inseparable, a place made more interesting and intriguing by people—indeed, a place that might not be distinctive at all were it not for the exploits of Elisha Mitchell, Big Tom Wilson, John Simcox Holmes, Locke Craig, and others. Acknowledging that relationship might help us accept certain changes induced by humans as integral parts of the Black Mountain environment.

Nature itself seems to be pointing us in that direction. In the mid-1990s, as part of North Carolina's Natural Heritage Program, officials surveyed various plant and animal communities in and around Mount Mitchell State Park. The study indicated that some of Appalachia's rarest native plants, including those normally confined to isolated mountain balds, seemed to be flourishing in clearings and other areas "disturbed" by human activity. Roan rattlesnake root, a plant botanists classify as "significantly rare" in the southern mountains, can in fact be found in abundance along the road leading to the radio towers on Clingman's Peak. Likewise New England cottontail rabbits, also listed as "significantly rare," seemed to be thriving around the park picnic area, due in part to the proliferation of brambles and shrub thickets amid the declining spruce-fir forests.[15] Were we to abandon the myth of Eden, we might learn to appreciate the irony, to admire this incipient wildness, and to regard rabbits and rattlesnake root (or for that matter yellow birch, rainbow trout, or any other species enhanced by human presence) with the same sense of wonder we instinctively reserve for an ancient stand of spruce-fir or a free-roaming black bear.[16]

But too many rabbits and too much rattlesnake root, like too many deer or rainbow trout, can quickly become a problem. Consequently, we cannot push road building or construction of recreational facilities too far. As a counterbalance to the ongoing development of Asheville and surrounding areas, state and federal agencies must continue to seek viable solutions to the decline of red spruce, Fraser fir, brook trout, and other unique native species—not in some wistful attempt to return to Eden or to re-create a landscape devoid of human influence, but because conifer forests and shaded mountain streams, like road cuts and forest clearings, have an essential element of wildness that visitors (including tourists and sportsmen) find desirable.

Though no one can say for sure, such efforts likely will require considerable human intervention or "gardening." An ideal plan for the future might involve various programs of reforestation and the reintroduction of wildlife, especially predators, something tried with mixed results elsewhere in North Carolina. In some respects this type of restoration might resemble the work of Progressive conservationists. But the new ethic would not attempt to wipe out or correct the past. It would not seek to eliminate Norway spruce or rainbow trout but, rather, to include them with indigenous species as part of a more diverse modern landscape. Those entrusted with this important task would go about their business always cognizant—indeed always wary—that human actions take place in a wild and often unpredictable world that we may never completely understand.

As appealing as it is, the gardener's ethic can carry us only so far. Any attempt at restoration will almost certainly fail unless managers can do something about air pollution, the adelgid infestation, and other causes of recent high-elevation forest decline. Environmentalists generally agree that reducing automobile use, cutting emissions from coal-fired industries and power plants, and turning to alternative sources of energy might lead to much cleaner air on Mount Mitchell, and elsewhere for that matter. They also believe that we need increased government regulation to ensure that various industries and automakers comply with more stringent controls. But given recent history and the country's insatiable appetite for fossil fuels, adopting such measures will require a fundamental reordering of American priorities and new ways of thinking about our wider relationship with the natural world.

Carolyn Merchant, an environmental historian and feminist, is another writer who has thought long and hard about the "complex and complicated" relationship between nature and human history. Asserting that the story of America has often been a tale of a male-dominated society exercising patriarchal power over feminine nature, she argues for a new "partnership ethic" to guide relations between people and the natural world. Less enthusiastic than Pollan about the idea of a garden (which implies "total domestication and control"), Merchant advocates a more ethical "nearly equal relationship" between humans and nature. Such a relationship would involve listening to many voices—including those of women and minorities—which might help alleviate some of the race and class prejudice that has marred the conservation story in the Blacks. And, Merchant believes, heeding those voices might help us recognize that we "have the potential to destroy life as we currently know it," especially through the use of pesticides, toxic chemicals, and unrestrained economic development.[17]

It is a simple but powerful proposition. Were we to take it to heart, everyone involved with air quality in southern Appalachia would have to concede that nature's needs for healthy forests are every bit as important as human needs for transportation and cheap energy. The Forest Service, the Tennessee Valley Authority, various local industries, and state agencies would have to act on concerns about airborne toxins even if absolute proof of a cause-and-effect relationship is not immediately forthcoming.

Some evidence, albeit fragmentary, indicates that such a partnership might go a long way toward reversing current trends of forest decline. Recent research conducted on Clingman's Dome (the 6,000-foot-plus peak in the Great Smoky Mountains named for Mitchell's antagonist) suggests that mature Fraser firs may indeed be developing some new natural defenses against the balsam woolly adelgid. Since the 1970s, managers there had tried various means of adelgid control, including the spraying of soaps and other biodegradable pesticides, a practice finally abandoned as ineffective in the summer of 2000. Within a year entomologists working in the area discovered that certain trees on the summit appear to be producing thicker bark, which makes them less prone to attack from the pest.[18] Scientific inquiry into this phenomenon is just beginning, and some experts insist that even if the trees develop resistance to the adelgid, they could still be prone to numerous other problems. But the implication seems clear: If people can do something to limit airborne toxins, nature might eventually take care of the adelgid.

The best argument for making nature an equal partner in our future Black Mountain endeavors, however, is far simpler and more self-serving. Until recently, western North Carolina had long been regarded as the healthiest part of the state—indeed one of the healthiest places in the nation—in which to live. But that is changing. The same toxins that hang suspended in the fog on Mount Mitchell also clog and contaminate human lungs. A growing body of evidence suggests that, over time, tiny sulfate particles borne in polluted air affect the body much like cigarette smoke. Chronic lung disease is on the rise throughout the southern mountains, and according to a recent study conducted by clean-air advocates, "Asheville ranked sixth-highest among U.S. cities in deaths attributable to fine [airborne] particles." In addition, increased mountain ozone levels may be responsible for growing allergy and asthma problems among western North Carolina's children. As one Buncombe County politician remarked in 2001, "This is a much bigger thing than some dead trees up on Mount Mitchell, as sad as that is."[19]

The area's economic health is suffering, too. Several lodges and hotels on the outskirts of Asheville, once renowned for scenic vistas, report that guests now routinely complain about hazy views. Focus groups cite air pollution as a major reason for not visiting the Smokies and other prime attractions. Just as in 1915, a few state lawmakers and others concerned with the future of mountain tourism have taken note of the problems. In 2001 several representatives introduced the Clean Smokestacks Act, a bill designed to tighten controls on local pollution and "give the state leverage in demanding similar improvements from TVA and elsewhere." But opposition from various manufacturing interests and fears about rising electricity costs kept the state house of representatives from acting on the legislation. At the end of 2001 the *Charlotte Observer* listed a "lack of political will and a failure of leadership" as the number one environmental problem facing North Carolina.[20]

Stung by such criticism, House Speaker Jim Black and Governor Mike

Easley pushed for compromise. Thanks to a provision that freezes utility rates for five years while power companies raise 70 percent of the cost of new antipollution technology, the legislature finally approved the Clean Smokestacks Act in June 2002. As adopted, the measure requires a 60 to 75 percent reduction in major pollutants from the state's coal-fired power plants over the next ten years. Though serious questions about enforcement remain, environmentalists have hailed the act as "a landmark bill" and an important first step toward reducing the effects of airborne pollutants on high mountain forests.[21]

We can—indeed we must—build on this new momentum. As equivocal and complicated as it is, the long narrative of human experience in the Black Mountains suggests that while our record is far from exemplary, people *can* be agents of positive change in the region. More than at any time in the past, scholars have an understanding of nature's inherent complexity and the problems involved in landscape management. We have new environmental ethics to guide future actions, practical plans that can help redefine our place in the natural world and develop pragmatic solutions to some of our most vexing problems. How will we use that information? Will we preserve the best qualities of the modern landscape and nurture the wildness that remains? Will we heed the lessons of the past and recognize nature as an equal partner as we write the next chapter of Black Mountain history? These are perhaps the toughest questions of all. The answers may well decide not only the fate of the East's highest mountains, but humanity's as well.

NOTES

PREFACE

1. Foster A. Sondley, *A History of Buncombe County, North Carolina* (1930; reprint, Spartanburg, S.C.: Reprint Co., 1970); S. Kent Schwarzkopf, *A History of Mount Mitchell and the Black Mountains: Exploration, Development, and Preservation* (Raleigh: N.C. Division of Archives and History, 1985); Jeff Lovelace, *Mount Mitchell: Its Railroad and Toll Road* (Johnson City, Tenn.: Overmountain Press, 1994).

2. Carolyn Merchant, ed., *Major Problems in American Environmental History, Major Problems in American History* (Lexington, Mass.: D.C. Heath, 1993), 1.

3. Ronald L. Lewis, *Transforming the Appalachian Countryside: Railroads, Deforestation, and Social Change in West Virginia, 1880–1920* (Chapel Hill: University of North Carolina Press, 1998); Donald Edward Davis, *Where There Are Mountains: An Environmental History of the Southern Appalachians* (Athens: University of Georgia Press, 2000); Daniel S. Pierce, *The Great Smokies: From Natural Habitat to National Park* (Knoxville: University of Tennessee Press, 2000); Margaret Lynn Brown, *The Wild East: A Biography of the Great Smoky Mountains* (Gainesville: University Press of Florida, 2000).

4. Donald Worster, quoted in Anne Matthews, "Out of the Wilderness," *Preservation*, January/February 1998, 64.

CHAPTER ONE

1. David T. Catlin, *A Naturalist's Blue Ridge Parkway* (Knoxville: University of Tennessee Press, 1984), 15; J. Wright Horton and Victor A. Zullo, "An Introduction to the Geology of the Carolinas," in *The Geology of the Carolinas*, ed. J. Wright Horton and Victor A. Zullo (Knoxville: University of Tennessee Press, 1991), 9; P. Albert Carpenter III, ed., *A Geologic Guide to North Carolina's State Parks*, North Carolina Geologic Survey, Bulletin 91 (Raleigh: North Carolina Geological Survey Section, Division of Land Resources, Dept. of Natural Resources and Community Development, 1989), 55–56; Frederick A. Cook, Larry D. Brown, and Jack E. Oliver, "The Southern Appalachians and the Growth of Continents," *Scientific American*, October 1980, 163–65.

2. Carpenter, *Geologic Guide*, 56. See also N.C. Department of Environment and Natural Resources, Division of State Parks, *The Geology of Mt. Mitchell State Park*, official park publication (n.p., n.d.).

3. Cook, Brown, and Oliver, "Southern Appalachians," 166–67; Thomas A. Caine and A. W. Mangum, *Soil Survey of the Mount Mitchell Area, North Carolina* (1902), reprint from Field Operations, Bureau of Soils, North Carolina Collection, University of North Carolina Library, Chapel Hill.

4. Paul Richard Saunders, "The Vegetational Impact of Human Disturbance on the Spruce-Fir Forests of the Southern Appalachian Mountains" (Ph.D. diss., Duke University, 1979), 25–26; N.C. Department of Environment and Natural Resources, *Geology of Mt. Mitchell State Park.*

5. "The Black Mountains," *Appalachian Voice*, Fall 1996, 1.

6. Caine and Mangum, *Soil Survey*, 262–63.

7. Peter S. White, Edward R. Buckner, J. Daniel Pittillo, and Charles V. Cogbill, "High-Elevation Forests: Spruce-Fir Forests, Northern Hardwoods Forests, and Associated Communities," in *Biodiversity of the Southeastern United States: Upland Terrestrial Communities*, ed. William H. Martin, Stephen G. Boyce, and Arthur C. Echternacht (New York: John Wiley, 1993), 308–9.

8. Jason Basil Deyton, "The Toe River Valley to 1865," *North Carolina Historical Review* 24 (October 1947): 424.

9. S. Kent Schwarzkopf, *A History of Mount Mitchell and the Black Mountains: Exploration, Development, and Preservation* (Raleigh: N.C. Division of Archives and History, 1985), frontispiece; Timothy Silver, *A New Face on the Countryside: Indians, Colonists, and Slaves in South Atlantic Forests, 1500–1800* (New York: Cambridge University Press, 1990), 10.

10. Jonathan L. Richardson, *Dimensions of Ecology* (Baltimore: Williams and Wilkins, 1977), 128–30; Silver, *New Face*, 13.

11. William H. MacLeish, *The Day before America: Changing the Nature of a Continent* (Boston: Houghton Mifflin, 1994), 32–33; Charles L. Matsch, *North America and the Great Ice Age* (New York: McGraw-Hill, 1976), 12–24, 80–81; Richard Foster Flint, *Glacial and Quaternary Geology* (New York: John Wiley, 1971), 1–10.

12. Hazel R. Delcourt and Paul A. Delcourt, "Late Quaternary History of the Spruce-Fir Ecosystem in the Southern Appalachian Mountain Region," in *The Southern Appalachian Spruce-Fir Ecosystem: Its Biology and Threats*, ed. Peter S. White, National Park Service, Southeast Region Research/Resource Management Report SER-71 (Gatlinburg, Tenn.: Uplands Field Research Laboratory, 1984), 30; MacLeish, *Day before America*, 37.

13. MacLeish, *Day before America*, 52; John E. Guilday, Paul W. Parmalee, and Harold W. Hamilton, *The Clark's Cave Bone Deposit and the Late Pleistocene Paleoecology of the Central Appalachian Mountains of Virginia*, Bulletin of Carnegie Museum of Natural History, no. 2 (Pittsburgh: Carnegie Museum of Natural History, 1977), 74–75, 79; John E. Guilday, Harold W. Hamilton, Elaine Anderson, and Paul W. Parmalee, *The Baker Bluff Cave Deposit, Tennessee, and the Late Pleistocene Faunal Gradient*, Bulletin of Carnegie Museum of Natural History, no. 11 (Pittsburgh: Carnegie Museum of Natural History, 1978), 57–61.

14. Guilday, Parmalee, and Hamilton, *Clark's Cave*, 79.

15. Delcourt and Delcourt, "Late Quaternary History," 30–32; Peter S. White and Charles V. Cogbill, "Spruce-Fir Forests of Eastern North America," in *Ecology and De-*

cline of Red Spruce in the Eastern United States, ed. Christopher Eagar and Mary Beth Adams, Ecological Studies 96 (New York: Springer-Verlag, 1992), 29.

16. Delcourt and Delcourt, "Late Quaternary History," 32; Peter S. White, "The Southern Appalachian Spruce-Fir Ecosystem: An Introduction," in White, *Southern Appalachian Spruce-Fir Ecosystem*, 6–9.

17. Ann Sutton and Myron Sutton, *Eastern Forests* (New York: Knopf, 1986), 84–85; John H. Davis, *Vegetation of the Black Mountains of North Carolina: An Ecological Study* (n.p., n.d.).

18. Stephen L. Stephenson, Andrew N. Nash, and Dean F. Stauffer, "Appalachian Oak Forests," in Martin, Boyce, and Echternacht, *Biodiversity of the Southeastern United States*, 275–78; Charles Hudson, *Knights of Spain, Warriors of the Sun: Hernando de Soto and the South's Ancient Chiefdoms* (Athens: University of Georgia Press, 1997), 190–91; Davis, *Vegetation of the Black Mountains*, 5–6; Robert J. Noyes, "Lands of Mount Mitchell Company, Mt. Mitchell Area," in files of Pisgah National Forest, Toecane Ranger District, Burnsville, N.C., 13.

19. E. C. Pielou, *After the Ice Age: The Return of Life to Glaciated North America* (Chicago: University of Chicago Press, 1991), 91–92; Daniel B. Botkin, *Discordant Harmonies: A New Ecology for the Twenty-first Century* (New York: Oxford University Press, 1990), 60.

20. Donald Worster, *Nature's Economy: A History of Ecological Ideas*, 2d ed. (New York: Cambridge University Press, 1994), 256–57. Much of what follows in this necessarily cursory discussion of the history of ecology derives from Worster's book.

21. Michael G. Barbour, "Ecological Fragmentation in the Fifties," in *Uncommon Ground: Rethinking the Human Place in Nature*, ed. William Cronon (New York: Norton, 1996), 234–35; Worster, *Nature's Economy*, 209–20; F. A. Clements, "Plant Succession: An Analysis of the Development of Vegetation," in *Ecological Succession*, ed. Frank B. Golley (Stroudsburg, Pa.: Dowden, Hutchinson, and Ross, 1977), 185.

22. Worster, *Nature's Economy*, 294–304.

23. E. P. Odum, "The Strategy of Ecosystem Development," in Golley, *Ecological Succession*, 278–82; Worster, *Nature's Economy*, 362–68.

24. Sutton and Sutton, *Eastern Forests*, 81.

25. H. A. Gleason, "The Individualistic Concept of the Plant Association," in Golley, *Ecological Succession*, 206; Barbour, "Ecological Fragmentation," 236–37; Worster, *Nature's Economy*, 391–92.

26. P. S. White and S. T. A. Pickett, "Natural Disturbance and Patch Dynamics: An Introduction," in *The Ecology of Natural Disturbance and Patch Dynamics*, ed. S. T. A. Pickett and P. S. White (Orlando, Fla.: Academic Press, 1985), xiii–xiv, 3–13; Worster, *Nature's Economy*, 392–94. As Worster points out, this more dynamic view of forest ecology was initially promoted by William H. Drury and Ian C. T. Nisbet in "Succession," an article published in the *Journal of the Arnold Arboretum* 54, no. 3 (1973): 331–68. I used the reprint in Golley, *Ecological Succession*, 287–324.

27. James Gleick, *Chaos: Making a New Science* (New York: Viking, 1987), 23, 309–11; Edward Lorenz, *The Essence of Chaos* (Seattle: University of Washington Press, 1993), 77–110; Worster, *Nature's Economy*, 405–12; Carolyn Merchant, "Reinventing Eden: Western Culture as a Recovery Narrative," in Cronon, *Uncommon Ground*, 156.

28. Worster, *Nature's Economy*, 411.

29. Indeed, the study of chaos began among meteorologists whose older computer models proved inadequate for predicting erratic weather. See Lorenz, *Essence of Chaos*, 77–110, and Gleick, *Chaos*, 11–31.

30. N.C. Department of Environment and Natural Resources, Division of Parks and Recreation, *Forest Decline on Mount Mitchell*, official park publication (n.p., 1994). Recent temperature data in this and following paragraphs provided by Lea Beazley, N.C. Department of Environment and Natural Resources, Division of Parks and Recreation, West District Office, Troutman, N.C.; Saunders, "Vegetational Impact," 24–25.

31. Monthly report of John Simcox Holmes, State Forester, January 1928, Board of Conservation and Development, Activities of the Board, 1925–35, box 3, folder: Forestry, 1928, N.C. Division of Archives and History, Raleigh; Beazley data.

32. "Temperature Here Is 57 as Snow Falls on Mt. Mitchell Summit," *Asheville Citizen-Times*, August 13, 1931; Beazley data.

33. *Asheville Citizen-Times*, February 10, 1978, Newspaper Clipping File, Pack Memorial Public Library, Asheville, N.C.; Beazley data.

34. The descriptions in the above paragraphs are drawn from White, Buckner, Pittillo, and Cogbill, "High-Elevation Forests"; Stephenson, Nash, and Stauffer, "Appalachian Oak Forests"; and my observations in the Black Mountains.

35. Lawrence S. Barden and Frank W. Woods, "Characteristics of Lightning Fires in the Southern Appalachian Forests," *Proceedings of the Tall Timbers Fire Ecology Conference* 3 (1972): 345–61; David H. Van Lear and Thomas A. Waldrop, *History, Uses, and Effects of Fire in the Appalachians*, U.S. Forest Service, Southeastern Forest Experiment Station General Technical Report SE-54 (Asheville, N.C.: U.S. Dept. of Agriculture, Forest Service, Southeastern Forest Experiment Station, 1989), 1, 11.

36. Louis S. Murphy, *The Red Spruce: Its Growth and Management*, U.S. Department of Agriculture Bulletin 554 (Washington, D.C.: U.S. Department of Agriculture, 1917), 22–23.

37. Clarence F. Korstian, "Perpetuation of Spruce on Cut-over and Burned Lands in the Higher Southern Appalachian Mountains," *Ecological Monographs* 7, no. 1 (January 1937): 132; White, "Southern Appalachian Spruce-Fir Ecosystem," 5; Saunders, "Vegetational Impact," 23–24.

38. Lea Beazley, "Threatened Forests: Why Are Spruce-Fir Trees on Mt. Mitchell Dying?," *Morganton (N.C.) News Herald*, June 4, 1996, Clipping File, William Leonard Eury Appalachian Collection, Appalachian State University, Boone, N.C.; Saunders, "Vegetational Impact," 25.

39. Paul Richard Saunders, "Recreational Impacts in the Southern Appalachian Spruce-Fir Ecosystem," in White, *Southern Appalachian Spruce-Fir Ecosystem*, 110; J. Daniel Pittillo, "Regional Differences of Spruce-Fir Forests of the Southern Blue Ridge South of Virginia," in White, *Southern Appalachian Spruce-Fir Ecosystem*, 75.

40. E. H. Frothingham, *Timber Growing and Logging Practice in the Southern Appalachian Region*, U.S. Department of Agriculture Technical Bulletin 250 (Washington, D.C.: U.S. Department of Agriculture, 1931), 42–43; Elbert L. Little, *Field Guide to North American Trees, Eastern Region* (New York: Knopf, 1994), 364, 504, 510; Saunders, "Vegetational Impact," 122–23; Marcus B. Simpson Jr., "Annotated Checklist of the

Birds of Mt. Mitchell State Park, North Carolina," *Journal of the Elisha Mitchell Scientific Society* 88, no. 4 (Winter 1972): 244–45.

41. Delcourt and Delcourt, "Late Quaternary History," 30–32.

42. Worster, *Nature's Economy*, 412.

43. White, "Southern Appalachian Spruce-Fir Ecosystem," 5; White, Buckner, Pittillo, and Cogbill, "High-Elevation Forests," 318–20.

44. J. Bruce Wallace, Jackson R. Webster, and Rex L. Lowe, "High Gradient Streams of the Appalachians," in *Biodiversity of the Southeastern United States: Aquatic Communities*, ed. Courtney T. Hackney, S. Marshall Adams, and William H. Martin (New York: John Wiley, 1992), 142–67; Catlin, *Naturalist's Parkway*, 106–7.

45. W. B. Willers, *Trout Biology: An Angler's Guide* (Madison: University of Wisconsin Press, 1981), 21–23.

46. Catlin, *Naturalist's Parkway*, 107; John O. Whitaker Jr., *The Audubon Society Field Guide to North American Mammals* (New York: Knopf, 1980), 589, 578, 457, 573, 574.

47. David S. Lee, John B. Funderburg Jr., and Mary K. Clark, *A Distributional Survey of North Carolina Mammals*, Occasional Papers of the North Carolina Biological Survey (Raleigh: North Carolina Biological Survey, N.C. State Museum of Natural History, 1982), 7–8, 63; William S. Powell, "Creatures of Carolina from Roanoke Island to Purgatory Mountain," *North Carolina Historical Review* 50 (April 1973): 159; Silver, *New Face*, 25–26; H. Trawick Ward, "The Bull in the North Carolina Buffalo," *North Carolina Archaeology*, no. 39 (1990): 19–30.

48. Fred M. Burnett, *This Was My Valley* (Charlotte: Heritage, 1960), 144; David S. Lee, "Unscrambling Rumors: The Status of the Panther in North Carolina," *Wildlife in North Carolina*, July 1977, 6–9; Whitaker, *Guide to Mammals*, 596, 539.

49. Samuel Botsford Buckley, "Mountains of North Carolina and Tennessee," *American Journal of Science and Arts* 27 (March 1850): 292; Whitaker, *Guide to Mammals*, 552.

50. Sutton and Sutton, *Eastern Forests*, 570.

51. Simpson, "Annotated Checklist of Birds," 244–51; J. S. Cairns, *List of Birds of Buncombe County, North Carolina* (Weaverville, N.C.: n.p., 1891), 6.

52. Donald W. Linzey, "Distribution Status of the Northern Flying Squirrel and the Northern Water Shrew in the Southern Appalachians," in White, *Southern Appalachian Spruce-Fir Ecosystem*, 199; John E. Cooper, ed., *Endangered and Threatened Plants and Animals of North Carolina*, Proceedings of a Symposium on Endangered and Threatened Biota of North Carolina (Raleigh: N.C. State Museum of Natural History, 1977), 423–24.

53. Andrew Jones and George F. Wilhere, "A Survey of Nonavian Vertebrates with Management Recommendations, Mount Mitchell State Park," School of the Environment, Duke University, May 1994, in files of N.C. Department of Environment and Natural Resources, Division of Parks and Recreation, Raleigh; Whitaker, *Guide to Mammals*, 351, 501; Linzey, "Distribution of Northern Flying Squirrel," 195–96; Sutton and Sutton, *Eastern Forests*, 440, 449, 466, 469, 470.

54. Joel Harp, "Itsy Bitsy Spider," *Friend of Wildlife: Journal of the North Carolina Wildlife Federation*, Winter 1996, 16; Delcourt and Delcourt, "Late Quaternary History," 30–32.

55. MacLeish, *Day before America*, 32–47; Richard White, "'Are You an Environmentalist or Do You Work for a Living?': Work and Nature," in Cronon, *Uncommon Ground*, 175–77.

56. William Cronon, introduction to Cronon, *Uncommon Ground*, 24–25, 36–37; Merchant, "Reinventing Eden," 140–56.

57. Botkin, *Discordant Harmonies*, 62; William Cronon, "The Trouble with Wilderness; or, Getting Back to the Wrong Nature," in Cronon, *Uncommon Ground*, 81.

58. Cronon introduction, 25.

CHAPTER TWO

1. For my descriptions of native life in this paragraph and those following I am indebted to David Moore, an archaeologist with the State Historic Preservation Office, N.C. Department of Cultural Resources, Raleigh. He gave me a tour of the Warren Wilson site and did his best to make the often technical world of archaeology accessible to a historian. Many of his insights appear throughout my discussion of the Warren Wilson site. I also benefited from "Views of the Past: 8000 Years of Native American History," a poster prepared by the Archaeology department of Warren Wilson College. Gwen Diehn's drawings, which appear on the poster, provided the visual basis for my descriptions of the landscape as it looked during the Archaic, Woodland, and Mississippian periods.

2. Roy S. Dickens Jr., *Cherokee Prehistory: The Pisgah Phase in the Appalachian Summit Region* (Knoxville: University of Tennessee Press, 1976), 9–10, 25; Charles Hudson, *The Southeastern Indians* (Knoxville: University of Tennessee Press, 1976), 44–55; William H. MacLeish, *The Day before America: Changing the Nature of a Continent* (Boston: Houghton Mifflin, 1994), 100–103.

3. Burton L. Purrington, "Ancient Mountaineers: An Overview of Prehistoric Archaeology of North Carolina's Western Mountain Region," in *The Prehistory of North Carolina: An Archaeological Symposium*, ed. Mark A. Mathis and Jeffrey J. Crow (Raleigh: N.C. Division of Archives and History, 1983), 110–31; Dickens, *Cherokee Prehistory*, 11–12; MacLeish, *Day before America*, 102.

4. Bennie Carlton Keel, "Woodland Phases of the Appalachian Summit Area" (Ph.D. diss., Washington State University, 1972), 22–24, 273–74.

5. Paul S. Gardner, "The Ecological Structure and Behavioral Implications of Mast Exploitation Strategies," in *People, Plants, and Landscapes: Studies in Paleoethnobotany*, ed. Kristen J. Gremillion (Tuscaloosa: University of Alabama Press, 1997), 162–72; Richard A. Yarnell, "Plant Remains from the Warren Wilson Site," in Dickens, *Cherokee Prehistory*, 217–24; Daniel L. Simpkins, "An Ethnobotanical Study of Plant Food Remains from the Warren Wilson Site (31Bn29), North Carolina: A Biocultural Approach" (M.A. thesis, University of North Carolina, Chapel Hill, 1984), 99–105; Keel, "Woodland Phases," 272–73; Richard A. Yarnell and M. Jean Black, "Temporal Trends Indicated by a Survey of Archaic and Woodland Plant Food Remains from Southeastern North America," *Southeastern Archaeology* 4, no. 2 (Winter 1985): 98–100, 103.

6. Hudson, *Southeastern Indians*, 56; MacLeish, *Day before America*, 120–24.

7. Bruce D. Smith, *The Emergence of Agriculture* (New York: Scientific American Library, 1995), 194–96; Bruce D. Smith, "Prehistoric Plant Husbandry in Eastern North

America," in *The Origins of Agriculture: An International Perspective*, ed. C. Wesley Cowan and Patty Jo Watson (Washington, D.C.: Smithsonian Institution Press, 1992), 102–3; C. Wesley Cowan and Patty Jo Watson, "Some Concluding Remarks," in Cowan and Watson, *Origins of Agriculture*, 209.

8. Yarnell, "Plant Remains," 220; Smith, *Emergence of Agriculture*, 186–89; Smith, "Prehistoric Plant Husbandry," 102. I got the idea for tasting *chenopodium* from MacLeish, *Day before America*, 125–26. It is indeed edible, but bitter.

9. Yarnell, "Plant Remains," 220; Simpkins, "Ethnobotanical Study," 99–102; Richard A. Yarnell, "Domestication of Sunflower and Sumpweed in Eastern North America," in *The Nature and Status of Ethnobotany*, ed. Richard I. Ford, Anthropological Papers, Museum of Anthropology, University of Michigan, no. 67 (Ann Arbor: Museum of Anthropology, 1978), 289, 296–97; Smith, *Emergence of Agriculture*, 190, 197.

10. Smith, "Prehistoric Plant Husbandry," 102–4; Smith, *Emergence of Agriculture*, 190–94, Richard A. Yarnell, "*Iva annua var. macrocarpa*: Extinct American Cultigen?," *American Anthropologist* 74 (June 1972): 335–41; Keel, "Woodland Phases," 17–19; Yarnell, "Plant Remains," 223.

11. For a representation of Woodland habitations on the Warren Wilson site, see Gwen Diehn's drawing on the poster "Views of the Past."

12. Smith, *Emergence of Agriculture*, 200; Charles Hudson, *Knights of Spain, Warriors of the Sun: Hernando de Soto and the South's Ancient Chiefdoms* (Athens: University of Georgia Press, 1997), 13–14; Paul A. Delcourt, Hazel R. Delcourt, Patricia A. Cridlebaugh, and Jefferson Chapman, "Holocene Ethnobotanical and Paleoecological Record of Human Impact on Vegetation in the Little Tennessee River Valley, Tennessee," *Quaternary Research* 25 (1986): 346–48.

13. Hudson, *Knights of Spain*, 13–15; Dickens, *Cherokee Prehistory*, 13–14, 19.

14. Delcourt, Delcourt, Cridlebaugh, and Chapman, "Holocene Ethnobotanical and Paleoecological Record," 331–33; Jefferson Chapman, Hazel R. Delcourt, and Paul A. Delcourt, "Strawberry Fields, Almost Forever," *Natural History*, September 1989, 51–54; Hudson, *Knights of Spain*, 24–30.

15. Dickens, *Cherokee Prehistory*, 16–18; Hudson, *Knights of Spain*, 194–96; Yarnell, "Plant Remains," 223.

16. Hudson, *Knights of Spain*, 194; Chapman, Delcourt, and Delcourt, "Strawberry Fields," 56–57; Dickens, *Cherokee Prehistory*, 19–68. For artists' representations of Pisgah villages, see Dickens, *Cherokee Prehistory*, 95, and the more detailed drawing by Gwen Diehn on the poster "Views of the Past."

17. Information regarding the capability of stone axes and firewood is drawn from my conversations with David Moore and my inspection of artifacts recovered at Warren Wilson and other sites.

18. Keel, "Woodland Phases," 272.

19. Delcourt, Delcourt, Cridlebaugh, and Chapman, "Holocene Ethnobotanical and Paleoecological Record," 333–35; Hazel R. Delcourt and Paul A. Delcourt, "Pre-Columbian Native American Use of Fire on Southern Appalachian Landscapes," *Conservation Biology* 11, no. 4 (August 1997): 1010–11; Chapman, Delcourt, and Delcourt, "Strawberry Fields," 56–57.

20. Paul A. Delcourt and Hazel R. Delcourt, "The Influence of Prehistoric Human-set Fires on Oak-Chestnut Forests in the Southern Appalachians," *Castanea* 63, no. 3

(September 1998): 37–45; Mark D. Abrams, "Fire and the Development of Oak Forests," *Bioscience* 42, no. 5 (May 1992): 349–52; Susan Power Bratton and Albert J. Meier, "The Recent Vegetation Disturbance History of the Chattooga River Watershed," *Castanea* 63, no. 3 (September 1998): 372–81.

21. Hudson, *Knights of Spain*, 14; Hudson, *Southeastern Indians*, 295; Timothy Silver, *A New Face on the Countryside: Indians, Colonists, and Slaves in South Atlantic Forests, 1500–1800* (New York: Cambridge University Press, 1990), 46–48; Gail E. Wagner, "'Their Women and Children do Continually Keepe it with Weeding': Late Prehistoric Women and Horticulture in Eastern America," in *The Influence of Women on the Southern Landscape*, Proceedings of the Tenth Conference on Restoring Southern Gardens and Landscapes, October 5–7, 1993, Winston-Salem, N.C. (Winston Salem: Old Salem, Inc., 1997), 26–29; Chapman, Delcourt, and Delcourt, "Strawberry Fields," 55. Recently some scholars have argued that North American natives did not clear and abandon fields on a large scale but instead sought to cultivate them permanently, allowing them to lie fallow every other year. If so, such fields rarely became reforested, which means that stands of very young timber might be comparatively rare along the Swannanoa; see William E. Doolittle, *Cultivated Landscapes of Native North America* (New York: Oxford University Press, 2000), 174–90. My descriptions of Indian fields and the valley are also drawn from observations of a Native American garden re-created on the Warren Wilson site.

22. Dickens, *Cherokee Prehistory*, 3–4; Hudson, *Knights of Spain*, 196–99. The link between Pisgah culture and the historic Cherokees is based in part on similarities in pottery made by both groups. As Hudson notes, such similarities are not "an infallible guide to the prehistoric distribution of Cherokee-speaking peoples" (*Knights of Spain*, 197). Moreover, Pisgah culture was changed by the Spanish explorers who traversed the mountains in the sixteenth century, and it is difficult to determine a clear line of descent between Pisgah people and historic Cherokees.

23. Robin A. Beck Jr., "From Joara to Chiaha: Spanish Exploration of the Appalachian Summit Area, 1540–1568," *Southeastern Archaeology* 16, no. 2 (Winter 1997): 163–65; Hudson, *Knights of Spain*, 190–94; Charles C. Mann, "1491," *Atlantic Monthly*, March 2002, 44–45. Historians and archaeologists still disagree about Hernando de Soto's exact route across the Appalachians. In describing his path I have chosen the most recent and, to my mind, the most reliable reconstructions of the route.

24. A Gentleman of Elvas, "True Relation of the Hardships Suffered by Governor Hernando De Soto & Certain Portuguese Gentlemen During the Discovery of the Province of Florida," trans. and ed. James Alexander Robertson, with footnotes and updates to Robertson's notes by John H. Hahn; Rodrigo Rangel, "Account of the Northern Conquest and Discovery of Hernando De Soto," ed. John E. Worth; David Bost, ed., and Charmion Shelby, trans., "La Florida by Garcilaso de la Vega, the Inca"; and John E. Worth, ed. and trans., "Relation of the Island of Florida by Luys Hernandez de Biedma," all in *The De Soto Chronicles: The Expedition of Hernando De Soto to North America in 1539–1543*, 2 vols., ed. Lawrence A. Clayton, Vernon James Knight Jr., and Edward C. Moore (Tuscaloosa: University of Alabama Press, 1993), 1:86, 281–82, 231–32, 2:316.

25. Richard White, "'Are You an Environmentalist or Do You Work for a Living?':

Work and Nature," in *Uncommon Ground: Rethinking the Human Place in Nature*, ed. William Cronon (New York: Norton, 1996), 176–77.

26. Rangel, "Account," 1:281.

27. Charles Hudson, *The Juan Pardo Expeditions: Exploration of the Carolinas and Tennessee, 1566–1568* (Washington, D.C.: Smithsonian Institution Press, 1990), 160.

28. Leland G. Ferguson, "Prehistoric Mica Mines in the Southern Appalachians," *South Carolina Antiquities* 6 (1974): 1; Jasper Leonidas Stuckey, *North Carolina: Its Geology and Mineral Resources* (Raleigh: N.C. Department of Conservation and Development, 1965), 410–12.

29. Gentleman of Elvas, "True Relation," 1:89–90; Hudson, *Knights of Spain*, 203.

30. Ferguson, "Prehistoric Mica Mines," 3–4; W. C. Kerr, quoted in Stuckey, *North Carolina*, 414–15.

31. Hudson, *Knights of Spain*, 398–410.

32. Hudson, *Juan Pardo Expeditions*, 10–46; Beck, "Joara to Chiaha," 165–88.

33. Paul Hoffman, trans., "The 'Long' Bandera Relation" and "The 'Short' Bandera Relation," in Hudson, *Juan Pardo Expeditions*, 265, 302.

34. Hudson, *Juan Pardo Expeditions*, 189–95. Reports of mining by explorers and tools purported to have been used by the Spanish surface from time to time in North Carolina. See, for example, Stuckey, *North Carolina*, 208. But to date there is no proof that de Soto or any other explorer did any mining around the Black Mountains.

35. Hudson, *Knights of Spain*, 417–26; Mann, "1491," 45.

36. Gentleman of Elvas, "True Relation," 1:83.

37. Hudson, *Knights of Spain*, 421–26; Silver, *New Face*, 77–81.

38. James H. Merrell, *The Indians' New World: Catawbas and Their Neighbors from European Contact through the Era of Removal* (Chapel Hill: University of North Carolina Press, 1989), 92–133; Jason Basil Deyton, "The Toe River Valley to 1865," *North Carolina Historical Review* 24 (October 1947): 436–37; Hudson, *Juan Pardo Expeditions*, 181–89; S. Kent Schwarzkopf, *A History of Mount Mitchell and the Black Mountains: Exploration, Development, and Preservation* (Raleigh: N.C. Division of Archives and History, 1985), 2; Tom Hatley, *The Dividing Paths: Cherokees and South Carolinians through the Era of Reconstruction* (New York: Oxford University Press, 1993), 4–7; Peter H. Wood, "The Changing Population of the Colonial South: An Overview by Race and Region, 1685–1790," in *Powhatan's Mantle: Indians in the Colonial Southeast*, ed. Peter H. Wood, Gregory A. Waselkov, and M. Thomas Hatley (Lincoln: University of Nebraska Press, 1989), 61–63.

39. John Lederer, *The Discoveries of John Lederer in three several Marches from Virginia, to the West of Carolina, And other parts of the Continent: Begun in March 1699 and ended in September 1670* (Readex Microprint Corporation, 1966), 23–24, 17–18.

40. Hatley, *Dividing Paths*, 17–41; Silver, *New Face*, 67–73.

41. "The Journeys of Needham and Arthur," in *The First Explorations of the Trans-Allegheny Region by the Virginians 1650–1674*, ed. Clarence Walworth Alvord and Lee Bidgood (Cleveland: Arthur H. Clark, 1912), 211–12; Deyton, "Toe River Valley," 429.

42. "The Discovery of the Ohio Waters," in Alvord and Bidgood, *First Explorations of the Trans-Allegheny Region*, 79–88.

43. Daniel H. Usner Jr., *Indians, Settlers, and Slaves in a Frontier Exchange Economy:*

The Lower Mississippi Valley before 1783 (Chapel Hill: University of North Carolina Press, 1992), 17–35; Hatley, *Dividing Paths*, 33–39.

44. Daniel H. Usner Jr., "The Deerskin Trade in French Louisiana," *Proceedings of the Tenth Meeting of the French Colonial Historical Society, April 12–14, 1984*, ed. Philip P. Boucher (Washington, D.C.: University Press of America, 1985), 76–81; Kathryn E. Holland Braund, *Deerskins and Duffels: The Creek Indian Trade with Anglo-America, 1685–1815* (Lincoln: University of Nebraska Press, 1993), 87–88.

45. Converse D. Clowse, *Economic Beginnings of South Carolina, 1670–1730* (Columbia: University of South Carolina Press, 1971), 256–57; Silver, *New Face*, 92–93; Usner, *Indians, Settlers, and Slaves*, 246–48; Hatley, *Dividing Paths*, 164.

46. Peter H. Wood, "The Impact of Smallpox on the Native Population of the Eighteenth Century South," *New York State Journal of Medicine* 87 (January 1987): 30–36; Silver, *New Face*, 73–88; Hatley, *Dividing Paths*, 164.

47. Wood, "Changing Population," 65. For detailed accounts of the wars, population decline, and the changing nature of Cherokee hunting, see Hatley, *Dividing Paths*, 119–40, 191–228.

48. Dale R. McCullough, "Lessons from the George Reserve, Michigan," in *White-Tailed Deer: Ecology and Management*, ed. Lowell K. Halls (Harrisburg, Pa.: Stackpole, 1984), 216–17; Richard E. McCabe and Thomas R. McCabe, "Of Slings and Arrows: An Historical Retrospection," in Halls, *White-Tailed Deer*, 60. On the link between Indian warfare and the recovery of deer populations, see Richard White, *The Roots of Dependency: Subsistence, Environment, and Social Change among the Choctaws, Pawnees, and Navajos* (Lincoln: University of Nebraska Press, 1983), chap. 3, and Silver, *New Face*, 93.

49. Hatley, *Dividing Paths*, 211–12.

50. Kim Derek Pritts, *Ginseng: How to Find, Grow, and Use America's Forest Gold* (Mechanicsburg, Pa.: Stackpole, 1995), 19–23; Alvar W. Carlson, "America's Botanical Drug Connection to the Orient," *Economic Botany* 40 (1986): 233–35.

51. Keith Thomas, *Man and the Natural World: A History of the Modern Sensibility* (New York: Pantheon, 1983), 225–27; Henry Savage Jr. and Elizabeth J. Savage, *André and François-André Michaux* (Charlottesville: University Press of Virginia, 1986), 3–6.

52. Thomas, *Man and the Natural World*, 65–67.

53. Savage and Savage, *Michaux*, 7–10, 53–77.

54. C. S. Sargent, ed., "Portions of the Journal of André Michaux, Botanist, written during his Travels in the United states and Canada, 1785–1796," *Proceedings of the American Philosophical Society* 26, no. 129 (January–July 1889): 54–55; Savage and Savage, *Michaux*, 101–2; Schwarzkopf, *History of Mount Mitchell*, 16–17.

55. Sargent, "Journal of Michaux," 62; Schwarzkopf, *History of Mount Mitchell*, 17.

56. Savage and Savage, *Michaux*, 106–7.

57. Sargent, "Journal of Michaux," 109–12; Schwarzkopf, *History of Mount Mitchell*, 17–18.

58. Savage and Savage, *Michaux*, 151–61; André Michaux, "Journal of Travels into Kentucky: July 15, 1793–April 11, 1796," in *Early Western Travels, 1748–1846*, 32 vols., ed. Reuben Gold Thwaites (New York: AMS Press, 1966), 3:56–58, 98–100.

59. Savage and Savage, *Michaux*, 161–80.

60. Schwarzkopf, *History of Mount Mitchell*, 19; Savage and Savage, *Michaux*, 65, 70;

Howard K. Kimbill, "Famous French Botanists Made 2 Trips to Mt. Mitchell," *Asheville Citizen-Times*, June 12, 1955.

61. Michaux, "Journal of Travels," 100; Savage and Savage, *Michaux*, 98–99.

62. Deyton, "Toe River Valley," 428–35; Schwarzkopf, *History of Mount Mitchell*, 5.

63. Wilma A. Dunaway, "Speculators and Settler Capitalists: Unthinking the Mythology about Appalachian Landholding, 1790–1860," in *Appalachia in the Making: The Mountain South in the Nineteenth Century*, ed. Mary Beth Pudup, Dwight B. Billings, and Altina L. Waller (Chapel Hill: University of North Carolina Press, 1995), 52–55; Deyton, "Toe River Valley," 435–36; A. R. Newsome, ed., "John Brown's Journal of Travel in Western North Carolina in 1795," *North Carolina Historical Review* 11 (October 1934): 284–313.

64. Deyton, "Toe River Valley," 433–34; Schwarzkopf, *History of Mount Mitchell*, 6–8; J. S. Otto and N. E. Anderson, "Slash-and-Burn Cultivation in the Highlands South: A Problem in Comparative Agricultural History," *Comparative Studies in Society and History* 24 (June 1982): 134–36.

65. John C. Inscoe, *Mountain Masters: Slavery and the Sectional Crisis in Western North Carolina* (Knoxville: University of Tennessee Press, 1989), 38–41; Newsome, "John Brown's Journal," 290; Deyton, "Toe River Valley," 454, 457.

66. Julius Rubin, "The Limits of Agricultural Progress in the Nineteenth-Century South," *Agricultural History* 49 (April 1975): 364–67; Inscoe, *Mountain Masters*, 14–15; Deyton, "Toe River Valley," 452.

67. Inscoe, *Mountain Masters*, 45–47; Deyton, "Toe River Valley," 451.

68. Otto and Anderson, "Slash-and-Burn Cultivation," 141–42.

69. Forrest McDonald and Grady McWhiney, "The Antebellum Herdsman: A Reinterpretation," *Journal of Southern History* 41 (February 1975): 158.

70. Grady McWhiney and Forrest McDonald, "Celtic Origins of Southern Herding Practices," *Journal of Southern History* 51 (February 1985): 165–81; Otto and Anderson, "Slash-and-Burn Cultivation," 136–37; John Fraser Hart, "Land Rotation in Appalachia," *Geographical Review* 67 (January 1977): 150–51.

71. Otto and Anderson, "Slash-and-Burn Cultivation," 142–43.

72. E. A. Johnson, "Effect of Farm Woodland Grazing on Watershed Values in the Southern Appalachian Mountains," *Journal of Forestry* 50 (1952): 110–11; H. H. Biswell and M. D. Hoover, "Appalachian Hardwood Trees Browsed by Cattle," *Journal of Forestry* 43 (1945): 675–76.

73. Kim Moreau Peterson, "Natural Origin and Maintenance of Southern Appalachian Balds: A Review of Hypotheses," in *Status and Management of Southern Appalachian Balds: Proceedings of a Workshop Sponsored by The Southern Appalachian Research/Resource Cooperative, November 5–7, 1980*, ed. Paul Richard Saunders (Cullowhee, N.C.: Western Carolina University, [1981?]), 7–15; Schwarzkopf, *History of Mount Mitchell*, 8–9.

74. Phil Gersmehl, "Factors Leading to Mountaintop Grazing in the Southern Appalachians," *Southeastern Geographer* 10 (April 1970): 67–72.

75. Ibid., 72; J. Daniel Pittillo, "Status and Dynamics of Balds in the Southern Appalachian Mountains," in Saunders, *Status and Management of Balds*, 44–48.

76. Edward Schell, "The Floral Values of Southern Appalachian Balds," in Saunders, *Status and Management of Balds*, 68–69.

77. Silver, *New Face*, 175–77; Liz Bomford, *The Complete Wolf* (New York: St. Martin's Press, 1993), 84; Deyton, "Toe River Valley," 437–38, 451.

78. Hudson, *Knights of Spain*, 188; Schwarzkopf, *History of Mount Mitchell*, 2; Scott Weidensaul, *Mountains of the Heart: A Natural History of the Appalachians* (Golden, Colo.: Fulcrum, 1994), 150.

79. Schwarzkopf, *History of Mount Mitchell*, 13; Elbert L. Little, *The Audubon Society Field Guide to American Trees: Eastern Region* (New York: Knopf, 1980), 278.

CHAPTER THREE

1. George H. Daniels, *American Science in the Age of Jackson* (New York: Columbia University Press, 1968), 18; Charles Phillips, ed., *A Memoir of the Rev. Elisha Mitchell, D.D., Late Professor of Chemistry, Mineralogy & Geology in the University of North Carolina: Together with the Tributes of Respect to his Memory, by Various Public Meetings and Literary Associations, and the Addresses Delivered at the Re-Interment of His Remains, by Rt. Rev. James H. Otey, D.D., Bishop of Tennessee, and Hon. David L. Swain, LL.D., President of the University, Chapel Hill* (Chapel Hill: J. M. Henderson, 1858), 5; Elgiva D. Watson, "Elisha Mitchell: A Connecticut Yankee in North Carolina," *Journal of the Elisha Mitchell Scientific Society* 100, no. 2 (1984): 43; Seymour H. Mauskopf, "Elisha Mitchell and European Science: The Case of Chemistry," *Journal of the Elisha Mitchell Scientific Society* 100, no. 2 (1984): 57.

2. Daniels, *Science in the Age of Jackson*, 34, 13–14; Rogers McVaugh, Michael R. McVaugh, and Mary Ayers, *Chapel Hill and Elisha Mitchell the Botanist* (Chapel Hill, N.C.: Botanical Garden Foundation, 1996), 4; Elgiva D. Watson, "Elisha Mitchell," 43.

3. Maria Mitchell to Hannah North, January 1, 1820, Elisha Mitchell Papers, Southern Historical Collection, Manuscripts Division, University of North Carolina, Chapel Hill; Michael McVaugh, "Elisha Mitchell's Library as an Index to His Scientific Interests," *Journal of the Elisha Mitchell Scientific Society* 100, no. 2 (1984): 55.

4. William Hooper to Elisha Mitchell, January 9, 1823, Mitchell Papers; McVaugh, McVaugh, and Ayers, *Chapel Hill and Elisha Mitchell*, 4–5, 9; Elgiva D. Watson, "Elisha Mitchell," 44, 46.

5. McVaugh, McVaugh, and Ayers, *Chapel Hill and Elisha Mitchell*, 6–28; Daniels, *Science in the Age of Jackson*, 114, 225; Elizabeth B. Keeney, *The Botanizers: Amateur Scientists in Nineteenth-Century America* (Chapel Hill: University of North Carolina Press, 1992), 32–3.

6. McVaugh, McVaugh, and Ayers, *Chapel Hill and Elisha Mitchell*, 10; McVaugh, "Elisha Mitchell's Library," 52–53.

7. Elisha Mitchell to Maria Mitchell, July 22, 1827, Mitchell Papers; Jasper Leonidas Stuckey, *North Carolina: Its Geology and Mineral Resources* (Raleigh: N.C. Department of Conservation and Development, 1965), 229–33; Walter H. Wheeler, "Elisha Mitchell as a Geologist," *Journal of the Elisha Mitchell Scientific Society* 100, no. 2 (1984): 61–63; McVaugh, "Elisha Mitchell's Library," 54.

8. Stuckey, *North Carolina*, 233; Kemp P. Battle, ed., *Diary of a Geological Tour by Dr. Elisha Mitchell in 1827 and 1828*, James Sprunt Historical Monograph No. 6 (Chapel Hill: University of North Carolina, 1905), 54; Elgiva D. Watson, "Elisha Mitchell," 46.

9. Roderick Nash, *Wilderness and the American Mind*, 3d ed. (New Haven: Yale University Press, 1982), 29, 30–31; William Cronon, "The Trouble with Wilderness; or, Getting Back to the Wrong Nature," in *Uncommon Ground: Rethinking the Human Place in Nature*, ed. William Cronon (New York: Norton, 1996), 70–71.

10. Battle, *Diary of a Geological Tour*, 34.

11. Harry L. Watson, *Jacksonian Politics and Community Conflict: The Emergence of the Second American Party System in Cumberland County North Carolina* (Baton Rouge: Louisiana State University Press, 1981), 49–51; Harry L. Watson, *An Independent People: The Way We Lived in North Carolina, 1770–1820* (Chapel Hill: University of North Carolina Press, 1983), 106–7.

12. Elgiva D. Watson, "Elisha Mitchell," 43.

13. Battle, *Diary of a Geological Tour*, 51–52.

14. John C. Inscoe, *Mountain Masters: Slavery and the Sectional Crisis in Western North Carolina* (Knoxville: University of Tennessee Press, 1989), 30–33; David L. Swain, "A Vindication of the Propriety of Giving the Name 'Mt. Mitchell,' to the Highest Peak of 'Black Mountain,'" in Phillips, *Memoir of Elisha Mitchell*, 79–80. In this speech, given June 16, 1858, on the occasion of the reinterment of Mitchell's body on the Blacks' highest peak, Swain noted that one of those who believed the East's tallest mountain lay in North Carolina was John C. Calhoun, who traveled extensively in the western part of the state. See also Elgiva D. Watson, "Elisha Mitchell," 47.

15. Charles Phillips, "Dr. Mitchell's Investigations among the Mountains of Yancey County," *North Carolina University Magazine* 7, no. 7 (March 1858): 294–95. The link between Swain's boosterism and Mitchell's explorations remains tenuous. Phillips suggests that Swain's interest in the west and his conversations with John C. Calhoun and others "doubtless . . . helped form [Mitchell's] resolution to visit so remarkable a district of our State." Although written in 1858 as part of an effort to prove Mitchell's claim to the high peak, Phillips's essay is one of the best-researched contemporary accounts of the professor's early explorations and, as such, provides the best evidence of Swain's interest and participation in Mitchell's explorations.

16. S. Kent Schwarzkopf, *A History of Mount Mitchell and the Black Mountains: Exploration, Development, and Preservation* (Raleigh: N.C. Division of Archives and History, 1985), 24; Perrin Wright, "Measuring Mitchell's Mountains," *Journal of the Elisha Mitchell Scientific Society* 108, no. 1 (Spring 1992): 2.

17. Schwarzkopf, *History of Mount Mitchell*, 20–34, 49–72, provides by far the best account of Mitchell's travels in the Blacks between 1835 and 1857. Throughout the following paragraphs I have relied extensively on Schwarzkopf's careful reconstruction of Mitchell's routes for determining the professor's whereabouts.

18. Wright, "Measuring Mitchell's Mountains," 1–10.

19. Phillips, "Mitchell's Investigations," 295; Wright, "Measuring Mitchell's Mountains," 2–3.

20. Elisha Mitchell, "The Mountains of Carolina," Mitchell Papers.

21. Elisha Mitchell, "The Mountains of Carolina," *Raleigh Register*, November 3, 1835.

22. *Raleigh Register*, November 10, 1835.

23. Schwarzkopf, *History of Mount Mitchell*, 30–31.

24. Elisha Mitchell to Maria Mitchell, July 7, 1844, Mitchell Papers; Thomas E. Jeffrey, "'A Whole Torrent of Mean and Malevolent Abuse': Party Politics and the Clingman-Mitchell Controversy," pt. 1, *North Carolina Historical Review* 70 (July 1993): 246.

25. Schwarzkopf, *History of Mount Mitchell*, 31–32; Jeffrey, "'Whole Torrent,'" pt. 1, 247 n. 17; "The Mitchell and Clingman Controversy," *Asheville Citizen-Times*, March 26, 1950.

26. Elisha Mitchell to Maria Mitchell, July 14, 1844, Mitchell Papers; Schwarzkopf, *History of Mount Mitchell*, 32.

27. Cronon, "Trouble with Wilderness," 73–75.

28. Elisha Mitchell to Thomas Clingman, *Asheville Highland Messenger*, January 24, 1845; Mitchell to Clingman, *Raleigh Register*, March 25, 1845; Jeffrey, "'Whole Torrent,'" pt. 1, 247–48; Schwarzkopf, *History of Mount Mitchell*, 32–33.

29. The sketch of Clingman's early life, education, and political career is drawn from Thomas E. Jeffrey, *Thomas Lanier Clingman: Fire-eater from the Carolina Mountains* (Athens: University of Georgia Press, 1998), 1–71, 133–40; Inscoe, *Mountain Masters*, 178–79; and Jeffrey, "'Whole Torrent,'" pt. 1, 243–45.

30. Thomas L. Clingman, *Selections from the Speeches and Writings of Hon. Thomas L. Clingman, of North Carolina, with Additions and Explanatory Notes* (Raleigh: John Nichols, 1877), 114–15.

31. Jeffrey, *Thomas Lanier Clingman*, 137–41.

32. Mitchell to Clingman, *Raleigh Register*, March 25, 1845; Thomas L. Clingman, *Measurements of the Black Mountain* (Washington, D.C.: G. S. Gideon, 1856), 6; Jeffrey, "'Whole Torrent,'" pt. 1, 248. Jeffrey's two-part article is the clearest and most thoroughly researched account of the Mitchell-Clingman feud. In the paragraphs that follow I have relied extensively on Jeffrey's work (esp. pp. 245–50) to develop a chronology of the dispute and to understand key points of debate between Mitchell and Clingman.

33. Thomas L. Clingman, "Topography of Black Mountain," in *Tenth Annual Report of the Board of Regents of the Smithsonian Institution* (Washington, D.C.: Cornelius Wendell, 1856), 299–300.

34. Ibid., 299; Schwarzkopf, *History of Mount Mitchell*, 73–78.

35. Jeffrey, "'Whole Torrent,'" pt. 1, 249; Elisha Mitchell to Joseph Henry, November 20, 1855, in Clingman, *Measurements*, 3–4; Schwarzkopf, *History of Mount Mitchell*, 54; Foster A. Sondley, *A History of Buncombe County, North Carolina* (1930; reprint, Spartanburg, N.C.: Reprint Co., 1977), 565.

36. Mitchell to Henry, November 20, 1855, and Thomas L. Clingman to Joseph Henry, December 1, 1855, in Clingman, *Measurements*, 3, 7.

37. Jeffrey, "'Whole Torrent,'" pt. 1, 242, 253.

38. Inscoe, *Mountain Masters*, 191–92; Thomas E. Jeffrey, "Thunder from the Mountains: Thomas Lanier Clingman and the End of Whig Supremacy in North Carolina," *North Carolina Historical Review* 56 (October 1979): 386–87.

39. Thomas L. Clingman to Joseph Henry, June 28, 1856, in Clingman, *Measurements*, 15; Jeffrey, "Thunder from the Mountains," 377–80.

40. Elisha Mitchell, *Black Mountain: Professor Mitchell's Reply to Mr. Clingman* (1856), North Carolina Collection, University of North Carolina Library, Chapel Hill, 7; Jeffrey, "'Whole Torrent,'" pt. 1, 254–56; Elgiva D. Watson, "Elisha Mitchell," 46–7.

41. Clingman, *Measurements*, 9–15.

42. Thomas Clingman to Thomas W. Atkin, *Asheville News*, October 22, 30, 1856; Jeffrey, "'Whole Torrent,'" pt. 1, 261–62; Thomas E. Jeffrey, "'A Whole Torrent of Mean and Malevolent Abuse': Party Politics and the Clingman-Mitchell Controversy," pt. 2, *North Carolina Historical Review* 70 (October 1993): 402.

43. Clingman, *Measurements*, 12; Mitchell, *Black Mountain*, 8.

44. Clingman recounted Mitchell's November 1856 letter in the *North Carolina Standard*, October 3, 1857; Jeffrey, "'Whole Torrent,'" pt. 1, 263–65.

45. Virginia Boone (great-granddaughter of Big Tom Wilson and daughter of Ewart Wilson), interview with author, Pensacola, N.C., January 5, 1996. See also Tim Silver, "Big Tom Wilson," *Wildlife in North Carolina*, November 1997, 19.

46. Harold E. Johnston, "'Big Tom' Wilson Leads Search for Mitchell," in *Common Times: Written and Pictorial History of Yancey County* (Burnsville, N.C.: Yancey Graphics, 1981), 22.

47. Interview with Boone.

48. The best description of nineteenth-century bear hunting in the Blacks is Fred M. Burnett, *This Was My Valley* (Charlotte: Heritage, 1960), 25–27.

49. Bill Sharpe, *A New Geography of North Carolina* (Raleigh: Sharpe, 1961), 1654.

50. Interview with Boone; Sharpe, *New Geography*, 1654.

51. Moses Ashley Curtis, quoted in A. Hunter Dupree, *Asa Gray, 1810–1888* (Cambridge: Harvard University Press, 1959), 96–97; "Moses Ashley Curtis," *American Journal of Science and Arts*, 3d ser., 5 (January–June 1873): 391; Charles F. Jenkins, "Asa Gray and His Quest for *Shortia Galacifolia*," *Bulletin of Popular Information of the Arnold Arboretum, Harvard University* 2, no. 3–4 (April 10, 1942): 13, 17–18.

52. Schwarzkopf, *History of Mount Mitchell*, 36–42.

53. "Trip to the Mountains," *North Carolina Standard* (Raleigh), September 30, 1857.

54. Charles Dudley Warner, *On Horseback: A Tour in Virginia, North Carolina, and Tennessee, with Notes of Travel in Mexico and California* (Boston: Houghton Mifflin, 1889), 87; "Trip to the Mountains."

55. Cecil D. Eby Jr., *Porte Crayon: The Life of David Hunter Strother, Writer of the Old South* (Chapel Hill: University of North Carolina Press, 1960), 68–70.

56. [Porte Crayon], "A Winter in the South, Third Paper," *Harper's New Monthly Magazine*, November 1857, 733–40; Eby, *Porte Crayon*, 93–95.

57. Clingman, *Measurements*, 12; Jeffrey, "'Whole Torrent,'" pt. 1, 262–64; Schwarzkopf, *History of Mount Mitchell*, 60–61.

58. Elisha Mitchell to Mary P. Ashe, June 1857, Mitchell Papers; Phillips, "Mitchell's Investigations," 318.

59. Johnston, "'Big Tom' Wilson Leads Search," 23; Schwarzkopf, *History of Mount Mitchell*, 61.

60. Johnston, "'Big Tom' Wilson Leads Search," 24; Judge David Schenck, "Big Tom Wilson Found Body of Elisha Mitchell in 1857," reprinted in *Asheville Citizen-Times*, March 26, 1950; Schwarzkopf, *History of Mount Mitchell*, 62–65; Jeffrey, "'Whole Torrent,'" pt. 1, 265.

61. Phillips, *Memoir of Elisha Mitchell*, 11–12; Johnston, "'Big Tom' Wilson Leads Search," 25.

62. Jeffrey, "'Whole Torrent,'" pt. 2, 405–9. Part 2 of Jeffrey's article covers the

efforts of Zebulon Vance and Charles Phillips to rehabilitate Mitchell's reputation and, in effect, rewrite the narrative of Black Mountain exploration. In the following paragraphs I quote and cite primary documents but have relied extensively on Jeffrey's provocative interpretation of the events following Mitchell's death.

63. Statements from the various witnesses and an account of William Wilson's recreation of the 1835 trip are included in Phillips, "Mitchell's Investigations," 305–10.

64. Phillips, "Mitchell's Investigations," 316–17; Jeffrey, "'Whole Torrent,'" pt. 2, 411.

65. For a summary of Clingman's arguments, see Clingman to T. W. Atkin, *North Carolina Standard*, October 3, 1857, and Jeffrey, "'Whole Torrent,'" pt. 2, 412–15.

66. Phillips, "Mitchell's Investigations," 293–301, 315.

67. Jeffrey, "'Whole Torrent,'" pt. 2, 415–16.

68. Ibid.; Phillips, "Mitchell's Investigations," 309–11.

69. Vance and Phillips did use Mitchell's estimates to prove their case (see Phillips, "Mitchell's Investigations," 313). But they also tried to use later measurements to prove that he had gone back on subsequent expeditions. On the reluctance of scholars to accept those numbers at face value, see Schwarzkopf, *History of Mount Mitchell*, 30.

70. Perrin Wright first suggested to me that if one takes weather into account, Mitchell's measurements might, in fact, be trustworthy. The interpretation of events in this and the following paragraphs, however, is my own.

71. Wright, "Measuring Mitchell's Mountains," 10. Much of this paragraph is also based on extensive conversations with Wright, his review of the manuscript, and a hike I took with him to Yeates Knob, August 15, 1999.

72. "Guyot's Measurement of the Mountains of Western North Carolina," in Clingman, *Speeches and Writings*, 141, 138, 143.

73. Jeffrey, "'Whole Torrent,'" pt. 2, 416–27.

74. Johnston, "'Big Tom' Wilson Leads Search"; Schenck, "Big Tom Wilson Found Body"; Silver, "Big Tom Wilson," 22.

75. Inscoe, *Mountain Masters*, 21–24.

76. Phillips, "Mitchell's Investigations," 309; Johnston, "'Big Tom' Wilson Leads Search," 24.

77. "Trip to the Mountains"; Schwarzkopf, *History of Mount Mitchell*, 42–43.

78. "Trip to the Mountains." On the endurance of the Vance-Phillips scheme of events, see Jeffrey, "'Whole Torrent,'" pt. 2, 241–42.

79. Schwarzkopf, *History of Mount Mitchell*, 42–43; Jeffrey, "'Whole Torrent,'" pt. 2, 427–29.

CHAPTER FOUR

1. John C. Inscoe and Gordon B. McKinney, *The Heart of Confederate Appalachia: Western North Carolina's Civil War* (Chapel Hill: University of North Carolina Press, 2000), 95, 105–6, 166–67 (I am indebted to Professors Inscoe and McKinney for allowing me to read their manuscript in advance of publication); Harold E. Johnston, "'Big Tom' Wilson Leads Search for Mitchell," in *Common Times: Written and Pictorial History of Yancey County* (Burnsville, N.C.: Yancey Graphics, 1981), 22; "Civil War Yancey: A County Divided, Microcosm of a Bitter Struggle," in *Common Times*, 26–27.

2. Gordon B. McKinney, "Women's Role in Civil War Western North Carolina,"

North Carolina Historical Review 69 (January 1992): 36–46; John C. Inscoe, "Coping in Confederate Appalachia: Portrait of a Mountain Woman and Her Community at War," *North Carolina Historical Review* 69 (October 1992): 388–413.

3. Paul D. Escott, *After Secession: Jefferson Davis and the Failure of Confederate Nationalism* (Baton Rouge: Louisiana State University Press, 1978); Phillip Shaw Paludan, *Victims: A True Story of the Civil War* (Knoxville: University of Tennessee Press, 1981), 68–69; Paul D. Escott, "Poverty and Governmental Aid for the Poor in Confederate North Carolina," *North Carolina Historical Review* 61 (October 1984): 465–66.

4. A. D. Childs to Zebulon Baird Vance, November 29, 1863, and S. J. Westall to Vance, January 17, 1864, in *The Papers of Zebulon Baird Vance*, ed. Gordon B. McKinney and Richard M. McMurray (Frederick, Md.: University Publications of America, 1987) (microfilm), reels 20, 21; McKinney, "Women's Role," 45.

5. Paludan, *Victims*, 80–81; McKinney, "Women's Role," 47; Escott, "Poverty and Governmental Aid," 464.

6. Escott, *After Secession*, 68–69.

7. J. W. McElroy to Zebulon Baird Vance, October 23, 1863, in McKinney and McMurray, *Papers of Vance*, reel 20.

8. Escott, "Poverty and Governmental Aid," 469–70, 477–80; Inscoe and McKinney, *Heart of Confederate Appalachia*, 197–98; McKinney, "Women's Role," 45.

9. Inscoe and McKinney, *Heart of Confederate Appalachia*, 124–25; "Civil War Yancey," 27; Richard Randolph, "Confederate Dilemma: North Carolina Troops and the Deserter Problem," *North Carolina Historical Review* 66 (January 1989): 61; Peter S. Bearman, "Desertion as Localism: Army Unit Solidarity and Group Norms in the U.S. Civil War," *Social Forces* 70 (December 1991): 329.

10. Fred M. Burnett, *This Was My Valley* (Charlotte: Heritage, 1960), 73.

11. William Hutson Abrams Jr., "The Western North Carolina Railroad, 1855–1894" (M.A. thesis, Western Carolina University, 1976), 34; S. Kent Schwarzkopf, *A History of Mount Mitchell and the Black Mountains: Exploitation, Development, and Preservation* (Raleigh: N.C. Division of Archives and History, 1985), 45; Charlotte Pyle and Michael P. Schafale, "Land Use History of Three Spruce-Fir Forest Sites in Southern Appalachia," *Journal of Forest History* 32 (January 1988): 11.

12. Walter H. Wheeler, "Elisha Mitchell as a Geologist," *Journal of the Elisha Mitchell Scientific Society* 100, no. 2 (1984): 61–64.

13. James H. Brown and Arthur C. Gibson, *Biogeography* (St. Louis: C. V. Mosby, 1983), 9–11; David Quammen, *The Song of the Dodo: Island Biogeography in an Age of Extinctions* (New York: Simon and Schuster, 1996), 17–20, 107–14; Peter J. Bowler, *Evolution: The History of an Idea* (Berkeley: University of California Press, 1984), 185–86.

14. E. D. Cope, "Observations on the Fauna of the Southern Alleghanies," *American Naturalist* 4 (1870): 394; Schwarzkopf, *History of Mount Mitchell*, 79–80; Quammen, *Song of the Dodo*, 86–96.

15. Cope, "Observations on Fauna," 402; Bowler, *Evolution*, 247–49, 289–300.

16. Roderick Nash, *Wilderness and the American Mind*, 3d ed. (New Haven: Yale University Press, 1982), 153–55; William Cronon, "The Trouble with Wilderness; or, Getting Back to the Wrong Nature," in *Uncommon Ground: Rethinking the Human Place in Nature*, ed. William Cronon (New York: Norton, 1996), 76–78.

17. Allen H. Bent, "The Mountaineering Clubs of America," *Appalachia* 14 (Decem-

ber 1916): 5–18; Nash, *Wilderness and the American Mind*, 154; Abrams, "Western North Carolina Railroad," 34–44.

18. A. E. Scott, "A Visit to Mitchell and Roan Mountains," *Appalachia* 4 (December 1884), 20, 16; Ellen Mitchell Summerell to Dr. J. J. Summerell, ca. 1878, typescript, Elisha Mitchell Papers, Southern Historical Collection, Manuscripts Division, University of North Carolina, Chapel Hill; "Climbing Mt. Mitchell Was Rugged in the 1870s," *Asheville Citizen-Times*, July 17, 1960; Charles Sloan Reid, "An Ascent of Mt. Mitchell," *Appalachia* 8 (July 1897): 232–35.

19. William Brewster, "An Ornithological Reconnaissance in Western North Carolina," *Auk* 3 (January 1886): 94; U.S. Forest Service, "Mt. Mitchell National Park Feasibility Study Wildlife Assessment," unpublished report, North Carolina Department of Environment and Natural Resources, Division of Parks and Recreation, Raleigh; Schwarzkopf, *History of Mount Mitchell*, 79.

20. "Notes and News," *Auk* 12 (July 1895): 315; *Asheville Daily Citizen*, June 12, 1895; Mark V. Barrow Jr., *A Passion for Birds: American Ornithology after Audubon* (Princeton: Princeton University Press, 1998), 20–33.

21. Stuart A. Marks, *Southern Hunting in Black and White: Nature, History, and Ritual in a Carolina Community* (Princeton: Princeton University Press, 1991), 45; John F. Reiger, *American Sportsmen and the Origins of Conservation*, 3d ed. (Corvallis: Oregon State University Press, 2001), 67–104.

22. Charles Dudley Warner, *On Horseback: A Tour in Virginia, North Carolina, and Tennessee, with Notes of Travel in Mexico and California* (Boston: Houghton Mifflin, 1889), 84; Phillips Russell, "A New Chapter in Mt. Mitchell's History," [1961], Clipping File, through 1975, North Carolina Collection, University of North Carolina Library, Chapel Hill.

23. Warner, *On Horseback*, 83; Schwarzkopf, *History of Mount Mitchell*, 80.

24. Warner, *On Horseback*, 88.

25. Johnston, "'Big Tom' Wilson Leads Search," 22; Hodge Mathes, [with] Explanation by Rev. M. C. Ray, *A Saga of the Carolina Hills: Being a True Story of the Naming of Mt. Mitchell*, Arthur C. Ray Papers, N.C. Division of Archives and History, Raleigh; Tim Silver, "Big Tom Wilson," *Wildlife in North Carolina*, November 1997, 18–23.

26. John Simcox Holmes, *Forest Conditions in Western North Carolina*, N.C. Geological and Economic Survey, Bulletin No. 23 (Raleigh: N.C. Geological and Economic Survey, 1911), 9–10.

27. Ronald L. Lewis, *Transforming the Appalachian Countryside: Railroads, Deforestation, and Social Change in West Virginia, 1880–1920* (Chapel Hill: University of North Carolina Press, 1998), 4–5; Ronald D. Eller, *Miners, Millhands, and Mountaineers: Industrialization of the Appalachian South, 1880–1930* (Knoxville: University of Tennessee Press, 1982), 99–102.

28. Chris Bolgiano, *The Appalachian Forest: A Search for Roots and Renewal* (Mechanicsburg, Pa.: Stackpole, 1998), 87; Eller, *Miners, Millhands, and Mountaineers*, 102–3; Harold K. Steen, *The U.S. Forest Service: A History* (Seattle: University of Washington Press, 1976), 64; Char Miller and V. Alaric Sample, "Gifford Pinchot and the Conservation Spirit," in *Breaking New Ground*, by Gifford Pinchot (Washington, D.C.: Island Press, 1998), xii–xiii.

29. John Simcox Holmes, "Autobiographical," Department of Conservation and De-

velopment, Activities of the Department, 1936–41, N.C. Division of Archives and History; Nash, *Wilderness and the American Mind*, 183; Steen, *U.S. Forest Service*, 64.

30. Pinchot, *Breaking New Ground*, 326; Holmes, "Autobiographical"; Stephen Fox, *The American Conservation Movement: John Muir and His Legacy* (Madison: University of Wisconsin Press, 1981), 109–15; Steen, *U.S. Forest Service*, 96–98.

31. Holmes, *Forest Conditions in Western North Carolina*, 48–49, 51–53.

32. Thomas D. Clark, *The Greening of the South: The Recovery of Land and Forest* (Lexington: University Press of Kentucky, 1984), 14; Pyle and Schafale, "Land Use History," 12; "Oral History Interview with Mr. Reuben B. Robertson and Mr. E. L. Demmon," Asheville, N.C., February 15, 1959, transcript, Forest History Society, Durham, N.C.

33. Pyle and Schafale, "Land Use History," 12; Jeff Lovelace, *Mount Mitchell: Its Railroad and Toll Road* (Johnson City, Tenn.: Overmountain Press, 1994), 1–9; Theresa Coletta, "'Timber!' Started Railroad Boom," in *Common Times*, 74; Schwarzkopf, *History of Mount Mitchell*, 82–85.

34. "Mount Mitchell Conquest," *Charlotte Observer*, June 1, 1913; "Mount Mitchell Railroad Thrown Open to Public," *Asheville Citizen-Times*, July 24, 1913; Clark, *Greening of the South*, 23–25; Paul Salstrom, *Appalachia's Path to Dependency: Rethinking a Region's Economic History, 1730–1940* (Lexington: University Press of Kentucky, 1994), 26.

35. "Mount Mitchell Railroad Thrown Open"; Pyle and Schafale, "Land Use History," 14; Lovelace, *Mount Mitchell*, 11–13; Schwarzkopf, *History of Mount Mitchell*, 93.

36. "Asheville Tourists on Trip over Strange Railroad Line," *Asheville Citizen-Times*, August 24, 1913; Schwarzkopf, *History of Mount Mitchell*, 86.

37. Samuel P. Hays, *Conservation and the Gospel of Efficiency: The Progressive Conservation Movement, 1890–1920* (New York: Atheneum, 1969), esp. 1–4, 122–74; Eller, *Miners, Millhands, and Mountaineers*, 113.

38. Edward L. Ayers, *The Promise of the New South: Life after Reconstruction* (New York: Oxford University Press, 1992), 413–19; George B. Tindall, "Business Progressivism: Southern Politics in the Twenties," *South Atlantic Quarterly* 62 (Winter 1963): 92–106.

39. "Inaugural Address of Governor Locke Craig," January 15, 1913, in *Public Letters and Papers of Locke Craig, Governor of North Carolina, 1913–1917*, ed. May F. Jones (Raleigh: Edwards and Broughton, 1916), 9, 11; "Mount Mitchell Railroad Thrown Open." On Craig's politics and reputation as a Progressive, see *Locke Craig for Governor: He States His Record on the Railroad Question—Reasons for His Nomination Given by Congressman W. T. Crawford*, 1907, campaign pamphlet, North Carolina Collection; T. W. Chambliss, "The Little Giant of the Blue Ridge," *Sky-Land Magazine*, June 1913, 22–24; and "Locke Craig: 'Little Giant of the Mountains,'" *Asheville Citizen-Times*, July 17, 1960.

40. Clark, *Greening of the South*, 26; Coletta, "'Timber!' Started Railroad Boom," 74; Schwarzkopf, *History of Mount Mitchell*, 85; Pyle and Schafale, "Land Use History," 12. See also Gerald Williams, "The Spruce Production Division," *Forest History Today*, Spring 1999, 2–10.

41. "Asheville Tourists on Trip over Strange Railroad Line"; Coletta, "'Timber!' Started Railroad Boom," 74; Virginia Boone (great-granddaughter of Big Tom Wil-

son and daughter of Ewart Wilson), interview with author, Pensacola, N.C., January 5, 1996; Schwarzkopf, *History of Mount Mitchell*, 85.

42. Pyle and Schafale, "Land Use History," 12; Schwarzkopf, *History of Mount Mitchell*, 85; Lovelace, *Mount Mitchell*, 9–10; "Large Lumber Deal at Black Mountain," *Asheville Citizen-Times*, October 8, 1913.

43. "Asheville Tourists on Trip over Strange Railroad Line"; "Large Lumber Deal at Black Mountain."

44. Robert J. Noyes, "Lands of Mount Mitchell Company, Mt. Mitchell Area," January 1917, in files of Pisgah National Forest, Toecane Ranger District, Burnsville, N.C., 3, 7–8, map; Pyle and Schafale, "Land Use History," 13.

45. Raymond Pullman, "Destroying Mt. Mitchell," *American Forestry*, February 1915, 86; Noyes, "Lands of Mount Mitchell Company," 5.

46. Pullman, "Destroying Mt. Mitchell," 83; Pyle and Schafale, "Land Use History," 12.

47. "Mt. Mitchell Fire Destroys Timber," *Asheville Citizen-Times*, July 8, 1914; J. S. Holmes, *Can Mt. Mitchell's Spruce Forests Be Saved?*, N.C. Geological and Economic Survey, Press Bulletin No. 135 (Raleigh: N.C. Geological and Economic Survey, 1914), 1.

48. J. S. Holmes, *Mount Mitchell State Forest: Opinions of Representative Men as Expressed at the Convention of the North Carolina Forestry Association, Raleigh, N.C., January 13, 1915*, N.C. Geological and Economic Survey, Press Bulletin No. 138 (Raleigh: N.C. Geological and Economic Survey, 1915), 3; Linda Flint McClelland, *Building the National Parks: Historic Landscape Design and Construction* (Baltimore: Johns Hopkins University Press, 1998), 53–54.

49. Nash, *Wilderness and the American Mind*, 16–81; Fox, *American Conservation Movement*, 139–47; Daniel S. Pierce, *The Great Smokies: From Natural Habitat to National Park* (Knoxville: University of Tennessee Press, 2000), 43–44.

50. Holmes, *Can Mt. Mitchell's Spruce Forests Be Saved?*, 1–4.

51. Pierce, *Great Smokies*, 41–43.

52. N. Buckner, Asheville Board of Trade, quoted in Pullman, "Destroying Mt. Mitchell," 88; Schwarzkopf, *History of Mount Mitchell*, 88.

53. On the possibility of condemnation proceedings, see Craig's speech to the N.C. Forestry Association, January 13, 1915, in Holmes, *Mount Mitchell State Forest*, 2–3; Locke Craig, "Mitchell's Peak Park," in Jones, *Letters and Papers of Locke Craig*, 217–18.

54. Craig to the N.C. Forestry Association, in Holmes, *Mount Mitchell State Forest*, 2.

55. Pullman, "Destroying Mt. Mitchell," 85, 89.

56. Craig, "Mitchell's Peak Park," 218; "Mount Mitchell Commission Preparing to Acquire Peak," *Asheville Citizen-Times*, April 12, 1915.

57. Noyes, "Lands of Mount Mitchell Company," map; "Mount Mitchell Commission Preparing to Acquire Peak"; Schwarzkopf, *History of Mount Mitchell*, 90.

58. Richard West Sellars, *Preserving Nature in the National Parks: A History* (New Haven: Yale University Press, 1997), 29, 38–39; Alfred Runte, *National Parks: The American Experience*, 3d ed. (Lincoln: University of Nebraska Press, 1997), 101–5.

59. [J. S. Holmes], "Special Forest Investigation: Chestnut Bark Disease," N.C. Geological and Economic Survey, *Biennial Report of the State Geologist, 1913–1914* (Raleigh: N.C. Geological and Economic Survey, 1915), 76–77.

60. Joseph R. Newhouse, "Chestnut Blight," *Scientific American*, July 1990, 106; G. F. Gravatt and R. P. Marshall, *Chestnut Blight in the Southern Appalachians*, U.S. Department of Agriculture, Department Circular 370 (Washington, D.C.: U.S. Department of Agriculture, 1926), 1; George H. Hepting, "Death of the American Chestnut," *Journal of Forest History* 72 (July 1974): 61–62; R. Kent Beattie and Jesse D. Diller, "Fifty Years of Chestnut Blight in America," *Journal of Forestry* 52 (1954): 323.

61. Hepting, "Death of the American Chestnut," 62; Newhouse, "Chestnut Blight," 106; Wilson B. Sayers, "The King Is Dead, Long Live the King," pt. 1, *American Forests* 77 (November 1971): 23.

62. Newhouse, "Chestnut Blight," 107; M. Ford Cochran, "Back from the Brink," *National Geographic*, February 1990, 134; Judy C. Treadwell, "American Chestnut, *Castanea dentata*," March 1996, <http://ncnatural.com/NCNatural/trees/chestnut/html>.

63. Hepting, "Death of the American Chestnut," 62–65; Newhouse, "Chestnut Blight," 106; Wilson B. Sayers, "The King Is Dead, Long Live the King," pt. 2, *American Forests* 77 (December 1971): 22–23, 38–41; G. F. Gravatt, "The Chestnut Blight in North Carolina," in *Chestnut and the Chestnut Blight in North Carolina*, by G. F. Gravatt, N.C. Geological and Economic Survey, Economic Paper No. 56 (Raleigh: N.C. Geological and Economic Survey, 1925), 15.

64. J. S. Holmes, *Scouting for Chestnut Blight in North Carolina*, N.C. Geological and Economic Survey, Press Bulletin No. 88 (Raleigh: N.C. Geological and Economic Survey, 1912); J. S. Holmes, "The Chestnut Bark Disease Which Threatens North Carolina," in *Proceedings of the Second Annual Convention of the North Carolina Forestry Association*, N.C. Geological and Economic Survey, Economic Paper No. 25 (Raleigh: N.C. Geological and Economic Survey, 1912), 14; J. S. Holmes to Haven Metcalf, April 1, 1915, and P. L. Buttrick to Holmes, November 23, 1912, Department of Conservation and Development, Economic and Geological Survey, Correspondence, 1912, N.C. Division of Archives and History; Emily W. B. Russell, "Pre-blight distribution of *Castanea dentata (Marsh.) Borkh*," *Bulletin of the Torrey Botanical Club*, April–June 1987, 185–86.

65. G. F. Gravatt and R. P. Marshall, *Chestnut Blight in the Southern Appalachians: The Natural Replacement of Blight-Killed Chestnut*, U.S. Department of Agriculture, Miscellaneous Circular No. 100 (Washington, D.C.: U.S. Department of Agriculture, 1927). Though this official report on the desirability of various "replacement trees" did not appear until the late 1920s, such speculation was part of the national debate over chestnut blight between 1910 and 1915. This report was based on studies done in Virginia and Maryland in 1924.

66. Sayers, "King Is Dead," pt. 1, 22–23; J. S. Holmes to J. L. Peters, September 27, 1915, and J. S. Holmes to Editor, *Charlotte Observer*, February 13, 1912, Department of Conservation and Development, Economic and Geological Survey, Correspondence, 1915, 1912, N.C. Division of Archives and History.

67. Hepting, "Death of the American Chestnut," 66; Sayers, "King Is Dead," pt. 2, 42, and pt. 1, 22.

68. Asbury F. Lever to J. S. Holmes, February 23, 1915, and Holmes to Haven Metcalf, April 1, 1915, Department of Conservation and Development, Economic and Geological Survey, Correspondence, 1915, N.C. Division of Archives and History; Treadwell, "American Chestnut"; Sayers, "King Is Dead," pt. 2, 40; E. Murray Bruner, "A Com-

prehensive Plan for the Marketing and Utilization of the Remaining Stand of Chestnut Necessitated by the Chestnut Blight Situation," in *Chestnut and the Chestnut Blight in North Carolina*, 18–23.

69. J. S. Holmes, foreword to *Chestnut and the Chestnut Blight in North Carolina*, ii; Gravatt, *Chestnut Blight in the Southern Appalachians*, USDA Circular 370 (1926), 5–6 (see maps); James R. Beavers, "Contemporary Forest/People Relationships in the Vicinity of Mount Mitchell," unpublished report, U.S. Forest Service, Asheville, N.C., December 1977, 12.

70. Noyes, "Lands of Mount Mitchell Company," map.

71. J. S. Holmes, *Organization of Co-operative Forest-Fire Protective Areas in North Carolina*, N.C. Geological and Economic Survey, Economic Paper No. 42 (Raleigh: N.C. Geological and Economic Survey, 1915), 8; Bruner, "Comprehensive Plan," 19–21.

CHAPTER FIVE

1. Ronald D. Eller, *Miners, Millhands, and Mountaineers: Industrialization of the Appalachian South, 1880–1930* (Knoxville: University of Tennessee Press, 1982), 117, 118–19; Southern Railway Company, *Camping on Mount Mitchell* (n.p., 1916), 6–7; N.C. Geological and Economic Survey, *Biennial Report of the State Geologist, 1915–16* (Raleigh: N.C. Geological and Economic Survey, 1917), 102; Kenneth B. Pomeroy and James G. Yoho, *North Carolina Lands: Ownership, Use, and Management of Forest and Related Lands* (Washington, D.C.: American Forestry Association, 1964), 211; Raymond Pullman, "Destroying Mount Mitchell," *American Forestry*, February 1915, 83; Stephen Pyne, *Fire in America: A Cultural History of Wildland and Rural Fire* (Princeton: Princeton University Press, 1982), 7, 155–56.

2. N.C. Geological and Economic Survey, *Biennial Report of the State Geologist, 1911–12* (Raleigh: N.C. Geological and Economic Survey, 1913), 58, 60–61; Fred M. Burnett, *This Was My Valley* (Charlotte: Heritage, 1960), 167–70.

3. J. S. Holmes to Henry Grinnell, March 9, 1915, Department of Conservation and Development, Economic and Geological Survey, Correspondence, 1915, N.C. Division of Archives and History, Raleigh; J. S. Holmes, "Sketch of State Forestry in North Carolina," Department of Conservation and Development, Activities of the Department, 1936–41, N.C. Division of Archives and History.

4. J. S. Holmes, *Organization of Co-Operative Forest Fire Protective Areas in North Carolina*, N.C. Geological and Economic Survey, Economic Paper No. 42 (Raleigh: N.C. Geological and Economic Survey, 1915), 27–31.

5. "Worst Forest Fire in Years Eating Its Way into City Watershed, with Thousand Men Battling to Save It," *Asheville Citizen-Times*, May 10, 1916; J. S. Holmes, "Collaborator's Annual Fire Report for the Calendar Year 1915, North Carolina," and J. S. Holmes to J. G. Peters, December 16, 1915, Department of Conservation and Development, Economic and Geological Survey, Correspondence, 1915, N.C. Division of Archives and History; Charlotte Pyle and Michael P. Schafale, "Land Use History of Three Spruce-Fir Forest Sites in Southern Appalachia," *Journal of Forest History* 32 (January 1988): 13.

6. Locke Craig, "Appointment of General Relief Committee and of Local Committees," in *Public Letters and Papers of Locke Craig, Governor of North Carolina, 1913–1917*,

ed. May F. Jones (Raleigh: Edwards and Broughton, 1916), 233; Jeff Lovelace, *Mount Mitchell: Its Railroad and Toll Road* (Johnson City: Overmountain Press, 1994), 10.

7. Reports of the State Forester to the State Geologist, May, June, August, November 1917, February, May 1918, Joseph Hyde Pratt Papers, Southern Historical Collection, Manuscripts Division, University of North Carolina, Chapel Hill; Pyle and Schafale, "Land Use History," 13.

8. Report of the State Forester to the State Geologist, August 1917, Pratt Papers; N.C. Geological and Economic Survey, *Biennial Report of the State Geologist, 1917–18* (Raleigh: N.C. Geological and Economic Survey, 1919), 41.

9. Report of the State Forester to the State Geologist, August 1917, Pratt Papers; N.C. Geological and Economic Survey, *Biennial Report of the State Geologist, 1921–22* (Raleigh: N.C. Geological and Economic Survey, 1923), 38.

10. Clarence F. Korstian, "Perpetuation of Spruce on Cut-over and Burned Lands in the Higher Southern Appalachian Mountains," *Ecological Monographs* 7, no. 1 (January 1937): 150; Paul Richard Saunders, "The Vegetational Impact of Human Disturbance on the Spruce-Fir Forests of the Southern Appalachian Mountains" (Ph.D. diss., Duke University, 1979), 126–27; "Mount Mitchell State Park, Natural Heritage Program Report," based on fieldwork conducted in the park on September 3–6, 1991, and observations of park volunteers during 1991–92, in files of N.C. Department of Environment and Natural Resources, Division of Parks and Recreation, Raleigh.

11. Paul M. Fink, *Backpacking Was the Only Way: A Chronicle of Camping Experiences in the Southern Appalachian Mountains* (Johnson City: Research Advisory Council, East Tennessee State University, 1975), 79, 80–81.

12. Robert J. Noyes, "Lands of Mount Mitchell Company, Mt. Mitchell Area," in files of Pisgah National Forest, Toecane District, Burnsville, N.C., 9–10, 6, 11.

13. Leon S. Minckler, "Reforestation in the Spruce Type in the Southern Appalachians," *Journal of Forestry* 43 (1945): 652–54.

14. Charles F. Speers, "Experimental Planting in Cutover Spruce-Fir in the Southern Appalachians: 50-Year Results," U.S. Forest Service Research Note SE-219, Southeastern Forest Experiment Station, Asheville, N.C., 1975, 2, 3–5.

15. Monthly Report of the State Forester, April 1929, Department of Conservation and Development, Division of Forestry, Forestry, 1926–35, box 4, folder: Monthly Reports of the State Forester, N.C. Division of Archives and History; Pyle and Schafale, "Land Use History," 14–15.

16. Planting Map, Mount Mitchell Project, Pisgah National Forest, in files of Pisgah National Forest.

17. E. E. Miller, "The Land beyond Kona: A Little Journey to Mount Mitchell, the Highest of the Appalachians," *American Forestry*, February 1923, 92–93.

18. Joseph Hyde Pratt to Lee A. Folger, Secretary, Gordon Motor Company, Richmond, Va., December 20, 1910, North Carolina Geological Survey Papers, Southern Historical Collection.

19. Harry Wilson McKown Jr. "Roads and Reform: The Good Roads Movement in North Carolina, 1885–1921" (M.A. thesis, University of North Carolina, Chapel Hill, 1972), 29–30; Howard Lawrence Preston, *Dirt Roads to Dixie: Accessibility and Modernization in the South, 1885–1935* (Knoxville: University of Tennessee Press, 1991), 5, 39.

20. Clark Howell to J. H. Pratt, October 7, 1910, and J. H. Pratt to L. W. Page, Janu-

ary 7, 1911, North Carolina Geological Survey Papers, Southern Historical Collection; McKown, "Roads and Reform," 37–39; Harley E. Jolley, *The Blue Ridge Parkway* (Knoxville: University of Tennessee Press, 1969), 11–13.

21. Henry McNair, "The Southern Appalachians from a Motor," *Travel*, May 1913, 65; J. H. Pratt to John L. Yates, October 26, 1910, N.C. Geological Survey Papers, Southern Historical Collection; J. H. Pratt to A. M. Griffin, October 7, 1915, Pratt Papers; McKown, "Roads and Reform," 38–39.

22. N. Buckner, Secretary, Asheville Board of Trade, to J. S. Holmes, Gov. Locke Craig, et al., May 8, 1915, Department of Conservation and Development, Economic and Geological Survey, Correspondence, 1915, N.C. Division of Archives and History; Robert E. Ireland, "Prison Reform, Road Building, and Southern Progressivism: Joseph Hyde Pratt and the Campaign for 'Good Roads and Good Men,'" *North Carolina Historical Review* 68 (April 1991): 143–44.

23. "Asheville Tourists on Trip over Strange Railroad Line," *Asheville Citizen-Times*, August 24, 1913; "Mount Mitchell Road to Open on Thursday," *Asheville Citizen-Times*, May 16, 1916; S. Kent Schwarzkopf, *A History of Mount Mitchell and the Black Mountains: Exploration, Development, and Preservation* (Raleigh: N.C. Division of Archives and History, 1985), 93–96.

24. Southern Railway Company, *Camping on Mount Mitchell*, 1916, 13–14; Schwarzkopf, *History of Mount Mitchell*, 94–96.

25. Lovelace, *Mount Mitchell*, 44. Specific information about the de facto segregation of North Carolina's state parks is difficult to come by. According to its minutes the Board of Conservation and Development discussed "Negro use of State Parks" at a meeting on July 1–2, 1956. The minutes state, "By tradition have always maintained separate parks for the two races. It is our opinion [the Board's] that in the event operation of such facilities results in the development of conditions unacceptable to the people of the state—close the facilities so affected" (Minutes, Board of Conservation and Development, Paul Kelly Papers, N.C. Division of Archives and History). This seems to have been the governing principle for North Carolina's state parks from their inception until the early 1960s. I am indebted to Michael Hill, Research Branch, N.C. Division of Archives and History, for this reference. On Jones Lake as a park for African Americans, see N.C. Department of Conservation and Development, *Tenth Biennial Report, 1942–44* (Raleigh: N.C. Department of Conservation and Development, 1945), 71.

26. See advertisement in *Asheville Citizen-Times*, May 14, 1916; Schwarzkopf, *History of Mount Mitchell*, 96–99; Lovelace, *Mount Mitchell*, 7.

27. "The One-Way Motor Road to the Summit of Mount Mitchell," *American Forestry*, February 1923, 96–97. Lovelace, *Mount Mitchell*, 44–45; Schwarzkopf, *History of Mount Mitchell*, 99–100.

28. "Report of Inspection of Toll Motor Roads to Mt. Mitchell," August 9, 1927, Board of Conservation and Development, 1925–35, folder: Forestry, 1927, N.C. Division of Archives and History.

29. James J. Flink, *The Automobile Age* (Cambridge: MIT Press, 1988), 145–47; Robert E. Ireland, *Entering the Auto Age: The Early Automobile in North Carolina, 1900–1930* (Raleigh: N.C. Division of Archives and History, 1990), 73–74; Daniel S. Pierce, "Boosters, Bureaucrats, Politicians, and Philanthropists: Coalition Building in the Establish-

ment of the Great Smoky Mountains National Park" (Ph.D. diss., University of Tennessee, 1995), 26.

30. J. S. Holmes, "A Forest Policy for North Carolina," *Journal of the Elisha Mitchell Scientific Society* 45, no. 1 (November 1929): 30; Paul S. Sutter, *Driven Wild: How the Fight against Automobiles Launched the Modern Wilderness Movement* (Seattle: University of Washington Press, forthcoming). I am indebted to Professor Sutter for allowing me to read his manuscript in advance of publication.

31. Ida Briggs Henderson, "Taking Care of Mount Mitchell," *The State*, July 17, 1935, 9–10; "Report of Inspection of Toll Motor Roads to Mt. Mitchell"; Virginia Boone (great-granddaughter of Big Tom Wilson and daughter of Ewart Wilson), interview with author, Pensacola, N.C., January 5, 1996; Tim Silver, "Big Tom Wilson," *Wildlife in North Carolina*, November 1997, 21.

32. Monthly Report of J. S. Holmes, State Forester, December 1928, Board of Conservation and Development, Activities of the Board, 1925–35, box 3, folder: Forestry, 1928, N.C. Division of Archives and History; "Report of Inspection of Toll Motor Roads to Mt. Mitchell." On the charges and countercharges during the dispute, see Perley's correspondence with Ewart Wilson and his attorneys in box 4, folder: N.C. Dept. of Conservation and Development, and box 6, folder: Wi–Wz, Fred A. Perley Papers, N.C. Division of Archives and History.

33. Jolley, *Blue Ridge Parkway*, 21–32.

34. Anne V. Mitchell, "Parkway Politics: Class, Culture, and Tourism in the Blue Ridge" (Ph.D. diss., University of North Carolina, 1997), 17–18, 25–31, 65, 147; Jolley, *Blue Ridge Parkway*, 39; Anthony J. Badger, *North Carolina and the New Deal* (Raleigh: N.C. Division of Archives and History, 1981), 7–12, 76–90.

35. A. E. Wilson to State Highway and Public Works Commission, April 3, 1936, Department of Conservation and Development, Activities of the Department, 1936–41, box 16, folder: Wilson Toll Road (Mount Mitchell), 1936–38, N.C. Division of Archives and History; Lovelace, *Mount Mitchell*, 43–45.

36. N.C. Department of Conservation and Development, *Sixth Biennial Report, 1937*, 56–57; "Mount Mitchell State Park Summit of Mount Mitchell, North Carolina," July 9, 1937, folder: State Parks, and J. S. Holmes, Memorandum to the Director, "Wilson Toll Road," May 25, 1938, folder: Wilson Toll Road (Mount Mitchell), 1936–38, both in Department of Conservation and Development, Activities of the Department, 1936–41, box 16, N.C. Division of Archives and History; "Operators Raise Toll Road Ante," *Raleigh News and Observer*, July 18, 1939.

37. N.C. Board of Conservation and Development, *Seventh Biennial Report, 1936–38* (Raleigh: N.C. Board of Conservation and Development, 1939), 68–69; Pyle and Schafale, "Land Use History," 15; Linda Flint McClelland, *Building the National Parks: Historic Landscape Design and Construction* (Baltimore: Johns Hopkins University Press, 1998), 5–10, 381–86.

38. Richard West Sellars, *Preserving Nature in the National Parks: A History* (New Haven: Yale University Press, 1997), 357; Susan L. Flader, *Thinking Like a Mountain: Aldo Leopold and the Evolution of an Ecological Attitude toward Deer, Wolves, and Forests* (Columbia: University of Missouri Press, 1974) 14; Harold K. Steen, *The U.S. Forest Service: A History* (Seattle: University of Washington Press, 1976), 115–18; Sutter, *Driven Wild*; Verne Rhoades, "The Recreational Development of the National Forests

of North Carolina," in *Forest Protection or Devastation? It is Up to North Carolina: Some of the Addresses Made at the Tenth Annual Meeting of the North Carolina Forestry Association*, Asheville, June 9–10, 1920 (Chapel Hill: Office of the Secretary, 1920), North Carolina Collection, University of North Carolina Library, Chapel Hill, 16–18.

39. Ruth More, "15,000 Persons Visited Place Last Summer," *Asheville Citizen-Times*, July 4, 1937; McClelland, *Building the National Parks*, 7–8.

40. Burnett, *This Was My Valley*, 173–74.

41. A. H. Marshall, "Forty Eight State Summits," *Appalachia* 21 (December 1936): 167; Hoyt McAfee, "Atop Mitchell's Peak," *Charlotte Observer*, May 30, 1937.

42. C. N. Mease, County Warden, Buncombe County, to J. K. Dixon, June 18, 1927, Wildlife Resources Commission, Game Correspondence, 1927–36, N.C. Division of Archives and History.

43. Undated newspaper article, Wildlife Resources Commission, Game Correspondence, 1927–36, N.C. Division of Archives and History.

44. *Hunting in North Carolina*, N.C. Department of Conservation and Development, Bulletin No. 36 (Raleigh: n.p., 1928), 11–12; Stuart A. Marks, *Southern Hunting in Black and White: Nature, History, and Ritual in a Carolina Community* (Princeton: Princeton University Press, 1991), 51–52.

45. *Hunting in North Carolina*, 11.

46. "An Act to Provide for the Protection and Conservation of Wild Birds and Animals" (North Carolina Game Law, 1927), in Wildlife Resources Commission, Game Refuge Correspondence, 1929–36, N.C. Division of Archives and History.

47. *Hunting in North Carolina*, 9–21; report by Wade H. Phillips to Governor A. W. McLean, June 1, 1927–June 30, 1928, Board of Conservation and Development, Activities of the Board, 1925–35, box 1, folder: Game, N.C. Division of Archives and History.

48. J. S. Holmes, *Mount Mitchell and Mitchell State Park: A Souvenir Dedicated to the General Federation of Women's Clubs* (Chapel Hill: n.p., 1919), North Carolina Collection, University of North Carolina Library, Chapel Hill, 5–7.

49. Apparently neither Holmes nor any other state official commented on such problems at the time, but the growing concern for diminishing fish populations is clearly evident in efforts to restock Black Mountain streams during the 1920s and 1930s. Those efforts are detailed below.

50. "Act to Provide for the Protection and Conservation of Wild Birds and Animals," and "Information Report on Mt. Mitchell Game Refuge," 1934, Wildlife Resources Commission, General Correspondence, folder: Game Refuges, N.C. Division of Archives and History.

51. Sellars, *Preserving Nature in the National Parks*, 24–25; U.S. Department of Agriculture, *Pisgah National Game Preserve: Regulations and Information for the Public*, Department Circular 161 (Washington, D.C.: U.S. Department of Agriculture, 1926), 1–2. I am indebted to Cheryl Oakes, librarian at the Forest History Society, Durham, N.C., for this reference.

52. Flader, *Thinking Like a Mountain*, 9–10. "Moses" is an appellation given Leopold by historian Donald Fleming, quoted in Philip Shabecoff, *A Fierce Green Fire: The American Environmental Movement* (New York: Hill and Wang, 1993), 90.

53. "Aldo Leopold and the Land Ethic: A Conversation with Susan Flader," *Environmental Review Newsletter* 3, no. 5 (May 1996) <http://www.igc.org/envreview/flader.

html>, 2–3, 7–8 (May 22, 2000); Aldo Leopold, *Game Management* (New York: Scribner's, 1948), 195–96; Flader, *Thinking Like a Mountain*, 25–26.

54. A. E. Ammons and Joe Scarborough, interview with author, Clyde, N.C., June 19, 1997; A. E. Ammons, telephone interview with author, June, 19, 2000; "Information Report on Mt. Mitchell Game Refuge, 1934."

55. U.S. Department of Agriculture, *Pisgah National Game Preserve*, 3; interview with Ammons and Scarborough; "Report on Game Received for Restocking the W.N.C. Game Refuges from 1928 to 1934," Wildlife Resources Commission, General Correspondence, N.C. Division of Archives and History; Carolyn Marlowe (daughter of C. N. Mease), interview with author, Burlington, N.C., August 13, 1997; "Deer Herd at Refuge Increasing," unidentified newspaper article dated 1932, in scrapbook kept by C. N. Mease, currently in possession of Carolyn Marlowe; "On the Parkway through Mitchell Game Refuge," *Charlotte Observer*, November 30, 1939.

56. "Report on Fish and Game Liberated on the Western North Carolina State Game Refuges to Date, January 1, 1933," Board of Conservation and Development, box 1, folder: Game Refuges, 1927–33, N.C. Division of Archives and History.

57. Perry Jones, "A Historical Study of the European Wild Boar in North Carolina" (M.A. thesis, Appalachian State Teachers College, 1957), 34, 52, 55–59; William L. Hamnett and David C. Thornton, *Tar Heel Wildlife* (Raleigh: N.C. Wildlife Resources Commission, 1953), 93, 67; U.S. Department of Agriculture, *Pisgah National Game Preserve*, 3.

58. "Aldo Leopold and the Land Ethic," 4; Sellars, *Preserving Nature in the National Parks*, 71–75.

59. Leopold, *Game Management*, 196; interview with Ammons and Scarborough.

60. "Report on Predatory's Killed On The Western North Carolina State Game Refuge's by the Refuge Warden's to Date" [*sic*], January 1, 1933, Board of Conservation and Development, box 1, folder: Game Refuges, 1927–33, N.C. Division of Archives and History; interview with Ammons and Scarborough. For an example of newspaper reports on the annual elimination of predators from game refuges, see "2,301 Predatory Birds and Animals Killed in Game Refuges in 1940," *Asheville Citizen-Times*, February 4, 1940.

61. "1934 Report on the Estimation of Fish and Game on the W.N.C. Game Refuges," Wildlife Resources Commission, General Correspondence, N.C. Division of Archives and History; interview with Ammons and Scarborough; "One-Eyed Elk to Eat in Refuge's Green Pastures," *Asheville Times*, January 19, 1939.

62. Interview with Ammons and Scarborough; "Federal Aid in Wildlife Restoration Act (Pittman-Robertson Act)," <http://www.fws.gov/laws/federal/summaries/pract.html> (June 21, 2001).

63. "Instructions for Deer Hunt Applicants, Mt. Mitchell Cooperative Wildlife Management Area, 1941," General Correspondence, 1937–43, box 1, and "Plans for Bear Hunts, 1940, North Carolina Cooperative Wildlife Management Areas of the Pisgah National Forest," General Correspondence, 1933–45, box 2, both in Wildlife Resources Commission, N.C. Division of Archives and History.

64. "Report on Fish and Game Liberated on the Western North Carolina State Game Refuges to Date January 1, 1933," and "Report on the Open Fishing Seasons on the Western North Carolina State Game Refuges to Date January 1, 1933," Board of Con-

servation and Development, box 1, folder: Game Refuges, 1927–33, N.C. Division of Archives and History; "Fish Management Plan for 1939, Mt. Mitchell Wildlife Area," Wildlife Resources Commission, box 1, General Correspondence, N.C. Division of Archives and History; interview with Ammons and Scarborough; "Refuge Trout Fishing Dates Are Announced," *Asheville Citizen-Times*, May 10, 1932.

65. "Report of Inland Fisheries Division, July 1, 1935–January 1, 1936" and "No Men Allowed in Angling Eve's Fishing Creek," State News Bureau, Raleigh, N.C., Wildlife Resources Commission, box 3, General Correspondence, 1933–45, N.C. Division of Archives and History; "Those Ladies Fished the Creek Hard, More Inland Angling Fans Than Ever," undated newspaper clipping in Mease scrapbook; "Fish Management Plan for 1939, Mt. Mitchell Wildlife Area."

66. "Report on Prosecutions on the Western North Carolina State Game Refuge's from January 1, 1930 to January 1, 1933," Board of Conservation and Development, box 1, folder: Game Refuges, 1927–33, N.C. Division of Archives and History.

67. J. S. Holmes, "Report of the Conservation Committee of the North Carolina Academy of Science," *Journal of the Elisha Mitchell Scientific Society* 53, no. 2 (December 1937): 214; J. S. Holmes, "Wildlife on Sate Parks," article requested by E. B. Floyd, ed., *N.C. Wildlife Conservation*, N.C. Department of Conservation and Development, box 26, Activities of the Department, folder: J. S. Holmes, State Forester, 1942, N.C. Division of Archives and History.

68. Untitled document [probably a press release ca. 1933–35], Board of Conservation and Development, Game Correspondence, box 1, folder: Game Refuges, 1927–33, N.C. Division of Archives and History; interview with Marlowe; "Hundreds Are Expected to Attend Dedication of Gilkey Memorial Park in Mount Mitchell," *Winston-Salem Journal*, October 13, 1940.

69. Douglas B. Starrett, "Mica Deposits of North Carolina," in *The Mining Industry in North Carolina during 1908, 1909, 1910*, ed. Joseph Hyde Pratt and H. M. Berry, N.C. Geological and Economic Survey, Economic Paper No. 23 (Raleigh: N.C. Geological and Economic Survey, 1911), 36–37; W. B. Phillips, "Mica Mining in North Carolina," *Journal of the Elisha Mitchell Scientific Society* 5, no. 2 (July–December 1888): 73–75.

70. Joseph Hyde Pratt and H. M. Berry, *The Mining Industry in North Carolina during 1913–17, Inclusive*, N.C. Geological and Economic Survey, Economic Paper No. 49 (Raleigh: N.C. Geological and Economic Survey, 1919), 93–94.

71. Jasper L. Stuckey, *The Mining Industry in North Carolina from 1946 through 1953*, N.C. Department of Conservation and Development, Economic Paper No. 66 (Raleigh: N.C. Department of Conservation and Development, 1955), 55–57.

72. Joseph Hyde Pratt, *The Mining Industry in North Carolina during 1911 and 1912*, N.C. Geological and Economic Survey, Economic Paper No. 34 (Raleigh: N.C. Geological and Economic Survey, 1914), 218; Schwarzkopf, *History of Mount Mitchell*, 82.

73. Noyes, "Lands of Mount Mitchell Company," 1; Pyle and Schafale, "Land Use History," 14; Schwarzkopf, *History of Mount Mitchell*, 82.

74. E. H. Frothingham, *Timber Growing and Logging Practice in the Southern Appalachian Region*, U.S. Department of Agriculture Technical Bulletin 250 (Washington, D.C.: U.S. Department of Agriculture, 1931), 49–50; *Chestnut and the Chestnut Blight in North Carolina*, N.C. Geological and Economic Survey, Economic Paper No. 56 (Raleigh: N.C. Geological and Economic Survey, 1925), 9.

75. C. F. Korstian, "The Tragedy of Chestnut," *Southern Lumberman*, December 20, 1924, 181; J. S. Holmes, *Scouting for Chestnut Blight in North Carolina*, N.C. Geological and Economic Survey, Press Bulletin No. 88 (Raleigh: N.C. Geological and Economic Survey, 1912), 2.

76. James R. Beavers, "Contemporary Forest/People Relationships in the Vicinity of Mount Mitchell," unpublished report, U.S. Forest Service, Asheville, N.C., December 1977, 12; M. Ford Cochran, "Back from the Brink," *National Geographic*, February 1990, 132–34.

77. "Our Mountain Forests," February 1933, N.C. Department of Conservation and Development, Activities of the Board, 1925–35, box 3, folder: Forestry, 1934, N.C. Division of Archives and History; Cochran, "Back from the Brink," 135; Beavers, "Contemporary Forest/People Relationships," 12.

78. Pomeroy and Yoho, *North Carolina Lands*, 22–23; "Oral History Interview with Mr. Reuben B. Robertson and Mr. E. L. Demmon," Asheville, N.C., February 15, 1959, transcript, Forest History Society, Durham, N.C., 10.

79. Beavers, "Contemporary Forest/People Relationships," 12; "Oral History Interview with Robertson and Demmon," 11–12; *The Story of Chestnut Extract* (n.p.: Champion Paper and Fibre Company, 1937), Reuben B. Robertson Papers, Special Collections, University of North Carolina, Asheville, 10–15.

80. Donald Edward Davis, *Where There Are Mountains: An Environmental History of the Southern Appalachians* (Athens: University of Georgia Press, 2000), 196.

81. Cochran, "Back from the Brink," 135; Joseph R. Newhouse, "Chestnut Blight," *Scientific American*, July 1990, 106–9; Steven L. Stephenson, Andrew N. Ash, and Dean F. Stauffer, "Appalachian Oak Forests," in *Biodiversity of the Southeastern United States: Upland Terrestrial Communities*, ed. William H. Martin, Stephen G. Boyce, and Arthur C. Echternacht (New York: John Wiley, 1993), 276–78.

82. Cochran, "Back from the Brink," 135; Newhouse, "Chestnut Blight," 109–11.

83. Korstian, "Tragedy of Chestnut," 182; Stephenson, Ashe, and Stauffer, "Appalachian Oak Forests," 275.

84. James M. Hill, "Wildlife Value of *Castanea dentata* Past and Present: The Historical Decline of the Chestnut and Its Future Use in Restoration of Natural Areas," *Proceedings of the International Chestnut Conference*, Morgantown, July 10–14, 1992 (Morgantown: West Virginia University Press, 1994), 191.

85. "Holmes Has Seen Wildlife Programs Make Rapid Headway," *Asheville Citizen-Times*, July 7, 1939, clipping in Mease scrapbook.

86. Davis, *Where There Are Mountains*, 196–97; Eller, *Miners, Millhands, and Mountaineers*, 122.

CHAPTER SIX

1. N.C. Department of Conservation and Development, *Tenth Biennial Report* (Raleigh: N.C. Department of Conservation and Development, 1944), 71.

2. Ibid. This report notes that a fire broke out in lumbering operations on the west side of the park, which suggests that loggers had stepped up efforts to harvest chestnut and other hardwoods. Richard West Sellars, *Preserving Nature in the National Parks: A History* (New Haven: Yale University Press, 1997), 150–55.

3. "Vast Changes Wrought on Mitchell in Last 25 Years," *Asheville Citizen-Times*, July 13, 1952; "Address of Governor Cherry before Meeting of Board of Conservation and Development at Ashevlle, N.C," press release, October 13, 1947, D. Hiden Ramsey Papers, Southern Historical Collection, University of North Carolina, Chapel Hill; N.C. Department of Conservation and Development, *Eleventh Biennial Report* (Raleigh: N.C. Department of Conservation and Development, 1946), 90; "Historic Occasion," *Raleigh News and Observer*, May 19, 1948.

4. "Vast Changes Wrought on Mitchell"; Charles F. Speers, "Experimental Planting in Cutover Spruce-Fir in the Southern Appalachians: 50-Year Results," U.S. Forest Service Research Note SE-219, Southeastern Forest Experiment Station, Asheville, N.C., 1975, 2.

5. Margaret Lynn Brown, *The Wild East: A Biography of the Great Smoky Mountains* (Gainesville: University Press of Florida, 2000), 175–76; Thomas C. Parramore, *Express Lanes and Country Roads: The Way We Lived in North Carolina, 1920–1970* (Chapel Hill: University of North Carolina Press, 1983), 50–53.

6. N.C. Department of Conservation and Development, *Eleventh Biennial Report*, 91; N.C. Department of Conservation and Development, *Twelfth Biennial Report* (Raleigh: N.C. Department of Conservation and Development, 1948), 83–84; "Extensive Plans for Developing Mount Mitchell Park Announced," *Greensboro Daily News*, 1946, Clipping File, North Carolina Collection, University of North Carolina Library, Chapel Hill.

7. "How Big Tom Got Its Name in Mount Mitchell State Park," *Asheville Citizen-Times*, April 10, 1949; "Inscriptions for Original Plaques and Markers for Mount Craig and Big Tom," Ramsey Papers; Fred M. Burnett, *This Was My Valley* (Charlotte: Heritage, 1960), 175–76.

8. "Work Is Launched on Picnic Grounds atop Mt. Mitchell," *Asheville Citizen-Times*, August 15, 1948; "Museum Is Latest Unit in Mt. Mitchell Park," *Asheville Citizen-Times*, June 2, 1953; N.C. Department of Conservation and Development, *Biennial Report, 1952–54* (Raleigh: N.C. Department of Conservation and Development, 1955), 76.

9. "Mount Mitchell Still Losing Its Majesty," *Winston-Salem Journal*, May 12, 1991; "More Funds Asked for Mt. Mitchell Park Improvements," *Asheville Citizen*, January 23, 1953; Paul Richard Saunders, "The Vegetational Impact of Human Disturbance on the Spruce-Fir Forests of the Southern Appalachian Mountains" (Ph.D. diss., Duke University, 1979), 50; R. Gage Smith, "Mount Mitchell State Park, Established 1915," Paul Kelly Papers, N.C. Division of Archives and History, Raleigh, 18.

10. "More Funds Asked for Mt. Mitchell Park Improvements"; N.C. Department of Conservation and Development, *Biennial Report, 1950–52* (Raleigh: N.C. Department of Conservation and Development, 1953), 57; "Legislative Move Endangers Mitchell Park Project," *Asheville Citizen-Times*, February 25, 1951.

11. N.C. Department of Conservation and Development, *Biennial Report, 1952–54*; Smith, "Mount Mitchell State Park," 23.

12. Linda Flint McClelland, *Building the National Parks: Historic Landscape Design and Construction* (Baltimore: Johns Hopkins University Press, 1998), 463, 466; Sellars, *Preserving Nature in the National Parks*, 149–50; Brown, *Wild East*, 176.

13. Smith, "Mount Mitchell State Park," 20–22; S. Kent Schwarzkopf, *A History of*

Mount Mitchell and the Black Mountains: Exploration, Development, and Preservation (Raleigh: N.C. Division of Archives and History, 1985), 106–7.

14. Thomas Ellis, telephone interview with author, February 20, 2001; John Drescher, *Triumph of Good Will: How Terry Sanford Beat a Champion of Segregation and Reshaped the South* (Jackson: University Press of Mississippi, 2000), xv–xvi; Howard E. Covington Jr. and Marion A. Ellis, *Terry Sanford: Politics, Progress, and Outrageous Ambitions* (Durham: Duke University Press, 1999), 277.

15. Interview with Ellis, February 20, 2001; Covington and Ellis, *Terry Sanford*, 277.

16. Interview with Ellis, February 20, 2001.

17. Burnett, *This Was My Valley*, 175.

18. Virginia Boone (great-granddaughter of Big Tom Wilson and daughter of Ewart Wilson), interview with author, Pensacola, N.C., January 5, 1996; Schwarzkopf, *History of Mount Mitchell*, 107.

19. Phillips Russell, "A New Chapter in Mt. Mitchell's History," *Raleigh News and Observer*, June 25, 1961; Thomas Ellis, telephone interview with author, February 22, 2001. My description of Ewart Wilson's inn and other facilities is based on a postcard provided by Virginia Boone.

20. Interview with Ellis, February 22, 2001.

21. Russell, "New Chapter"; interview with Ellis, February 22, 2001.

22. Russell, "New Chapter."

23. Interview with Ellis, February 20, 2001; Schwarzkopf, *History of Mount Mitchell*, 108–9.

24. Sellars, *Preserving Nature in the National Parks*, 186–90.

25. "Fire Destroys Mt. Mitchell Park Lodge," *Asheville Citizen-Times*, July 6, 1949; A. E. Ammons and Joe Scarborough, interview with author, Clyde, N.C., June 19, 1997; Carolyn Marlowe (daughter of C. N. Mease), interview with author, Burlington, N.C., August 13, 1997.

26. Frank B. Barrick, *Hunting in North Carolina* (Raleigh: N.C. Wildlife Resources Commission, 1972), 1; interview with Marlowe; interview with Ammons and Scarborough.

27. William L. Hamnett and David C. Thornton, *Tar Heel Wildlife* (Raleigh: N.C. Wildlife Resources Commission, 1953), 58; Richard F. Harlow and Zelbert F. Palmer, "Clearcutting in Coordinated Deer-Timber Management in the Southern Appalachians," *Wildlife in North Carolina*, December 1967, 14–15.

28. Paul W. Hirt, *A Conspiracy of Optimism: Management of the National Forests since World War Two* (Lincoln: University of Nebraska Press, 1994), 245–51.

29. Harlow and Palmer, "Clearcutting in Coordinated Deer-Timber Management," 15; Charles E. Hill, "Mt. Mitchell Deer Movement," *Wildlife in North Carolina*, November 1966, 20.

30. Interview with Ammons and Scarborough.

31. Harlow and Palmer, "Clearcutting in Coordinated Deer-Timber Management," 15.

32. Hirt, *Conspiracy of Optimism*, 294.

33. Interview with Ammons and Scarborough.

34. Ibid. In all likelihood efforts to rid Mount Mitchell's streams of rough fish paral-

leled similar programs instituted in the Great Smoky Mountains National Park. See Brown, *Wild East*, 186–88.

35. Interview with Ammons and Scarborough; N.C. Wildlife Resources Commission, *Managed Hunts: Regulations—Schedules—Information on Public Wildlife Management Areas for Deer—Bear—Boar—Wild Hog—Squirrel—Grouse—Raccoon—Dove—Quail—Rabbit—Waterfowl* (Raleigh: N.C. Wildlife Commission, Game Division, 1960).

36. Interview with Ammons and Scarborough.

37. James R. Beavers, "Contemporary Forest/People Relationships in the Vicinity of Mount Mitchell," unpublished report, U.S. Forest Service, Asheville, N.C., December 1977, 19–20.

38. Aldo Leopold, *A Sand County Almanac with Essays on Conservation from Round River* (New York: Oxford University Press, 1966), 259; Thomas R. Dunlap, *Saving America's Wildlife* (Princeton: Princeton University Press, 1988), 103–4.

39. Barrick, *Hunting in North Carolina*, 1; Dunlap, *Saving America's Wildlife*, 104–5.

40. Barrick, *Hunting in North Carolina*, 26; interview with Ammons and Scarborough.

41. Harold Warren, "Mt. Mitchell: The Crusade for a Park," *Charlotte Observer*, October 31, 1976; Schwarzkopf, *History of Mount Mitchell*, 109.

42. Warren, "Mt. Mitchell"; John Opie, *Nature's Nation: An Environmental History of the United States* (Ft. Worth: Harcourt Brace, 1998), 435–44; Samuel P. Hays, *A History of Environmental Politics since 1945* (Pittsburgh: University of Pittsburgh Press, 2000), 17–73.

43. Warren, "Mt. Mitchell."

44. Beavers, "Contemporary Forest/People Relationships," 14–22; Ed Spears, "Park Change in Trouble," *Asheville Citizen-Times*, April 17, 1977; Schwarzkopf, *History of Mount Mitchell*, 110–11.

45. Charles F. Speers, "The Balsam Woolly Aphid in the Southeast," *Journal of Forestry* 56 (1958): 515–16; Charles W. Dull, James D. Ward, H. Daniel Brown, George W. Ryan, William H. Clerke, and Robert J. Uhler, *Evaluation of Spruce and Fir Mortality in the Southern Appalachian Mountains*, U.S. Department of Agriculture, Forest Service, Southern Region, Protection Report R8-PR13 (Atlanta: U.S. Department of Agriculture, Forest Service, Southern Region, 1988), 39.

46. Garrett A. Smathers, *Balsam Woolly Aphid (Adelges piceae)*, Research/Resource Management Information Bulletin No. 1 (Washington, D.C.: U.S. Department of the Interior, National Park Service, 1978).

47. Ibid.; Whiteford L. Baker, *Eastern Forest Insects*, U.S. Department of Agriculture, Forest Service Miscellaneous Publication 1175 (Washington, D.C.: U.S. Department of Agriculture, Forest Service, 1972), 88–89; Jill R. Sidebottom, "The Balsam Woolly Adelgid (Aphid)," Christmas Tree Notes, N.C. Cooperative Extension Service, 1993, updated 1995, <http://www.ces.ncsu.edu/nreos/forest/xmas/CTN020.html> (June 8, 2001).

48. Southern Appalachian Research/Resource Management Cooperative, "Fraser Fir and the Balsam Woolly Aphid: A Problem Analysis," with a summary of information by Kristine D. Johnson, September 30, 1980, and "Status of the Balsam Woolly Aphid in the Southern Appalachians," Minutes of a Meeting of Representatives of

Land-Managing Agencies, November 18, 1958, Southeastern Forest Experiment Station, Asheville, N.C., 2, both in files of N.C. Department of Environment and Natural Resources, Division of Forest Resources, Raleigh; Gene D. Amman, "Seasonal Biology of the Balsam Woolly Aphid on Mt. Mitchell," *Journal of Economic Entomology* 55 (1962), 96–98; Peter S. White and Charles V. Cogbill, "Spruce-Fir Forests of Eastern North America," in *Ecology and Decline of Red Spruce in the Eastern United States*, ed. Christopher Eagar and Mary Beth Adams, Ecological Studies 96 (New York: Springer-Verlag, 1992), 33.

49. Christopher Eagar, "Review of the Biology and Ecology of the Balsam Woolly Aphid in Southern Appalachian Spruce-Fir Forests," in *The Southern Appalachian Spruce-Fir Ecosystem: Its Biology and Threats*, ed. Peter S. White, National Park Service, Southeast Region Research/Resource Management Report SER-71 (Gatlinburg, Tenn.: Uplands Field Research Laboratory, 1984), 37.

50. W. P. Nagel, "Status of the Balsam Woolly Aphid in the Southeast in 1958, with Special Reference to the Infestations on Mt. Mitchell, N.C., and Adjacent Lands," U.S. Department of Agriculture, Forest Service, Southeastern Forest Experiment Station, Asheville, N.C., July 1959, 2.

51. Southern Appalachian Research/Resource Management Cooperative, "Fraser Fir and the Balsam Woolly Aphid," 8, 12; Nagel, "Status of the Balsam Woolly Aphid," 3.

52. Southern Appalachian Research/Resource Management Cooperative, "Fraser Fir and the Balsam Woolly Aphid," 8–9.

53. Edmund Russell, *War and Nature: Fighting Humans and Insects with Chemicals from World War I to* Silent Spring (New York: Cambridge University Press, 2001), esp. 184–203.

54. R. P. Teulings, "Synopsis of File Information and History of Aphid Control at Mount Mitchell, 1958–1981," May 1982, in files of N.C. Department of Environment and Natural Resources, Division of Forest Resources, Raleigh; "Lindane (Isotox) Chemical Profile, 4/85," <http://pmep.cce.cornell.edu/profiles/ins . . . hylpara/lindane/insect-prof-lindane.html> (June 7, 2001).

55. Pest Control Forester Maxwell to State Forester Winkworth, "Balsam Woolly Aphid—Control Activities, Mt. Mitchell State Park," June 26, 1967, in files of N.C. Department of Environment and Natural Resources, Division of Forest Resources, Raleigh; Teulings, "Synopsis of File Information"; Southern Appalachian Research/Resource Management Cooperative, "Fraser Fir and the Balsam Woolly Aphid," 41; "Balsam Woolly Adelgid, *Adelges piceae*," <http://www.forestpests.org/southern/Insects/blsmwlly.htm> (June 8, 2001).

56. Conley L. Moffett, "Impact of Selected Insecticides on the Biota at Mount Mitchell," Progress Report MTM-2, in files of N.C. Department of Environment and Natural Resources, Division of Forest Resources, Raleigh.

57. Southern Appalachian Research/Resource Management Cooperative, "Fraser Fir and the Balsam Woolly Aphid," 40–41.

58. Opie, *Nature's Nation*, 414–15; M. D. Jackson, T. J. Sheets, and C. L. Moffett, "Persistence and Movement of BHC in a Watershed, Mount Mitchell State Park, North Carolina, 1967–72," *Pesticides Monitoring Journal* 8 (December 1974): 202–5; Teulings, "Synopsis of File Information."

59. Jackson, Sheets, and Moffett, "Persistence and Movement of BHC," 204–5; Teul-

ings, "Synopsis of File Information"; "Lindane (Isotox) Chemical Profile 4/85"; "About Lindane: Excerpts from the book *The Best Control* by Stephen Tvedten," <http://www.safe2use.com/poisons-pesticides/pesticides/lindane/lindane.htm> (June 7, 2001).

60. Teulings, "Synopsis of File Information"; Sidebottom, "Balsam Woolly Adelgid (Aphid)."

61. Speers, "Experimental Planting in Cutover Spruce-Fir," 5; Southern Appalachian Research/Resource Management Cooperative, "Fraser Fir and the Balsam Woolly Aphid," 42.

62. White and Cogbill, "Spruce-Fir Forests of Eastern North America," 33.

63. Charles E. Little, *The Dying of the Trees: The Pandemic in America's Forests* (New York: Viking, 1995), 45; Robert Ian Bruck, interview with author, Raleigh, N.C., June 5, 1995; Lawrence S. Earley, "Trouble on Mount Mitchell," *Wildlife in North Carolina*, December 1984, 12–13; Bruce Henderson, "A Dying Forest: International Team Explores Mount Mitchell," *Charlotte Observer*, June 29, 1984.

64. N.-H. Lin and V. K. Saxena, "Interannual Variability in Acidic Deposition on the Mt. Mitchell Area Forest," *Atmospheric Environment* 25A, no. 2 (1991), 517; Earley, "Trouble on Mount Mitchell," 13–14.

65. Little, *Dying of the Trees*, 48; Robert Ian Bruck, "Forest Decline Syndromes in the Southeastern United States," in *Air Pollution's Toll on Forests and Crops*, ed. James J. MacKenzie and Mohamed T. El-Ashry (New Haven: Yale University Press, 1989), 134–84.

66. Bruck, "Forest Decline Syndromes." For less technical explanations of Bruck's findings, see Earley, "Trouble on Mount Mitchell," 12–15; Little, *Dying of the Trees*, 47–52; and *The Search for Clean Air*, prod. Hugh Morton, PBS (1994), videocassette.

67. Jon R. Luoma, "Acid Murder No Longer a Mystery," *Audubon*, November 1988. On the publicity generated by Bruck's work, see Little, *Dying of the Trees*, 41, and Earley, "Trouble on Mount Mitchell," 11.

68. "Mount Mitchell Still Losing Its Majesty."

69. The various effects of air pollution on forest decline are detailed in Bruck, "Forest Decline Syndromes," 134–84. For a less technical explanation of the different theories, see Jon R. Luoma, "Forests Are Dying, but Is Acid Rain Really to Blame?," *Audubon*, March 1987, 46–48.

70. David R. Peart, N. S. Nicholas, Shepard M. Zedaker, Margaret M. Miller-Weeks, and Thomas G. Siccama, "Conditions and Recent Trends in High-Elevation Red Spruce Populations," in Eagar and Adams, *Ecology and Decline of Red Spruce*, 181; N. S. Nicholas, "Spruce-Fir Decline," in *Threats to Forest Health in the Southern Appalachians, Workshops at Chattanooga, Tennessee, Asheville, North Carolina, and Roanoke, Virginia*, ed. Carole Ferguson and Pamela Bowman (Washington, D.C.: U.S. Department of Agriculture, Forest Service, 1994), 34; Jean Muhlbaier Dasch, "Hydrological and Chemical Inputs to Fir Trees from Rain and Clouds during a One-Month Study at Clingman's Peak, NC," *Atmospheric Environment* 22, no. 10 (1988): 2255–62.

71. Edward R. Cook and Shepard M. Zedaker, "The Dendroecology of Red Spruce Decline," in Eagar and Adams, *Ecology and Decline of Red Spruce*, 211.

72. Interview with Bruck, June 5, 1995. For one of Bruck's many statements about the difficulty of determining cause and effect, see Jim Conrad, "An Acid-Rain Trilogy:

A Very Special Look at 'Forest Death' from the Black Forest to Mt. Mitchell—and a Philosophy for Change," *American Forests* 93 (November–December 1987): 78.

73. On the political nature of the debate and the apparent reluctance of the state park to include air pollution as part of the explanation for forest decline, see Earley, "Trouble on Mount Mitchell," 12–15, and Little, *Dying of the Trees*, 38–52. The Forest Service position is best illustrated by Ferguson and Bowman, *Threats to Forest Health*, and Dull et al., *Evaluation of Spruce and Fir Mortality*. In the former the effects of air pollution are noted but downplayed in relation to other factors. In the latter the emphasis is on the extent of damage to spruce and fir and the changing composition of high-elevation forests.

74. Earley, "Trouble on Mount Mitchell," 12. For Shepard Zedaker's affiliation and academic interests, see <http://www.cnf.vt.edu/forestry/faculty/zedaker.html>. In 1994 Niki Nicholas listed her affiliation as "Forest Ecologist, Tennessee Valley Authority" (see Ferguson and Bowman, *Threats to Forest Health*, 31); Jean Muhlbaier Dasch ("Hydrological and Chemical Inputs") listed her affiliation as "Environmental Science Department, General Motors Research Laboratories, Warren, MI."

75. For one environmentalist's reading of the differences, see Harvard Ayers, "The Cry of the Mountains," in *An Appalachian Tragedy: Air Pollution and Tree Death in the Eastern Forests of North America*, ed. Harvard Ayers, Jenny Hager, and Charles E. Little (San Francisco: Sierra Club Books, 1998), 2–3.

76. Interview with Bruck; Joseph E. Barnard, *NAPAP Report 16: Changes in Forest Health and Productivity in the United States and Canada, Acid Deposition: State of Science and Technology* (Washington, D.C.: Government Printing Office, 1990), 126; Adela Backiel, "Research Service Issue Brief for Congress 86031: Acid Rain, Air Pollution, and Forest Decline," updated October 12, 1990, <http://www.cnie.org/nle/nrgen-15.html>, 5 (June 12, 2001); Little, *Dying of the Trees*, 42–44.

77. Little, *Dying of the Trees*, 41; Tom Sieg, "Report on Acid Rain Downplays Dying Trees," *Winston-Salem Journal*, September 9, 1990.

78. Hays, *History of Environmental Politics since 1945*, 116–18, 111–12, 125–28; Opie, *Nature's Nation*, 449; Little, *Dying of the Trees*, 41–42.

79. Sieg, "Report on Acid Rain," C9; "Clouds' Effects May Hurt N.C. Mountain," *Robesonian* (Lumberton, N.C.), July 18, 1993.

80. Viney P. Aneja, "Organic Compounds in Cloud Water and Their Deposition at a Remote Continental Site," *Journal of Air and Waste Management Association* 43 (September 1993): 1240–43; "Pollution in Mountains Traced Closer to Home," *Charlotte Observer*, December 7, 2001; Little, *Dying of the Trees*, 48; "Clouds' Effects May Hurt N.C. Mountain."

81. "Mount Mitchell Still Losing Its Majesty"; Little, *Dying of the Trees*, 55; N.C. Department of Environment and Natural Resources, Division of Parks and Recreation, *Forest Decline on Mount Mitchell*, official park publication (n.p., 1994).

82. Bruck, "Forest Decline Syndromes," 186; Nicholas, "Spruce-Fir Decline," 34; Robert L. Anderson, "How People, Pests, and the Environment Have Changed, and Continue to Change the Southern Appalachian Forest Landscape," in Ferguson and Bowman, *Threats to Forest Health*, 3.

83. Bruck, "Forest Decline Syndromes," 135–46; Little, *Dying of the Trees*, 39, 179–81; Ayers, Hager, and Little, *Appalachian Tragedy*, 77.

84. Bruck, quoted in Little, *Dying of the Trees*, 47; Wes Humphries, Andy Kilpatrick, Michael Long, Ford Mauney, Morgan Williams, and George Venitsanos, "Forest Decline on Mount Mitchell," unpublished report, Environmental Management, April 17, 1995, in files of N.C. Department of Environment and Natural Resources, Division of Parks and Recreation, Raleigh.

85. Andrew Jones and George F. Wilhere, "A Survey of Nonavian Vertebrates with Management Recommendations, Mount Mitchell State Park," School of the Environment, Duke University, May 1994; David A. Adams and John S. Hammond, "Changes in Forest, Bird, and Small Mammal Populations at Mount Mitchell, North Carolina, 1959/62 and 1985," *Journal of the Elisha Mitchell Scientific Society* 107, no. 1 (Spring 1991): 10–11; Kerry N. Rabenold, "Birds of Appalachian Spruce-Fir Forests: Dynamics of Habitat-Island Communities," in White, *Southern Appalachian Spruce-Fir Ecosystem*, 182–83.

86. Joel Harp, "Itsy Bitsy Spider," *Friend of Wildlife: Journal of the North Carolina Wildlife Federation*, Winter 1996, 16.

87. *Search for Clean Air*; John Hoeffel, "Environmental Group Puts N.C. on Danger List," *Winston-Salem Journal*, December 21, 1995; Little, *Dying of the Trees*, esp., 38, 54–55; Ayers, Hager, and Little, *Appalachian Tragedy*, frontispiece, 59–61.

88. "Mount Mitchell Still Losing Its Majesty."

89. Ibid.

CONCLUSION

1. Douglas M. Orr Jr. and Alfred W. Stuart, *The North Carolina Atlas: Portrait for a New Century* (Chapel Hill: University of North Carolina Press, 2000), 130; "July 1, 2000 Certified Population Estimates," N.C. State Data Center, <http://census.nc.us/census_body.html> (October 19, 2001).

2. "Mount Mitchell Golf Club," <http://www.insidenc.com/mountain/Mtn.Mitchell Golf.htm> (July 10, 2001).

3. William Cronon, "The Trouble with Wilderness; or, Getting Back to the Wrong Nature," in *Uncommon Ground: Rethinking the Human Place in Nature*, ed. William Cronon (New York: Norton, 1996), 73; Alfred Runte, *National Parks: The American Experience*, 3d ed. (Lincoln: University of Nebraska Press, 1997), 33–47.

4. National Geodetic Survey, Mount Mitchell GPS Project, January 14, 1994. My thanks to Perrin Wright for providing me with a copy of the data.

5. The classic example of this kind of declensionist interpretation is Ronald Eller, *Miners, Millhands, and Mountaineers: Industrialization of the Appalachian South, 1880–1930* (Knoxville: University of Tennessee Press, 1982). More recently some historians of Appalachia have taken a longer view of the region's history by describing how land speculation, slaveholding, and other economic activities in preindustrial Appalachia had already incorporated various regions into the "capitalist world system." See, for example, Paul Salstrom, *Appalachia's Path to Dependency: Rethinking a Region's Economic History, 1730–1940* (Lexington: University Press of Kentucky, 1994). Salstrom suggests that Appalachia's agriculture and economy were already in trouble before the arrival of lumbermen. Even so, I would argue, the narrative is still one of declension. The question is simply, When did capitalism begin and decline set in? For a sum-

mary of the trends in preindustrial Appalachian historiography, see Dwight B. Billings, Mary Beth Pudup, and Altina L. Waller, "Taking Exception with Exceptionalism: The Emergence and Transformation of Historical Studies of Appalachia," in *Appalachia in the Making: The Mountain South in the Nineteenth Century*, ed. Mary Beth Pudup, Dwight B. Billings, and Altina L. Waller (Chapel Hill: University of North Carolina Press, 1995), 1–24.

6. William Cronon, "A Place for Stories: Nature, History, and Narrative," *Journal of American History* 78, no. 4 (March 1992): 1347–48.

7. Paul Schullery, *Searching for Yellowstone: Ecology and Wonder in the Last Wilderness* (Boston: Houghton Mifflin, 1997), 61.

8. See, for examples, Eller, *Miners, Millhands, and Mountaineers*, 6–8, 110–11, and Ronald L. Lewis, *Transforming the Appalachian Countryside: Railroads, Deforestation, and Social Change in West Virginia, 1880–1920* (Chapel Hill: University of North Carolina Press, 1998), 15–18, 265–68, 277–78.

9. This trend, which for the Blacks I defined as a manifestation of Murphy's Law, is well known to students of environmental management. One of the clearest statements of the problem comes from forester and historian Nancy Langston, who encountered it during a study of Forest Service policy in Oregon's Blue Mountains. As Langston writes, "All attempts to manage are attempts to tell a story about how the land ought to be." In the Blue Mountains, she concludes, "every time a manager tried to fix one problem, the solution created a worse problem elsewhere" (Nancy Langston, *Forest Dreams, Forest Nightmares: The Paradox of Old Growth in the Inland West* [Seattle: University of Washington Press, 1995], 297, 296).

10. Orr and Stuart, *North Carolina Atlas*, 414–15.

11. To my knowledge no one has yet completed a systematic study of the decline of brook trout in the Black Mountains. But extensive work on competition between fish species in the Great Smoky Mountains clearly demonstrates this trend. See Margaret Lynn Brown, *The Wild East: A Biography of the Great Smoky Mountains* (Gainesville: University Press of Florida, 2000), 290–92.

12. Cronon, "Trouble with Wilderness," 83–86; Michael Pollan, *Second Nature: A Gardener's Education* (New York: Dell, 1991), 212.

13. Carolyn Merchant, "Reinventing Eden: Western Culture as a Recovery Narrative," in Cronon, *Uncommon Ground*, 132–59.

14. Pollan, *Second Nature*, 225, 230–31.

15. "Mount Mitchell State Park," N.C. Natural Heritage Program, in files of N.C. Department of Environment and Natural Resources, Division of Parks and Recreation, Raleigh, 5, 13, 22–23.

16. Cronon, "Trouble with Wilderness," 90.

17. Merchant, "Reinventing Eden," 158–59.

18. "Researchers Study Failing Firs, Look for Signs of Hope," *Watauga Democrat*, August 10, 2001.

19. Bruce Henderson, "Asheville's Health in Air Pollution's Clutches," *Charlotte Observer*, July 14, 2001.

20. Ibid.; Preston Howard Jr., "N.C. Clean Air Bill Isn't Fair," *Charlotte Observer*, November 29, 2001; "Environmental Woe: The 10 Worst Problems in N.C.," *Charlotte Observer*, December 16, 2001.

21. April Bethea, "Air Quality Should Improve," *Raleigh News and Observer*, June 19, 2002; "Smokestacks Bill OK, a Strong Victory for All North Carolinians," *Asheville Citizen-Times*, June 26, 2002; "Environmental Defense Praises House Vote on Smokestacks Bill," <http//www.environmentaldefense.org/pressrelease.cfm?Content ID=2099> (August 9, 2002).

INDEX